Multivariate Approximation

MATHEMATICAL RESEARCH

Volume 101

Multivariate Approximation

Recent Trends and Results

Proceedings of the 2nd International Conference
on Multivariate Approximation Theory
held at Witten-Bommerholz, Germany,
September 29 – October 4, 1996

edited by
Werner Haußmann
Kurt Jetter
Manfred Reimer

Akademie Verlag

Editors:

Prof. Dr. Werner Haußmann, Gerhard-Mercator-University of Duisburg
Prof. Dr. Kurt Jetter, University of Hohenheim
Prof. Dr. Manfred Reimer, University of Dortmund

With 11 figures

1st edition

Die Deutsche Bibliothek – CIP-Einheitsaufnahme

Multivariate approximation : recent trends and results ; proceedings
of the 2nd International Conference on Multivariate Approximation Theory,
held at Witten-Bommerholz, Germany, September 29 – October 4, 1996 /
ed. by Werner Haußmann ... – Berlin : Akad. Verl., 1997
  (Mathematical research ; Vol. 101)
  ISBN 3-05-501770-6

ISSN 0138-3019

© Akademie Verlag GmbH, Berlin 1997
Akademie Verlag is a member of WILEY-VCH.

Printed on non-acid paper.
The paper used corresponds to both the U.S. standard ANSI Z.39.48 – 1984
and the European standard ISO TC 46.

Printing: GAM Media GmbH, Berlin
Bookbinding: Druckhaus „Thomas Müntzer" GmbH, Bad Langensalza

Printed in the Federal Republic of Germany

Akademie Verlag GmbH
Mühlenstraße 33–34
D-13187 Berlin
Federal Republic of Germany

Preface

This volume consists of the Proceedings of the 2nd International Conference on Multivariate Approximation Theory at Bommerholz, Germany. The conference took place during the week of September 29 through October 4, 1996, and was hosted by "Haus Bommerholz", the guest–house of the University of Dortmund, located in the beautiful countryside south of Dortmund. It was attended by 48 participants from 14 countries. The program included 10 invited key lectures and 23 contributed talks. Most of these lectures were written up in a form suitable for publication, and appear here in print after careful peer–refereeing.

The scientific scope of this conference was directed towards the following topics which combine classical aspects of Approximation Theory with new and modern methods in Applied Mathematics: Multiscale Techniques, Approximation by Solutions of Differential Equations, and Geometric Design. We are happy that we were able to gather leading experts of these areas of Mathematics, as well as quite a number of promising young researchers. The idea of restricting talks to these subjects, with possible applications in science and engineering, proved a valuable stimulus to interesting discussions. Last but not least, the hospitality and the warm atmosphere of the house itself provided an excellent environment for lively interactions and scientific progress.

The editors are grateful to all who have contributed to the success of the conference and the production of this Proceedings volume. In particular, we are deeply indebted to the Deutsche Forschungsgemeinschaft whose funding made the conference possible. Our thanks also go to the authors and referees for their efforts to guarantee publications of high scientific standard, to the Akademie–Verlag for their kind cooperation and for publishing these Proceedings in the Mathematical Research series, and to the sponsors of this volume which are listed separately. Finally, we would especially like to thank the conference secretary, Ms. Naujokat, and the staff of Haus Bommerholz, for all their help.

<table>
<tr><td>May 1, 1997</td><td>Werner Haußmann</td></tr>
<tr><td></td><td>Kurt Jetter</td></tr>
<tr><td></td><td>Manfred Reimer</td></tr>
</table>

Acknowledgment

The production of this volume was sponsored by the following distinguished companies and trusts of the city of Dortmund:

Vereinigte Elektrizitätswerke Westfalen AG

Westfälische Hypothekenbank AG

Continentale Versicherung

Signal Versicherungen

Volkswohl Bund Versicherungen

We are grateful for their generous support.

The Editors

Table of Contents

Participants 9

Scientific Program of the Conference 15

Weights and Angles of Quadrature Rules on the Sphere S^{n-1}
    S. Blunck 19

Local Box Dimension of Fractal Functions
    Z. Ciesielski 27

Trigonometric Approximation in Multivariate Periodic Hilbert Spaces
    F.–J. Delvos 35

Harmonic Approximation and Boundary Behaviour
    S. J. Gardiner 45

Properties of Bivariate Refinable Spline Pairs
    T. N. T. Goodman 63

Polar Cone Splines
    R. Gormaz 83

Discrepancies of Point Sequences on the Sphere and Numerical Integration
    P. J. Grabner, B. Klinger and R. F. Tichy 95

Regularity of Multivariate Hermite Interpolation
    H. Hakopian 113

Extremizers for the Multivariate Landau–Kolmogorov Inequality
    O. Kounchev 123

Nested Sequences of Triangular Finite Element Spaces

  A. Le Méhauté 133

Charge Distribution of Points on the Sphere and Corresponding
Cubature Formulae

  U. Maier and J. Fliege 147

Solving of Algebraic Equations — An Interplay of Symbolical
and Numerical Methods

  H. M. Möller 161

An Orthonormal Bivariate Algebraic Polynomial Basis for $C(I^2)$
of Low Degree

  J. Prestin and F. Sprengel 177

Tensor Product Splines on Refined Grids in S–Convex Interpolation

  J. W. Schmidt and M. Walther 189

Quasi–Balayage and A Priori Estimates for the Laplace Operator I

  H. S. Shapiro 203

Quasi–Balayage and A Priori Estimates for the Laplace Operator II

  H. S. Shapiro 231

Interpolation and Hyperinterpolation on the Sphere

  I. H. Sloan 255

Interpolation and Wavelets on Sparse Gauß–Chebyshev Grids

  F. Sprengel 269

The General Structure of Multivariate Affine Frames

  J. Stöckler 287

Spectral Sythesis with Wavelet Methods

  G. Zimmermann 303

Participants

David Armitage, *Department of Pure Mathematics, Queen's University, Belfast BT7 1NN, Northern Ireland*
email: d.armitage@qub.ac.uk

Peter Binev, *Institute of Mathematics, Bulgarian Academy of Sciences, Sofia 1090, Bulgaria*
email: binev@fmi.uni-sofia.de

Heribert Blum, *Fachbereich Mathematik, Universität Dortmund, D–44221 Dortmund, Germany*
email: blum@math.uni-dortmund.de

Sönke Blunck, *Mathematisches Seminar, Universität Kiel, Ludewig–Meyn–Str. 4, D–24098 Kiel, Germany*
email: nms78@rz.uni-kiel.d400.de

Len Bos, *Department of Mathematics and Statistics, University of Calgary, Calgary, Alberta T2N 1N4, Canada*
email: lpbos@math.ucalgary.ca

Martin Buhmann, *Fachbereich Mathematik, Universität Dortmund, D–44221 Dortmund, Germany*
email: mdb@math.uni-dortmund.de

Zbigniew Ciesielski, *Mathematical Institute PAN, ul 23 Abrahama 18, 81–825 Sopot, Poland*
email: matzc@halina.univ.gda.pl

Franz–Jürgen Delvos, *Fachbereich 6 – Mathematik, Universität–GH–Siegen, Hölderlinstraße 3. D–57076 Siegen, Germany*

Uwe Depczynski, *Institut für Angewandte Mathematik und Statistik,*
Universität Hohenheim, D–70593 Stuttgart, Germany
email: depczyns@uni-hohenheim.de

Jörg Fliege, *Fachbereich Mathematik, Universität Dortmund,*
D–44221 Dortmund, Germany
email: fliege@math.uni-dortmund.de

Stephen J. Gardiner, *University College Dublin,*
Dublin 4, Ireland
email: gardiner@acadamh.ucd.ie

Tim N. T. Goodman, *Department of Mathematics, University of Dundee,*
Dundee DD1 4HN, Scotland
email: tgoodman@mcs.dundee.ac.uk

Raúl Gormaz, *Departamento de Ing. Matemática, Universidad de Chile,*
Casilla 170/3, Correo 3, Santiago, Chile
email: rgormaz@dim.uchile.cl

Peter J. Grabner, *Institut für Mathematik A, Technische Universität Graz,*
Steyrergasse 30, A–8010 Graz, Austria
email: grabner@weyl.math.tu-graz.ac.at

Hakop A. Hakopian, *Department of Mathematics, Yerevan State University,*
Alex Manoukian 1, Yerevan 375049, Armenia
email: instmath@pnas.sci.am

Werner Haußmann, *Fachbereich Mathematik, Universität Duisburg,*
D–47048 Duisburg, Germany
email: haussmann@math.uni-duisburg.de

Kurt Jetter, *Institut für Angewandte Mathematik und Statistik,*
Universität Hohenheim, D–70593 Stuttgart, Germany
email: kjetter@uni-hohenheim.de

Rong–Qing Jia, *Department of Mathematics, University of Alberta,*
Edmonton T6G 2G1, Canada
email: jia@xihu.math.ualberta.ca

Lavi Karp, *Department of Mathematics, University of Stockholm,*
S–10691 Stockholm, Sweden
email: karp@matematik.su.se

John Klinkhammer, *Fachbereich Mathematik, Universität Duisburg,*
D–47048 Duisburg, Germany
email: klinkhammer@math.uni-duisburg.de

Hermann König, *Mathematisches Seminar, Universität Kiel,*
Ludewig–Meyn–Str. 4, D–24098 Kiel, Germany
email: nms22@rz.uni-kiel.d400.de

Ognyan Kounchev, *Institute of Mathematics, Bulgarian Academy of*
Sciences, Acad. G. Bonchev Str. 8, BG–1113 Sofia, Bulgaria
email: kounchev@bgearn.acad.bg

Peter Kubach, *Fachbereich Mathematik, FernUniversität Hagen,*
D–58084 Hagen, Germany
email: peter.kubach@fernuni-hagen.de

Pierre–Jean Laurent, *LMC – IMAG, Université Joseph Fourier,*
BP 53 X, F–38041 Grenoble, France
email: pjl@imag.fr

Alain Le Méhauté, *Département de Mathématiques, Université de Nantes,*
2 Rue de la Houssinière, F–44072 Nantes Cedex, France
email: alm@namath.univ-nantes.fr

Rudolph A. Lorentz, *Gesellschaft für Mathematik und Datenverarbeitung,*
Schloß Birlinghoven, D–53757 St. Augustin, Germany
email: lorentz@gmd.de

Detlef Mache, *Fachbereich Mathematik, Universität Dortmund,*
D–44221 Dortmund, Germany
email: mache@math.uni-dortmund.de

Ulrike Maier, *Fachbereich Mathematik, Universität Dortmund,*
D–44221 Dortmund, Germany
email: umaier@math.uni-dortmund.de

Charles A. Micchelli, *Thomas J. Watson Research Center, IBM Corporation,*
Yorktown Heights, NY 10598, USA
email: cam@watson.ibm.com

H. Michael Möller, *Fachbereich Mathematik, Universität Dortmund,*
D–44221 Dortmund, Germany
email: moeller@math.uni-dortmund.de

Wilhelm Niethammer, *Institut für Praktische Mathematik, Universität (TH)*
Karlsruhe, Englerstraße 2, D–76128 Karlsruhe, Germany
email: niethammer@math.uni-karlsruhe.de

Günther Nürnberger, *Lehrstuhl für Mathematik IV, Universität Mannheim,*
D–68131 Mannheim, Germany
email: nuernberger@math.uni-mannheim.de

Jürgen Prestin, *Fachbereich Mathematik, Universität Rostock,*
Universitätsplatz 1, D–18055 Rostock, Germany
email: prestin@mathematik.uni-rostock.d400.de

Manfred Reimer, *Fachbereich Mathematik, Universität Dortmund,*
D–44221 Dortmund, Germany
email: reimer@math.uni-dortmund.de

Robert Schaback, *Institut für Numerische und Angewandte Mathematik,*
Universität Göttingen, D–37083 Göttingen, Germany
email: schaback@math.uni-goettingen.de

Andrea Schenk, *Fachbereich Mathematik, Universität Dortmund,*
D–44221 Dortmund, Germany

Jochen W. Schmidt, *Institut für Numerische Mathematik, Technische Univer-*
sität Dresden, D–01062 Dresden, Germany
email: jschmidt@math.tu-dresden.de

Peter Schroeder, *Fachbereich Mathematik, FernUniversität Hagen,*
D–58084 Hagen, Germany
email: peter.schroeder@Fernuni-Hagen.de

Kathi Selig, *Institut für Angewandte Mathematik und Statistik,*
Universität Hohenheim, D–70593 Stuttgart, Germany
email: selig@uni-hohenheim.de

Harold S. Shapiro, *Department of Mathematics, Royal Institute of Technology,*
S–10044 Stockholm, Sweden
email: shapiro@math.kth.se

Ian H. Sloan, *School of Mathematics, University of New South Wales,*
Sydney, Australia 2052
email: I.sloan@unsw.edu.au

Frauke Sprengel, *Fachbereich Mathematik, Universität Rostock,*
D–18051 Rostock, Germany
email: frauke.sprengel@mathematik.uni-rostock.de

Gabriele Steidl, *Fachbereich Mathematik, Universität Mannheim,*
D–68131 Mannheim, Germany
email: steidl@kiwi.math.uni-mannheim.de

Joachim Stöckler, *Institut für Angewandte Mathematik und Statistik,*
Universität Hohenheim, D–70593 Stuttgart, Germany
email: stockler@uni-hohenheim.de

Ralf Tenberg, *Fachbereich Mathematik, Universität Dortmund,*
D–44221 Dortmund, Germany
email: tenberg@math.uni-dortmund.de

Robert F. Tichy, *Institut für Mathematik A, Technische Universität Graz,*
Steyrergasse 30, A–8010 Graz, Austria
email: tichy@weyl.math.tu-graz.ac.at

Hans–Jörg Wenz, *Fachbereich Mathematik, Universität Duisburg,*
D–47048 Duisburg, Germany
email: wenz@informatik.uni-duisburg.de

Georg Zimmermann, *Mathematisches Institut, Universität Wien,*
Strudlhofgasse 4, A–1090 Wien, Austria
email: gzim@tyche.mat.univie.ac.at

Scientific Program of the Conference

Monday, September 30:

Opening Session: Chairman *M. Reimer*

9:00 — Opening of the Conference

9:30 – 10:30 — *I. H. Sloan*: Interpolation, Hyperinterpolation and Spherical Designs

Morning Session: Chairman *G. Nürnberger*

11:00 – 11:30 — *F.–J. Delvos*: Multivariate Periodic Hilbert Spaces

11:30 – 12:00 — *J. Klinkhammer*: Remarks on the Pompeiu Problem

12:00 – 12:30 — *L. Bos*: On Kergin Interpolation

Afternoon Session: Chairman *H. König*

16:00 – 17:00 — *P. J. Grabner*: Numerical Integration, Discrepancy and Spherical Designs

17:00 – 17:30 — *R. F. Tichy*: Polynomial Discrepancy

17:30 – 18:00 — *U. Maier*: Distribution of Points on the Sphere and Corresponding Cubature Formulae

18:00 – 18:30 — *S. Blunck*: Angles and Weights of Cubature Formulas for Polynomials on S^{n-1}

Tuesday, October 1:

Morning Session: Chairman *K. Jetter*

9:00 – 10:00	*T. N. T. Goodman*: Pairs of Refinable Bivariate Spline Functions
10:30 – 11:00	*Z. Ciesielski*: Fractal Functions and Schauder Bases
11:00 – 11:30	*J. W. Schmidt*: Tensor Product Splines in S-Convexity Preserving Interpolation
11:30 – 12:00	*R. Schaback*: Solving Partial Differential Equations Using Radial Basis Functions
12:00 – 12:30	*R. A. Lorentz*: Nonexistence of Prewavelets in $H^s(\mathbb{R}^d)$

Afternoon Session: Chairman *P.–J. Laurent*

16:00 – 17:00	*G. Steidl*: Fast Fourier Transforms of Different Kind
17:00 – 17:30	*M. D. Buhmann*: Asymptotically Optimal Approximation and Numerical Solutions of PDEs
17:30 – 18:00	*R. Gormaz*: Polyhedral Polar Splines
18:00 – 18:30	*P. Binev*: L_p-Approximations by Sigmoidal Functions

Wednesday, October 2:

Morning Session: Chairman *W. Niethammer*

9:00 – 10:00	*R.–Q. Jia*: Multiple Refinable Functions and Multiple Wavelets
10:30 – 11:00	*J. Prestin*: Multivariate Polynomial Bases
11:00 – 11:30	*J. Stöckler*: Multivariate Wavelets and Frames
11:30 – 12:00	*I. H. Sloan*: Interpolation on the Sphere – a Conjecture Proved

Afternoon Session: Chairman *J. W. Schmidt*

16:00 – 17:00   *H. Hakopian*: Regularity of Multivariate Hermite
Interpolation

17:00 – 17:30   *F. Sprengel*: Interpolation and Wavelets on Sparse
Gauß–Chebyshev Grids

17:30 – 18:00   *G. Zimmermann*: Spectral Synthesis with Wavelet
Methods

18:00 – 18:30   *A. Le Méhauté*: Nested Sequences of Finite Elements

Thursday, October 3:

Morning Session: Chairman *D. H. Armitage*

9:00 – 10:00   *C. A. Micchelli*: Regularity of Multiwavelets

10:30 – 11:30   *H. S. Shapiro*: Approximate Balayage and Some
Applications

11:30 – 12:00   *L. Karp*: Approximation of Harmonic Functions
in the L^1–Norm and the Cauchy Problem for the
Laplace Operator

Friday, October 4:

Morning Session: Chairman *Z. Ciesielski*

9:00 – 10:00   *S. J. Gardiner*: Harmonic Approximation and
Boundary Behaviour

10:30 – 11:00   *D. H. Armitage*: Best Harmonic L^1–Approximants
to Subharmonic Functions

11:00 – 11:30   *P. Kubach*: Direct and Inverse Mean Value Properties
of Harmonic Functions with Respect to Ellipsoids

11:30 – 12:00   *O. I. Kounchev*: Extremizers of Multivariate Landau–
Kolmogorov Inequalities are Polysplines

Afternoon Session: Chairman *W. Haußmann*

13:30 – 14:30   *H. M. Möller*: Symbolic Preprocessing for Solving Systems of Algebraic Equations

14:30   Closing of the Conference

Weights and Angles of Quadrature Rules on the Sphere S^{n-1}

S. Blunck

Abstract

Let $\mu : Y \to \mathbb{R}_+$ be a quadrature rule on the unit sphere S^{n-1} of $\mathbb{K}^n$ for all polynomials P of a certain degree $2k$ of homogeneity. We study the relationship between the weights $\mu(y)$ and the angles in the support Y. In the case of at most k angles we show that the points have to be equiweighted and equiangled, i.e. every $y \in Y$ has the same weight $\mu(y) = |Y|^{-1}$ and the same number of neighbours at a certain angle to y. In the general case we get similar results if we impose regularity conditions on the angle matrix $\left(|\langle y, y' \rangle|^2 \right)$.

Introduction

By weighting the nodes of a quadrature rule one tries to balance irregularities in their distribution over the sphere. One concept to describe such irregularities is to compare the positions of the nodes to each other. So for every node y we look at the number $d(y, \alpha)$ of neighbours in a certain angle α to y. If all these angle multiplicities are independent of y the nodes seem well–distributed so that it is natural to give the same weight to all of them. Conversely, if all weights are equal then apparently differences in the $d(y, \alpha)$ cannot be balanced so that they have to be equal, too.

This concept is investigated here.

For that purpose let $n, k \in \mathbb{N}$ and $\mathbb{K} \in \{\mathbb{R}, \mathbb{C}\}$. By S^{n-1} we denote the unit sphere of $\mathbb{K}^n$ and by σ the (rotation invariant, normalized) Haar measure on S^{n-1}.

Definition 0.1 *Let $Y \subset S^{n-1}$ be finite. A map $\mu : Y \to \mathbb{R}_+$ is called a* quadrature rule of index $2k$ *if*

$$\int_{S^{n-1}} P(x) \, d\sigma(x) \; = \; \sum_{y \in Y} \mu(y) \, P(y)$$

Multivariate Approximation: Recent Trends and Results; W. Haußmann, K. Jetter and M. Reimer (eds.)
Mathematical Research, Vol. 101, pp. 19–26, ISBN 3-05-501770-6
© Akademie-Verlag, Berlin 1997

holds for all homogeneous polynomials P of degree $2k$ in n variables [8] and

$$\sum_{y \in Y} \mu(y) \;=\; 1 \,.$$

An efficient criterion for establishing a quadrature rule is the following (cf. [6] for the real case, [8]).

Lemma 0.2 *Let $\mu : Y \to \mathrm{I\!R}_+$ be such that $\sum_{y \in Y} \mu(y) = 1$. Then*

$$\sum_{y,y' \in Y} \mu(y)\,\mu(y')\,|\langle y, y'\rangle|^{2k} \;\geq\; \int |x_1|^{2k}\, d\sigma(x) \;=:\; \gamma(n, 2k)\,.$$

Equality holds iff $\mu : Y \to \mathrm{I\!R}_+$ is a quadrature rule of index $2k$.

We now introduce the quantities we use for describing the angle properties of a support Y [4].

Definition 0.3

(a) $A(Y) := \{\, |\langle y, y'\rangle|^2 \mid y, y' \in Y\,,\ y \neq y' \,\}$ *is called the* angle-set *of* Y .
(b) $d(y_0, \alpha) := |\{\, y \in Y \mid |\langle y, y_0\rangle|^2 = \alpha \,\}|$.

We say that a quadrature rule $\mu : Y \to \mathrm{I\!R}_+$ is *equiweighted* (*equiangled*) if $\mu(y)$ (each $d(y, \alpha)$) is independent of $y \in Y$.
We always assume that Y does not contain collinear vectors or, equivalently, $1 \notin A(Y)$.

1   Supports with Few Angles

The next theorem characterizes that a set Y with at most k angles supports a quadrature rule of index $2k$. In particular it turns out that in this case the quadrature rule has to be equiangled and equiweighted.

Proposition 1.1 *Let $|A(Y)| \leq k$, $A := A(Y) \cup \{1\}$. Let q_α be the polynomial with leading coefficient 1 whose set of zeroes is $A \setminus \{\alpha\}$ and $Q_\alpha := q_\alpha(\,|\langle \cdot, e_1\rangle|^2\,)$. Then the following statements are equivalent :*

(a) *There exists a quadrature rule $\mu : Y \to \mathrm{I\!R}_+$ of index $2k$.*

(b) *For all $\alpha \in A$ we have*

$$d(y_0, \alpha) \;=\; \frac{|Y|}{q_\alpha(\alpha)} \int Q_\alpha \, d\sigma$$

independently of $y_0 \in Y$, and

$$\sum_{\alpha \in A} \frac{\alpha^k}{q_\alpha(\alpha)} \int Q_\alpha \, d\sigma \;=\; \gamma(m, 2k) . \tag{1.1}$$

In this case we have $\mu \equiv |Y|^{-1}$.

Proof: Choose a rotation U such that $U(y_0) = e_1$.

(a)$\Rightarrow$(b): Since $\deg(q_\alpha) = |A(Y)| \leq k$ the quadrature rule is exact for the Q_α.
So the identity

$$\int Q_1 \, d\sigma \;=\; \int Q_1 \circ U \, d\sigma \;=\; \sum_{y \in Y} \mu(y) Q_1(Uy) \;=\; \mu(y_0) q_1(1)$$

shows that μ is constant. Hence we obtain

$$\int Q_\alpha \, d\sigma \;=\; \sum_{y \in Y} \mu(y) Q_\alpha(Uy) \;=\; |Y|^{-1} \sum_{y \in Y} q_\alpha(|\langle y, y_0 \rangle|^2)$$

$$=\; |Y|^{-1} d(y_0, \alpha) q_\alpha(\alpha) \ .$$

This yields

$$\sum_{y, y' \in Y} \mu(y) \mu(y') |\langle y, y' \rangle|^{2k} \;=\; |Y|^{-1} \sum_{y \in Y} |\langle y, y_0 \rangle|^{2k}$$

$$=\; |Y|^{-1} \sum_{\alpha \in A} d(y_0, \alpha) \alpha^k$$

$$=\; \sum_{\alpha \in A} \frac{\alpha^k}{q_\alpha(\alpha)} \int Q_\alpha \, d\sigma$$

and we can appeal to Lemma 0.2.

(b)$\Rightarrow$(a): Define $\mu \equiv |Y|^{-1}$. By repeating the last computation we see that

$$\sum_{y, y' \in Y} \mu(y) \mu(y') |\langle y, y' \rangle|^{2k} \;=\; \sum_{\alpha \in A} \frac{\alpha^k}{q_\alpha(\alpha)} \int Q_\alpha \, d\sigma$$

and we appeal again to the same Lemma. $\square$

Remark 1.2 *In case of* $|A(Y)| = k$, *(1.1) always holds.*

Proof : The polynomial

$$p_1(t) \quad := \quad \sum_{\alpha \in A} \frac{\alpha^k}{q_\alpha(\alpha)} \, q_\alpha(t)$$

has degree at most k and interpolates $p_2(t) := t^k$ at the points of A. Since $|A| = k+1$ we conclude $p_1 = p_2$.

By definition $\gamma(n, 2k) = \int p_2(|x_1|^2) \, d\sigma(x)$ and (1.1) follows from

$$\sum_{\alpha \in A} \frac{\alpha^k}{q_\alpha(\alpha)} \int Q_\alpha \, d\sigma \quad = \quad \int \sum_{\alpha \in A} \frac{\alpha^k}{q_\alpha(\alpha)} \, q_\alpha(|x_1|^2) \, d\sigma(x)$$

$$= \quad \int p_1(|x_1|^2) \, d\sigma(x)$$

$$= \quad \int p_2(|x_1|^2) \, d\sigma(x). \qquad \qquad \square$$

In general a quadrature rule with $k+1$ angles is not equiweighted ([10], 9.26, $(n, k) = (6, 2)$). But still there is a strong connection between equiangled- and equiweightedness also in the case of supports with $k+1$ angles :

Proposition 1.3 *Let* $|A(Y)| = k+1$, $A := A(Y)$. *Let* q_α *and* Q_α *be as in Proposition* 1.1.

Then the following statements are equivalent :

(a) *There exists a constant quadrature rule* $\mu : Y \to \mathbb{R}_+$ *of index* $2k$.

(b) *For all* $\alpha \in A$ *we have*

$$d(y_0, \alpha) \quad = \quad \frac{|Y|}{q_\alpha(\alpha)} \int Q_\alpha \, d\sigma - \frac{q_\alpha(1)}{q_\alpha(\alpha)}$$

independently of $y_0 \in Y$.

This is shown by combining the technique of 1.1 and the interpolation argument of 1.2 .

Observe that the propositions 1.1 and 1.3 exclude many sets $Y \subset S^{n-1}$ with at most $k+1$ angles from supporting a quadrature rule just by its angle properties.

For only the angles and not the concrete points of Y are involved in the formulas for the multiplicities $d(y,\alpha)$. And whenever one of their right–hand sides evaluates to a non–integer then Y does not support a quadrature rule .

This can be used in order to prove in a simple way parts of the results in [1] and [2] about the non–existence of so–called tight quadrature rules ([6, 9]) where the number of points reaches a general lower bound. In this case the angles of a potential support Y are known, they have to be zeroes of certain k–th order polynomials ([5, 7, 8]). And indeed for many values of (n,k) our formulas yield non–integer values for the $d(y,\alpha)$ and thus show that a tight quadrature rule does not exist [3].

2   The General Case

Now we investigate the connection between equal weights and equal angle multiplicities without restrictions on $|A(Y)|$.

One approach might be to replace the annihilator polynomial q_α of degree $|A(Y)|$ by a k–th order approximating polynomial and to show with the method of 1.1 that equal weights imply equiangledness.

This idea is not pursued here.

We show that under certain regularity conditions on the angle matrix $(\,|\langle y, y'\rangle|^2\,)$ equiangledness implies equal weights.

Definition 2.1 *Let $N \in \mathrm{I\!N}$. A matrix $A \in \mathrm{I\!K}^{N\times N}$ is called a* circulant *if it has the following form:*

$$A \;=\; \begin{pmatrix} a_0 & a_1 & a_2 & \cdots & a_{N-1} \\ a_{N-1} & a_0 & a_1 & \cdots & a_{N-2} \\ a_{N-2} & a_{N-1} & a_0 & \cdots & a_{N-3} \\ \vdots & \vdots & \vdots & \ddots & \vdots \\ a_1 & a_2 & a_3 & \cdots & a_0 \end{pmatrix}. \qquad (2.1)$$

Theorem 2.2 *Let $\mu : Y \to \mathrm{I\!R}_+$ be a quadrature rule of index $2k$ and $N := |Y|$.*

(a) *If there exists an indexing $Y = \{y_0,\ldots,y_{N-1}\}$ such that $A := (\,|\langle y_i, y_j\rangle|^2\,) \in \mathrm{I\!K}^{N\times N}$ is circulant, then Y supports a constant quadrature rule of index $2k$.*

(b) *If $\mu : Y \to \mathrm{I\!R}_+$ is equiangled and $A := (\,|\langle y, y'\rangle|^2\,) \in \mathrm{I\!K}^{N\times N}$ is non–singular, then $\mu : Y \to \mathrm{I\!R}_+$ is equiweighted.*

Proof : (a) Setting $a_j := |\langle y_0, y_j \rangle|^{2k}$ for $j = 0, \ldots, N-1$ gives A exactly the form (2.1) . Using

$$\gamma(n, 2k) \;=\; \int |\langle \cdot, y_j \rangle|^{2k} \, d\sigma \;=\; \sum_{i=0}^{N-1} \mu(y_i) \, |\langle y_i, y_j \rangle|^{2k}$$

$$=\; \sum_{i=0}^{N-1} \mu(y_i) \, a_{(j-i) \bmod N}^{k} \;=\; \sum_{r=0}^{N-1} \mu(y_{(j-r) \bmod N}) \, a_r^{k}$$

for $j = 0, \ldots, N-1$ we find

$$\gamma(n, 2k) \;=\; \frac{1}{N} \sum_{j=0}^{N-1} \sum_{r=0}^{N-1} \mu(y_{(j-r) \bmod N}) \, a_r^{k}$$

$$=\; \frac{1}{N} \sum_{r=0}^{N-1} a_r^{k} \sum_{j=0}^{N-1} \mu(y_{(j-r) \bmod N}) \;=\; \frac{1}{N} \sum_{r=0}^{N-1} a_r^{k} \, .$$

We finish the proof via Lemma 0.2 :

$$\sum_{y, y' \in Y} \frac{1}{N} \frac{1}{N} \, |\langle y, y' \rangle|^{2k} \;=\; \frac{1}{N^2} \sum_{j=0}^{N-1} \sum_{i=0}^{N-1} |\langle y_i, y_j \rangle|^{2k}$$

$$=\; \frac{1}{N^2} \sum_{j=0}^{N-1} \sum_{r=0}^{N-1} a_r^{k} \;=\; \gamma(n, 2k) \, .$$

(b) Let $Y = \{y_1, \ldots, y_n\}$ be arbitrarily indexed . Then $\widetilde{A} := (\, |\langle y_i, y_j \rangle|^2 \,)$ is non–singular and $\Sigma := \sum_{\alpha \in A(Y)} \alpha \, d(y_j, \alpha)$ is independent of $j \in \mathrm{I\!N}_{\leq n}$. Let

$$f : \mathrm{I\!R}^N \to \mathrm{I\!R} \, , \; (x_1, \ldots, x_N) \mapsto \sum_{i,j=1}^{N} x_i \, x_j \, |\langle y_i, y_j \rangle|^2 \, ,$$

$$g : \mathrm{I\!R}^N \to \mathrm{I\!R} \, , \; (x_1, \ldots, x_N) \mapsto (\, \sum_{i=1}^{N} x_i \,) - 1 \, ,$$

$$S := \{ \, x \in \mathrm{I\!R}^N \, | \, g(x) = 0 \, \} \, .$$

For $x_0 := (\mu(y_1), \ldots, \mu(y_N)) \in \mathrm{I\!R}^N$ we have $x_0 \in M := \{ x \in S \, | \, f(x) = \gamma(n, 2) \}$ since $\mu : Y \to \mathrm{I\!R}_+$ obviously provides a quadrature rule of index 2. Therefore it suffices to show that $M = \{v\}$ holds for $v := (\frac{1}{N}, \ldots, \frac{1}{N}) \in \mathrm{I\!R}^N$. For that purpose let $x \in M$. According to Lemma 0.2 the vector x is a minimum point of f on S. This implies the existence of a Lagrange multiplyer $\lambda \in \mathbb{R}$ such that $\frac{\partial}{\partial x_j} f(x) - 2\lambda \frac{\partial}{\partial x_j} g(x) = 0$ for all $j \in \mathbb{N}_{\leq N}$, i.e. such that

$$\sum_{i=1}^{N} x_i \, |\langle y_i, y_j \rangle|^2 \; = \; \lambda \quad \text{for } j = 1, \ldots, N \, .$$

Thus $\lambda \; = \; (\sum_{j=1}^{N} x_j)\lambda \; = \; \sum_{j=1}^{N} x_j \sum_{i=1}^{N} x_i \, |\langle y_i, y_j \rangle|^2 \; = \; f(x) \; = \; \gamma(n,2)$. Hence $\tilde{A} x = b$ holds for $b := (\gamma(n,2), \ldots, \gamma(n,2)) \in \mathbb{R}^N$. On the other hand we have for $j = 1, \ldots, N$:

$$(\tilde{A} v)_j \; = \; \frac{1}{N} \sum_{i=1}^{N} |\langle y_i, y_j \rangle|^2 \; = \; \frac{1}{N} \left(1 + \sum_{\alpha \in A(Y)} \alpha \, d(y_j, \alpha) \right) \; = \; \frac{1}{N} \, (1 + \Sigma) \, .$$

For $c := \frac{1+\Sigma}{N} \frac{1}{\gamma(n,2)}$ this means $\tilde{A} v = c b = c \tilde{A} x = \tilde{A}(c x)$, and our assumption on $\tilde{A}$ yields $v = c x$. Since $1 = \sum x_i = \frac{1}{c} \sum v_i = \frac{1}{c}$ we conclude $c = 1$ and $v = x$. $\qquad \square$

So in the situation of (a) an equal weighting of the nodes y is one possible choice while in the situation of (b) it is the only.

Examples of quadrature rules with circulant angle matrices are provided in [7, Example 2] and [10, (8.7)].

In the proof of (b) we only need the weaker condition that $\sum_{\alpha \in A(Y)} \alpha \, d(y, \alpha)$ is independent of $y \in Y$. Conversely, this is always implied by equal weights.

Remark 2.3 *If the quadrature rule $\mu : Y \to \mathbb{R}_+$ of index $2k$ is equiweighted then*

$$\sum_{\alpha \in A(Y)} \alpha \, d(y_0, \alpha) \; = \; \frac{|Y| - n}{n} \quad \textit{independently of } y_0 \in Y \, .$$

Proof: We have

$$\frac{1}{n} \; = \; \gamma(n,2) \; = \; \int |\langle \cdot, y_0 \rangle|^2 \, d\sigma \; = \; |Y|^{-1} \sum_{y \in Y} |\langle y, y_0 \rangle|^2$$

$$= \; |Y|^{-1} (1 + \sum_{\alpha \in A(Y)} \alpha \, d(y_0, \alpha) \,),$$

where we have used polar coordinates to prove the first identity. $\qquad \square$

References

[1]   E. Bannai, R. M. Damerell: *Tight spherical designs I*, J. Math. Soc. Japan **31** (1979), 199–207.

[2]   E. Bannai, R. M. Damerell: *Tight spherical designs II*, J. London Math. Soc. (2) **21** (1980), 13–30.

[3]   S. Blunck: *Isometrische Einbettungen von euklidischen Räumen in l_p-Räume*, Diplomarbeit, Kiel 1995.

[4]   P. Delsarte, J. M. Goethals, J. J. Seidel: *Bounds for systems of lines, and Jacobi polynomials*, Philips Res. Repts. **30** (1975), 91–105.

[5]   P. Delsarte, J. M. Goethals, J. J. Seidel: *Spherical codes and designs*, Geom. Ded. **6** (1977), 363–388.

[6]   J. M. Goethals, J. J. Seidel: *Spherical designs*, Proc. Symp. Pure Math. **34** (1979), 255–272.

[7]   S. G. Hoggar: *t-designs in projective spaces*, Europ. J. Comb. **3** (1982), 233–254.

[8]   H. König: *Isometric imbeddings of Euclidean spaces into finite dimensional l_p-spaces*, Banach Center Publ. **34** (1995), 79–87.

[9]   Yu. Lyubich, L. Vaserstein: *Isometric imbeddings between classical Banach spaces, cubature formulas, and spherical designs*, Geom. Dedicata **47** (1993), 327–362.

[10]   B. Reznick: *Sums of even powers of real linear forms*, Memoirs Amer. Math. Soc. **463** (1992).

Address:

Sönke Blunck
Mathematisches Seminar
Universität Kiel
Ludewig–Meyn–Str. 4
D–24098 Kiel
Germany

Local Box Dimension of Fractal Functions

Z. Ciesielski

Abstract

For a given real valued function f over a cube in R^d the local box (entropy) dimension of its graph is discussed. Sufficient conditions for both the lower and upper dimension in terms of the coefficients of the expansion of f in the Schauder diamond basis are given.

1 Introduction

In recent years more and more attention is paid in mathematical papers to *fractal functions*. We call a bounded $f : I^d \to R^d$, $I = [0,1]$, a *fractal* function over a compact $Q \subset I^d$, by definition, if its graph $\Gamma(f, Q) = \{(\underline{t}, f(\underline{t})) : \underline{t} \in Q\}$ has box dimension satisfying the inequalities $d < \dim_b(\Gamma(f, Q)) \le d + 1$. To define this dimension we start with its lower and upper versions. For a given $\delta > 0$ let $N(\delta)$ denote the minimal number of cubes with edge of size δ needed to cover $\Gamma(f, Q)$. Now, by definition, the upper box dimension is equal to

$$\overline{\dim}_b(\Gamma(f, Q)) := \overline{\lim}_{\delta \to 0} \frac{\log N(\delta)}{\log \frac{1}{\delta}},$$

and the lower box dimension to

$$\underline{\dim}_b(\Gamma(f, Q)) := \underline{\lim}_{\delta \to 0} \frac{\log N(\delta)}{\log \frac{1}{\delta}}.$$

We are now ready to define the upper and lower box dimension of the graph of a bounded $f : I^d \to R$ at a given point $t \in I^d$, respectively. Namely, they are given by the following formulae

$$\overline{\dim}_b(f; t) := \overline{\lim} \ \overline{\dim}_b(\Gamma(f, Q(t, \varepsilon))), \tag{1.1}$$

$$\underline{\dim}_b(f; t) := \underline{\lim} \ \underline{\dim}_b(\Gamma(f, Q(t, \varepsilon))), \tag{1.2}$$

where the $\overline{\lim}$ and $\underline{\lim}$ are taken over all cubes $Q(t, \varepsilon)$, with edges of length 2ε parallel to the corresponding coordinate axes, containing in its interior t

Multivariate Approximation: Recent Trends and Results; W. Haußmann, K. Jetter and M. Reimer (eds.)
Mathematical Research, Vol. 101, pp. 27–33, ISBN 3–05–501770–6
© Akademie–Verlag, Berlin 1997

and shrinking to t with $\varepsilon \to 0$. For more on related definitions we refer to [2] and [5].

Our aim is to describe some subclasses of continuous functions f defined over I^d for which the local dimensions (1.1) and (1.2) can be estimated or computed. This can be done using corresponding results on the dimension $\dim_b(\Gamma(f, Q))$ from [1]. The proofs are constructive and the criteria are given in terms of the behavior of the coefficients of the expansion of f into the Schauder diamond basis over I^d.

In Section 2 we describe the construction of the Schauder diamond basis and recall the basic result on $\dim_b(\Gamma(f, I^d))$ from [1]. Section 3 contains the main result on the local box dimension.

2   The Diamond Basis Over I^d

To define the Schauder diamond basis over I^d we start with the function $\psi(t) = \max[0, 1 - |t|]$ and with the set D of all dyadic points in I. Define $D_0 = \{0, 1\}$, $D_k = \{\frac{2j-1}{2^k} : j = 1, \ldots, 2^{k-1}\}$ for $k = 1, 2, \ldots$. Thus

$$D = \bigcup_{k \geq 0} D_k,$$

and the Schauder functions over I are defined as follows

$$\phi_\tau(t) = \psi(2^k(t - \tau)) \quad \text{for} \quad \tau \in D_k, \ k = 0, 1, \ldots$$

For the Schauder functions over I^d it is convenient to introduce $C_0 = D_0$, $C_k = C_{k-1} \cup D_k$. Then

$$C_k^d = C_{k-1}^d \cup D_{k,d},$$

where

$$D_{k,d} = \{\underline{\tau} = (\tau_1, \ldots, \tau_d) \in C_k^d : \exists_i \ \tau_i \in D_k\} \text{ and } D_{0,d} = D_0^d.$$

Now, define

$$\phi_{\underline{\tau}}(\underline{t}) = \prod_{i \in \mathcal{D}} \psi(2^k(t_i - \tau_i)) \text{ for } \underline{\tau} \in D_{k,d}, \ k = 0, 1, \ldots \tag{2.1}$$

The system is called *the diamond* or *multi–affine* (cf. [3],[4]) basis in the Banach space $C(I^d)$. Some of its properties proved in [1] we mention here. For some constant C, which depends on the dimension, we have for $1 \leq p \leq \infty$, the inequalities

$$\frac{1}{C} \cdot B_{k,p} \leq \| \sum_{\underline{\tau} \in D_{k,d}} b_{\underline{\tau}} \cdot \phi_{\underline{\tau}} \|_p \leq C \cdot B_{k,p}, \tag{2.2}$$

where the L^p-norm $\| \cdot \|_p$ is taken over I^d and

$$B_{k,p} = \left(\frac{1}{|D_{k,d}|} \sum_{\underline{\tau} \in D_{k,d}} |b_{\underline{\tau}}|^p \right)^{\frac{1}{p}}, \tag{2.3}$$

with $|D_{k,d}|$ denoting the cardinality of $D_{k,d}$.

According to [3], we know the system of linear functionals over $C(I^d)$, which is biorthogonal to $(\phi_{\underline{\tau}}(\underline{t}), \ \underline{\tau} \in D^d)$; for given $f \in C(I^d)$ and $\underline{\tau} \in D^d$ the corresponding functionals are defined as follows:

$$b_{\underline{\tau}}(f) = f(\underline{\tau}) \quad \text{for} \quad \underline{\tau} \in D_{0,d},$$

and for $k \geq 1$

$$b_{\underline{\tau}}(f) = \frac{1}{2^d} \sum_{\varepsilon \in \{-1,1\}^d} (f(\underline{\tau}) - f(\underline{\tau}^\varepsilon)) \quad \text{for} \quad \underline{\tau} \in D_{k,d}, \tag{2.4}$$

where $\underline{\tau}^\varepsilon = (\tau_1^\varepsilon, \ldots, \tau_d^\varepsilon)$ with

$$\tau_i^\varepsilon = \begin{cases} \tau_i + \varepsilon_i \cdot 2^{-k} & \text{if } \tau_i \in D_k; \\ \tau_i & \text{if } \tau_i \in C_{k-1}. \end{cases}$$

It is convenient to introduce the finite dimensional projections in $C(I^d)$

$$R_k(f) = \sum_{\underline{\tau} \in D_{k,d}} b_{\underline{\tau}}(f) \cdot \phi_{\underline{\tau}}.$$

The fact that $(\phi_{\underline{\tau}}(\underline{t}), \ \underline{\tau} \in D^d)$ is a Schauder basis in $C(I^d)$ can now be stated as follows: for each $f \in C(I^d)$ the series

$$\sum_{k=0}^{\infty} R_k(f)$$

converges to f in the maximum norm. It now follows by (2.2) and (2.3) that

$$\|R_k(f)\|_p \sim \left(\frac{1}{|D_{k,d}|} \sum_{\underline{\tau} \in D_{k,d}} |b_{\underline{\tau}}(f)|^p\right)^{\frac{1}{p}}.$$

Now, let us recall the criterions for box dimension of a graph over the cube I^d as proved in [1]:

Theorem 2.1   *Let $0 < \alpha \le \beta \le 1$ and let the function f be given on I^d by the Schauder series*

$$f = \sum_k \sum_{\underline{\tau} \in D_{k,d}} b_{\underline{\tau}} \cdot \phi_{\underline{\tau}}. \tag{2.5}$$

If

$$\max_{D_{k,d}} |b_{\underline{\tau}}| = O(\frac{1}{2^{\alpha k}}), \tag{2.6}$$

then

$$\overline{\dim}_b(\Gamma(f, I^d)) \le d + 1 - \alpha.$$

Moreover, if for some $C > 0$, for large k

$$\frac{1}{|D_{k,d}|} \sum_{D_{k,d}} |b_{\underline{\tau}}| \ge C \cdot \frac{1}{2^{\beta k}}, \tag{2.7}$$

then

$$\underline{\dim}_b(\Gamma(f, I^d)) \ge d + 1 - \beta.$$

As it follows from [3], the criterion (2.6) corresponds to the well–known global Hölder condition and (2.7) to the anti–Hölder condition criterion in the terminology of [5]. Now, for any cube $Q \subset I^d$ with dyadic rational vertices and edges parallel to the axes, the system of functions $(\phi_{\underline{\tau}} : \underline{\tau} \in D^d \cap Q)$, restricted to Q, becomes a Schauder basis in $C(Q)$. Theorem 2.1 remains true over Q for this new basis. Moreover, let $k(Q)$ be the largest k such that there is a vertex of Q which is in $D_{k,d}$, then the basic expansion of $f \in C(Q)$

$$f = \sum_{\underline{\tau} \in D^d \cap Q} c_{\underline{\tau}}(f) \cdot \phi_{\underline{\tau}}, \tag{2.8}$$

has the property that for $k > k(Q)$ and $\underline{\tau} \in D_{k,d} \cap Q$ we have the equality $c_{\underline{\tau}}(f) = b_{\underline{\tau}}(f)$ with the last term given by formula (2.4). This property of the diamond basis is important for investigating the local dimensions.

3 The Main Result on Local Box Dimension

We are interested now in a situation over I^d in which the exponents α and β appearing in the lower and upper estimates of $b_{\underline{\tau}}$ in Theorem 2.1 depend explicitly on $\underline{\tau}$. More precisely, we are going to treat functions represented by (2.5) with coefficients satisfying, for some positive finite constants C_1, C_2 and for large k, the inequalities:

$$|b_{\underline{\tau}}| \le C_1 \cdot \frac{k^\gamma}{2^{\alpha(\underline{\tau}) \cdot k}} \quad \text{for} \quad \underline{\tau} \in D_{k,d}, \tag{3.1}$$

with $\gamma \in R$, $\alpha(\cdot) : I^d \to (0,1]$ and

$$|b_{\underline{\tau}}| \ge C_2 \cdot \frac{k^\delta}{2^{\beta(\underline{\tau}) \cdot k}} \quad \text{for} \quad \underline{\tau} \in D_{k,d}, \tag{3.2}$$

with $\delta \in R$ and $\beta(\cdot) : I^d \to (0,1]$.

Theorem 3.1 *Let $f \in C(I^d)$ be given by (2.5). Then, condition (3.1) implies*

$$\overline{\dim}_b(f;t) \le d + 1 - \alpha(t), \tag{3.3}$$

at each point t of lower semi–continuity of $\alpha(\cdot)$ and (3.2) implies

$$\underline{\dim}_b(f;s) \ge d + 1 - \beta(s), \tag{3.4}$$

at each point s of upper semi–continuity of $\beta(\cdot)$.

Proof: It should be clear that in the definitions (1.1) and (1.2) one may assume without loss of generality that the cubes $Q(t, \varepsilon)$ have rational dyadic vertices. Now, for $t_0 \in I^d$ being a point of lower semi–continuity $\alpha(\cdot)$ we have by the very definition that for small $\eta > 0$ there is ε_0 such that

$$\alpha(t) \ge \alpha(t_0) - \eta > 0 \quad \text{for} \quad t \in Q(t_0, \varepsilon_0),$$

where $Q := Q(t_0, \varepsilon_0)$ is a fixed small cube containing t_0 in its interior . This and (3.1) imply that

$$|b_{\underline{\tau}}| \le C_1 \cdot \frac{k^\gamma}{2^{\alpha \cdot k}} \quad \text{for} \quad \underline{\tau} \in D_{k,d} \cap Q,$$

with $\alpha = \alpha(t_0) - \eta$. Now, application of Theorem 2.1 to the restricted diamond basis to the cube Q gives for ε such that $Q(t_0, \varepsilon) \subset Q$

$$\overline{\dim}_b(\Gamma(f, Q(t_0, \varepsilon))) \le \overline{\dim}_b(\Gamma(f, Q)) \le d + 1 - \alpha(t_0) + \eta,$$

and consequently by (1.1), in case $t = t_0$, inequality (3.3) follows.

For the proof of (3.4) let s_0 be a point of upper semi–continuity for $\beta(\cdot)$. This implies that for given $\eta > 0$ there is an ε_0 such that

$$\beta(t) \leq \beta(s_0) + \eta \quad \text{for} \quad t \in Q(s_0, \varepsilon), 0 < \varepsilon < \varepsilon_0,$$

where $Q := Q(s_0, \varepsilon)$ is again a fixed small cube containing in its interior s_0. This and (3.2) give

$$|b_{\underline{\tau}}| \geq C_2 \cdot \frac{k^\delta}{2^{\beta \cdot k}} \quad \text{for} \quad \underline{\tau} \in D_{k,d} \cap Q,$$

where $\beta = \beta(s_0) + \eta$. Averaging the last inequality we obtain

$$\frac{1}{|D_{k,d} \cap Q|} \sum_{\underline{\tau} \in D_{k,d} \cap Q} |b_{\underline{\tau}}| \geq C_2 \cdot \frac{k^\delta}{2^{\beta \cdot k}},$$

which means that (2.7) is satisfied for the restricted to Q diamond basis. Therefore the second part of Theorem 2.1 implies that for $Q(s_0, \varepsilon) \subset Q(s_0, \varepsilon_0)$ we have

$$\underline{\dim}_b(\Gamma(f, Q(s_0, \varepsilon))) \geq d + 1 - \beta(s_0) - \eta,$$

and this completes the proof of (3.4). $\square$

Corollary 3.2 *If $\alpha(\cdot) = \beta(\cdot)$ and $\alpha(\cdot)$ is continuous at $t_0 \in I^d$, then under the assumptions of Theorem 3.1*

$$\dim_b(\Gamma(f, t_0)) = d + 1 - \alpha(t_0).$$

Corollary 3.3 *If in Theorem 3.1 $\alpha(\cdot) = \beta(\cdot)$ and $\alpha(\cdot)$ is Riemann integrable over I^d, then*

$$\dim_b(\Gamma(f, t)) = d + 1 - \alpha(t) \quad \text{for almost all} \quad t \in I^d.$$

Here almost all means with respect to Lebesgue measure.

Acknowledgement. The author would like to thank Dr. Anna Kamont for a number of helpful discussions.

References

[1] Z. Ciesielski: *Fractal functions and Schauder bases,* Comp. Math. Appl. **30** (1995), 283–291.

[2] K. Falconner: *Fractal Geometry,* Mathematical Foundations and Applications, Wiley, Chichester 1990.

[3] J. Ryll: *Schauder bases for the space of continuous functions on an n–dimensional cube,* Comment. Math. Prace Mat. **27** (1973), 201–213.

[4] Z. Semadeni: *Schauder Bases in Banach Spaces of Continuous Functions,* Lecture Notes Math. **918**, Springer, Berlin 1982.

[5] C. Tricot: *Curves and Fractal Dimension,* Springer, New York 1995.

Address:

ZBIGNIEW CIESIELSKI
Instytut Matematyczny PAN
ul. 23 Abrahama 18
Sopot
Poland

Trigonometric Approximation in Multivariate Periodic Hilbert Spaces

F.–J. Delvos

Abstract

The concept of periodic Hilbert spaces was introduced by Babuška and Prager to study universally optimal approximation of linear functionals. We extend their construction to the multivariate case and investigate approximation by trigonometric polynomials in these spaces.

Introduction

The concept of univariate periodic Hilbert spaces was introduced in Babuška [1, 2] and further investigated in Prager [6] and Delvos [3, 4, 5]. Minimum norm interpolation in these spaces was compared with trigonometric interpolation which turned out to be a universal approximation method for the whole class of these spaces. It is the objective of this paper to introduce multivariate periodic Hilbert spaces with special emphasis on two constructions based on univariate periodic Hilbert spaces. In these spaces trigonometric approximation is investigated.

1 Periodic Hilbert Spaces

As in the univariate case, a multivariate periodic Hilbert space is related to a positive *defining sequence* $D \in l_1(\mathbf{Z}^n)$. Let $m(\mathbf{Z}^n)$ be the linear space of all "measurable" sequences. We consider sequences $F \in m(\mathbf{Z}^n)$ satisfying

$$F/D^{\frac{1}{2}} \in l_2(\mathbf{Z}^n).$$

Since

$$D^{\frac{1}{2}} \in l_2(\mathbf{Z}^n)$$

we have

$$\|F\|_1 \leq \left\| F/D^{\frac{1}{2}} \right\|_2 \|D\|_1^{\frac{1}{2}}.$$

Multivariate Approximation: Recent Trends and Results; W. Haußmann, K. Jetter and M. Reimer (eds.)
Mathematical Research, Vol. 101, pp. 35–44, ISBN 3–05–501770–6
© Akademie–Verlag, Berlin 1997

These sequences form a subspace $l_1^D(\mathbf{Z}^n)$ of the convolution algebra $l_1(\mathbf{Z}^n)$. Let $\mathbf{T}$ denote the univariate torus. Recall that $l_1(\mathbf{Z}^n)$ is isomorphic to the multiplicative Wiener algebra $A(\mathbf{T}^n)$ via the *finite Fourier transform* of $f \in A(\mathbf{T}^n)$:

$$F_k := \left(\frac{1}{2\pi}\right)^n \int_{[-\pi,\pi]^n} f(t)\exp(-i(k,t))dt, \ k \in \mathbf{Z}^n.$$

Any $f \in A(\mathbf{T}^n)$ possesses an absolutely convergent Fourier series representation

$$f(x) = \sum_{k \in \mathbf{Z}^n} F_k e_k(x), \quad e_k(x) = \exp(i(x,k)).$$

$A(\mathbf{T}^n)$ is a Banach algebra with respect to the norm

$$\|f\|_a := \sum_{k \in \mathbf{Z}^n} |F_k|.$$

Let $C(\mathbf{T}^n)$ denote the Banach algebra of continuous complex-valued periodic functions equipped with the maximum norm

$$\|f\|_\infty := \max\{|f(t) : t \in \mathbf{T}^n\}.$$

$A(\mathbf{T}^n)$ is a subalgebra of $C(\mathbf{T}^n)$ satisfying

$$\|f\|_\infty \leq \|f\|_a$$

with equality if the finite Fourier transform F of f is nonnegative.

The *periodic Hilbert space* $H_D(\mathbf{T}^n)$ with defining sequence $D \in l_1(\mathbf{Z}^n)$ is the set of functions $f \in A(\mathbf{T}^n)$ satisfying $F \in l_1^D(\mathbf{Z}^n)$. Its inner product is defined by

$$(f,g)_D := \sum_{k \in \mathbf{Z}^n} F_k \overline{G_k}/D_k.$$

Any periodic Hilbert space $H_D(\mathbf{T}^n)$ is contained in the Wiener algebra and the following relations hold:

$$H_D(\mathbf{T}^n) \subseteq A(\mathbf{T}^n) \subseteq C(\mathbf{T}^n), \quad \|f\|_\infty \leq \|f\|_a \leq \|f\|_D \|D\|_1^{\frac{1}{2}}.$$

The *generating function* of the periodic Hilbert space $H_D(\mathbf{T}^n)$ is the Fourier series of its defining sequence D :

$$d(x) = \sum_{k \in \mathbf{Z}^n} D_k e_k(x), \quad e_k(x) = \exp(i(x,k)).$$

The inner product of $H_D(\mathbf{T}^n)$ is translation invariant:

$$(f(\cdot - a), g(\cdot - a))_D = (f,g)_D, \quad a \in \mathbf{R}^n.$$

Since

$$f(x) = (f, d(\cdot - x))_D \,,$$

any periodic Hilbert space is a *reproducing kernel* Hilbert space.

Before considering multivariate constructions of periodic Hilbert spaces, we consider two typical univariate examples.

Let $D_k = 2/(1 + k^2)$, $(k \in \mathbf{Z})$. Then its generating function is the periodic *exponential spline*

$$d(x) = \frac{\pi}{\sinh(\pi)} \cosh(\pi - |x|), \quad -\pi < x < \pi.$$

The associated periodic Hilbert space is the univariate *periodic Sobolev space* $W^1(\mathbf{T})$.

Choosing as defining sequence $D_k = \exp(-|k|), (k \in \mathbf{Z})$, the generating function is the *Poisson kernel*

$$d(x) = \sinh(1)/(\cosh(1) - \cos(x)),$$

and the periodic Hilbert space is a Hilbert space of *periodic holomorphic* functions.

The *anisotropic* periodic Hilbert space $H_{DP}(\mathbf{T}^n)$ is constructed from a univariate periodic Hilbert space $H_D(\mathbf{T})$ by choosing the generating sequence

$$D_k^P = D_{k_1} \ldots D_{k_n}, \quad (k_1, \ldots, k_n) \in \mathbf{Z}^n$$

$H_{DP}(\mathbf{T}^n)$ is a *tensor product* periodic Hilbert space :

$$H_{DP}(\mathbf{T}^n) = H_D(\mathbf{T}) \otimes \ldots \otimes H_D(\mathbf{T}).$$

For the defining univariate sequence $D_k = 2/(1 + k^2)$ we obtain the tensor product periodic Sobolev space

$$H_{DP}(\mathbf{T}^n) = W^{(1,\ldots,1)}(\mathbf{T}^n).$$

If the defining univariate sequence is given by $D_k = \exp(-|k|)$, we obtain a tensor product periodic Hilbert space of holomorphic functions

$$H_{D^P}(\mathbf{T}^n) = H^{(1,\ldots,1)}(\mathbf{T}^n).$$

The *isotropic* periodic Hilbert space $H_{DS}(\mathbf{T}^n)$ is constructed from a univariate periodic Hilbert space $H_D(\mathbf{T})$ by choosing the generating sequence

$$D_k^S = (D_{k_1}^{-1} + \ldots + D_{k_n}^{-1})^{-n}.$$

Since

$$(D_{k_1}^{-1} + \ldots + D_{k_n}^{-1})^{-n} \leq n^{-n} D_{k_1} \ldots D_{k_n}$$

we obtain a defining sequence. Note that $H_{DS}(\mathbf{T}^n)$ is contained in $H_{D^P}(\mathbf{T}^n)$.

For the defining univariate sequence $D_k = 2/(1+k^2)$ we obtain the multivariate periodic Sobolev space

$$H_{DS}(\mathbf{T}^n) = W^n(\mathbf{T}^n).$$

If the defining univariate sequence is given by $D_k = \exp(-|k|)$ we obtain a multivariate periodic Hilbert space of holomorphic functions

$$H_{DS}(\mathbf{T}^n) = H^n(\mathbf{T}^n).$$

2   Approximation by Fourier Partial Sums

Let $b = (b_1,\ldots,b_n) \in \mathbf{Z}_+^n$ and $[-b,b] = [-b_1,b_1] \times \ldots \times [-b_n,b_n]$. The *Fourier partial sum* projector is defined on the Wiener algebra $f \in A(\mathbf{T}^n)$ by

$$S_b(f)(x) = \sum_{k \in \mathbf{Z}^n} \chi_{[-b,b]}(k) F_k e_k(x).$$

S_b is a continuous projector on $A(\mathbf{T}^n)$ in view of

$$\|S_b(f)\|_\infty \leq \|S_b(f)\|_a \leq \|f\|_a .$$

Moreover, the *qualitative convergence* result holds:

$$\|f - S_b(f)\|_\infty \leq \|f - S_b(f)\|_a \to 0, \ |b| \to \infty. \tag{2.1}$$

The range of S_b is denoted by $A_b(\mathbf{T}^n)$. It is a tensor product space of univariate trigonometric polynomials:

$$A_b(\mathbf{T}^n) = A_{b_1}(\mathbf{T}) \otimes \ldots \otimes A_{b_n}(\mathbf{T}).$$

S_b is an *orthogonal* projector on $H_D(\mathbf{T}^n)$. For fixed x and b we define the linear functional

$$L_x^b(f) = f(x) - S_b(f)(x), \quad f \in H_D(\mathbf{T}^n).$$

Since $H_D(\mathbf{T}^n)$ is a reproducing kernel Hilbert space we have

$$f(x) - S_b(f)(x) = (f - S_b(f), d(\cdot - x) - S_b(d(\cdot - x)))_D \,.$$

Thus, the norm of L_x^b is given by

$$\left\| L_x^b \right\| = \|d - S_b(d)\|_D = \|d - S_b(d)\|_a^{\frac{1}{2}} \tag{2.2}$$

in view of $S_b(d(\cdot - x)) = S_b(d)(\cdot - x))$ and the translation invariance of the norm of $H_D(\mathbf{T}^n)$. Thus we obtain

Proposition 1. *Let $f \in H_D(\mathbf{T}^n)$. Then we have*

$$\|f - S_b(f)\|_\infty \leq \|f - S_b(f)\|_D \, \|d - S_b(d)\|_a^{\frac{1}{2}} \,. \tag{2.3}$$

Before considering applications in multivariate periodic Hilbert spaces we discuss two univariate examples. First we choose $D_k = 2/(1 + k^2)$. In this case $\|d - S_b(d)\|_a = O(b^{-1})$ which implies

$$\|f - S_b(f)\|_\infty = o(b^{-\frac{1}{2}})), \quad f \in W^1(\mathbf{T}). \tag{2.4}$$

For $D_k = \exp(-|k|)$, we have $\|d - S_b(d)\|_a = O(\exp(-b))$ and therefore

$$\|f - S_b(f)\|_\infty = o(\exp(-\frac{b}{2})), \quad f \in H^1(\mathbf{T}). \tag{2.5}$$

In the multivariate case it is convenient to assume $d(0) = 1$ and

$$b_j = \underline{b}, \quad j = 1, \ldots, n.$$

We first consider the anisotropic Hilbert space $H_{D^P}(\mathbf{T}^n)$ with generating function $d^P(x_1, \ldots, x_n) = d(x_1) \ldots d(x_n)$. Note that

$$1 - \chi_{[-b,b]}(k) \leq \sum_{j=1}^{n} (1 - \chi_{[-\underline{b},\underline{b}]}(k_j)). \tag{2.6}$$

Then

$$\left\| d^P - S_b(d^P) \right\|_{D^P}^2 \le \sum_{j=1}^{n} \sum_{k_j \in \mathbf{Z}} (1 - \chi_{[-\underline{b},\underline{b}]}(k_j)) D_{k_j} = n \left\| d - S_{\underline{b}}(d) \right\|_D^2 .$$

Thus Proposition 1 yields

Proposition 2. *Let* $f \in H_{D^P}(\mathbf{T}^n)$. *Then we have*

$$\| f - S_b(f) \|_\infty = o(\| d - S_{\underline{b}}(d) \|_a^{\frac{1}{2}}). \tag{2.7}$$

For $D_k = 2/(1 + k^2)$ we get

$$\| f - S_b(f) \|_\infty = o(\underline{b}^{-\frac{1}{2}}), \quad f \in W^{(1,\ldots,1)}(\mathbf{T}^n). \tag{2.8}$$

Similarly, for $D_k = \exp(-|k|)$ we have

$$\| f - S_b(f) \|_\infty = o(\exp(-\tfrac{1}{2}\underline{b})), \quad f \in H^{(1,\ldots,1)}(\mathbf{T}^n). \tag{2.9}$$

In the anisotropic case, the error estimate is mainly determined by the univariate error. This is different from the more complicated case of isotropic periodic Hilbert spaces $H_{DS}(\mathbf{T}^n)$. In view of (2.6) we have

$$\left\| d^S - S_b(d^S) \right\|_a = \sum_{k \in \mathbf{Z}^n} (1 - \chi_{[-b,b]}(k)) D_k^S$$

$$\le \sum_{j=1}^{n} \sum_{k \in \mathbf{Z}^n} (1 - \chi_{[-\underline{b},\underline{b}]}(k_j)) D_k^S = n \sum_{k \in \mathbf{Z}^n} (1 - \chi_{[-\underline{b},\underline{b}]}(k_1)) D_k^S .$$

Let

$$J := \sum_{k \in \mathbf{Z}^n} (1 - \chi_{[-\underline{b},\underline{b}]}(k_1)) D_k^S = \sum_{k \in \mathbf{Z}^n} (1 - \chi_{[-\underline{b},\underline{b}]}(k_1))(D_{k_1}^{-1} + \ldots + D_{k_n}^{-1})^{-n} .$$

This series J can be written as a sum of series of the following type

$$J_r := \sum_{k \in \mathbf{Z}^n} \prod_{s=1}^{r} (1 - \chi_{[-\underline{b},\underline{b}]}(k_s)) \prod_{s=r+1}^{n} \chi_{[-\underline{b},\underline{b}]}(k_s)(D_{k_1}^{-1} + \ldots + D_{k_n}^{-1})^{-n}$$

and similar series which are obtained from permutations of $k_2, \ldots, k_n$. Now

$$(D_{k_1}^{-1} + \ldots + D_{k_n}^{-1})^{-n} \leq ((D_{k_1}^{-1} + \ldots + D_{k_r}^{-1})^{-r})^{\frac{n}{r}} \leq (D_{k_1} \ldots D_{k_r})^{\frac{n}{r}}$$

implies

$$J_r = O(\underline{b}^{n-r}(\sum_{k \in \mathbf{Z}}(1 - \chi_{[-\underline{b},\underline{b}]}(k))D_k^{\frac{n}{r}})^r), \ r = 1, \ldots, n,$$

and we have proved

Proposition 3. *Let $f \in H_{DS}(\mathbf{T}^n)$. Then*

$$\|f - S_{\underline{b}}(f)\|_\infty = o\left(\left(\sup_{1 \leq r \leq n} \underline{b}^{n-r}\left[\sum_{k \in \mathbf{Z}}(1 - \chi_{[-\underline{b},\underline{b}]}(k))D_k^{\frac{n}{r}}\right]^r\right)^{\frac{1}{2}}\right). \qquad (2.10)$$

We first apply Proposition 3 to $D_k = 2/(1 + k^2)$. Then

$$\underline{b}^{n-r}(\sum_{k \in \mathbf{Z}}(1 - \chi_{[-\underline{b},\underline{b}]}(k))D_k^{n/r})^r = O(\underline{b}^{n-r}(\int_{\underline{b}}^\infty x^{-2n/r}dx)^r)$$

$$= O(\underline{b}^{n-r}(\underline{b}^{1-2n/r})^r) = O(\underline{b}^{-n}),$$

and we have

$$\|f - S_{\underline{b}}(f)\|_\infty = o(\underline{b}^{-\frac{n}{2}}), \quad f \in W^n(\mathbf{T}^n). \qquad (2.11)$$

Assume next $D_k = \exp(-|k|)$. Then

$$\underline{b}^{n-r}(\sum_{k \in \mathbf{Z}}(1 - \chi_{[-\underline{b},\underline{b}]}(k))D_k^{n/r})^r$$

$$= O(\underline{b}^{n-r}(\int_{\underline{b}}^\infty \exp(-\frac{n}{r}x)dx)^r) = O(\underline{b}^{n-1}\exp(-n\underline{b})),$$

and we get

$$\|f - S_{\underline{b}}(f)\|_\infty = o(\underline{b}^{\frac{n-1}{2}}\exp(-n\frac{\underline{b}}{2})), \quad f \in H^n(\mathbf{T}^n). \qquad (2.12)$$

3   Periodization and Interpolation

Assume $f \in A(\mathbf{T}^n)$, i.e.,

$$f(x) = \sum_{k \in \mathbf{Z}^n} F_k e_k(x), \quad F \in l_1(\mathbf{Z}^n).$$

Given $b \in \mathbf{N}^n$ define $M_j = 2b_j + 1$, $j = 1, \ldots, n$. The $M - periodization$ of $F \in l_1(\mathbf{Z}^n)$ is defined by

$$F_k^M = \sum_{r \in \mathbf{Z}^n} F_{k+rM}, \quad k \in \mathbf{Z}^n.$$

Then we have

$$f(\frac{2\pi}{M}j) = \sum_{k \in [-b,b] \cap \mathbf{Z}^n} F_k^M e_k(\frac{2\pi}{M}j), \quad j \in [-b,b].$$

The trigonometric polynomial related to the $M-$periodization of F,

$$T_b(f)(x) = \sum_{k \in [-b,b] \cap \mathbf{Z}^n} F_k^M e_k(x),$$

interpolates f uniquely at the points $\frac{2\pi}{M}j$, $j \in [-b,b]$:

$$T_b(f)(\frac{2\pi}{M}j) = f(\frac{2\pi}{M}j), \quad j \in [-b,b].$$

T_b is a bounded projector on $A(\mathbf{T}^n)$ in view of

$$\|T_b(f)\|_a \leq \|f\|_a.$$

The *trigonometric interpolation projector* T_b provides a natural approximation of the projector S_b. Since

$$S_b(f)(x) = \sum_{k \in \mathbf{Z}^n} \chi_{[-b,b]}(k) F_k e_k(x),$$

$$T_b(f)(x) = \sum_{k \in \mathbf{Z}^n} \chi_{[-b,b]}(k) F_k^M e_k(x),$$

we have

$$|T_b(f)(x) - S_b(f)(x)| \leq \sum_{k \in [-b,b]} \left| \sum_{r \in \mathbf{Z}^n} F_{k+rM} - F_k \right| = \|S_b(f) - T_b(f)\|_a$$

$$= \sum_{k\in[-b,b]} \left| \sum_{r\in\mathbf{Z}^n, r\neq 0} F_{k+rM} \right| \leq \sum_{k\in[-b,b]} \sum_{r\in\mathbf{Z}^n, r\neq 0} |F_{k+rM}| = \|f - S_b(f)\|_a \ .$$

This implies

Proposition 4. *Let $f \in A(\mathbf{T}^n)$. Then*

$$\|T_b(f) - S_b(f)\|_\infty \leq \|T_b(f) - S_b(f)\|_a \leq \|f - S_b(f)\|_a \,,$$

$$\|f - T_b(f)\|_\infty \leq \|f - T_b(f)\|_a \leq 2\|f - S_b(f)\|_a \ .$$

Note that for $f \in H_D(\mathbf{T}^n)$

$$\|f - S_b(f)\|_a = \sum_{k\in[-b,b]} \sum_{r\in\mathbf{Z}^n, r\neq 0} |F_{k+rM}| = \sum_{k\in\mathbf{Z}^n} (1 - \chi_{[-b,b]}(k))\sqrt{D_k}\,|F_k|\,/\sqrt{D_k}$$

$$\leq \Big(\sum_{k\in\mathbf{Z}^n} (1 - \chi_{[-b,b]}(k))\,|F_k|^2\,/D_k \Big)^{\frac{1}{2}} \Big(\sum_{k\in\mathbf{Z}^n} (1 - \chi_{[-b,b]}(k))D_k \Big)^{\frac{1}{2}} .$$

Thus, we have shown

Proposition 5. *Let $f \in H_D(\mathbf{T}^n)$. Then*

$$\|f - S_b(f)\|_a \leq \|f - S_b(f)\|_D\,\|d - S_b(d)\|_a^{\frac{1}{2}} \ .$$

In view of Propositions 4 and 5, the error estimates for S_b as established in section 2 hold equally for T_b. In particular, we have in the anisotropic case:

Proposition 6. *Let $f \in H_{D^P}(\mathbf{T}^n)$. Then*

$$\|f - T_b(f)\|_\infty = o(\|d - S_{\underline{b}}(d)\|_a^{\frac{1}{2}}).$$

As special cases we obtain (see (2.8) and (2.9))

$$\|f - T_b(f)\|_\infty = o(\underline{b}^{-\frac{1}{2}}), \ f \in W^{(1,\dots,1)}(\mathbf{T}^n),$$

$$\|f - T_b(f)\|_\infty = o(exp(-\frac{1}{2}\underline{b})), \ f \in H^{(1,\dots,1)}(\mathbf{T}^n).$$

In the isotropic case the interpolatory counterpart of Proposition 3 reads

Proposition 7. *Let $f \in H_{D^s}(\mathbf{T}^n)$. Then we have*

$$\|f - T_b(f)\|_\infty = o\Big(\big(\sup_{1 \le r \le n} \underline{b}^{n-r}[\sum_{k \in \mathbf{Z}}(1 - \chi_{[-\underline{b},\underline{b}]}(k))D_k^{\frac{n}{r}}]^r\big)^{\frac{1}{2}}\Big).$$

As special cases we get (see (2.11) and (2.12))

$$\|f - T_b(f)\|_\infty = o(\underline{b}^{-\frac{n}{2}}), \quad f \in W^n(\mathbf{T}^n),$$

$$\|f - T_b(f)\|_\infty = o(\underline{b}^{\frac{n-1}{2}} \exp(-\frac{n}{2}\underline{b})), \quad f \in H^n(\mathbf{T}^n).$$

References

[1]   I. Babuška: *Über universal optimale Quadraturformeln I*, Apl. Math. **13** (1968), 304–308.

[2]   I. Babuška: *Über universal optimale Quadraturformeln II*, Apl. Math. **13** (1968), 388–404.

[3]   F.-J. Delvos: *Approximation by optimal periodic interpolation*, Apl. Math. **35** (1990), 451–457.

[4]   F.-J. Delvos: *Approximation properties of periodic interpolation by translates of one function*, Modélisation Mathematique et Analyse Numérique **28** (1994), 177–188 .

[5]   F.-J. Delvos: *Mean square approximation by optimal periodic interpolation*, Apl. Math. **40** (1995), 451–457.

[6]   M. Prager: *Universally optimal approximation of functionals*, Apl. Math. **24** (1979), 406–420.

Address:

FRANZ–JÜRGEN DELVOS
University of Siegen
FB Mathematik I
Hölderinstraße 3
D–57068 Siegen
Germany

Harmonic Approximation and Boundary Behaviour

S. J. Gardiner

Abstract

The theory of uniform approximation by harmonic functions on non–compact sets has advanced substantially in the last few years. This paper presents an introduction to the subject, including some very recent results and also a sample application concerning the boundary cluster sets of harmonic functions.

1   Introduction

The classical approximation theorem of Weierstrass, which says that a continuous function on a closed bounded interval can be uniformly approximated by polynomials, has the following elegant generalization [5].

Carleman's Theorem. *Given any continuous function $f : \mathbb{R} \to \mathbb{C}$ and any continuous function $\epsilon : \mathbb{R} \to (0, 1]$, there is an entire function g such that*

$$|g(x) - f(x)| \le \epsilon(x) \quad (x \in \mathbb{R}).$$

The two new elements here are that the approximation takes place on an unbounded set, and that better–than–uniform approximation is achievable. Both of these themes will occur in this survey, but we will primarily be interested in the ability to approximate on non–compact sets. Such results are interesting not just because they require new methods and display new features, but also because it is precisely the non–compactness of the set on which the approximation takes place that renders them so useful in applications.

Among many contributions to the theory of holomorphic approximation on non–compact sets, we highlight papers by Alice Roth [30, 31], Arakelyan [1, 2], Nersesyan [26] and Stray [34]. A helpful account of some of these results may be found in the book by Gaier [13]. This paper is concerned with approximation by harmonic functions on such sets, and here the pioneering work was done by

Multivariate Approximation: Recent Trends and Results; W. Haußmann, K. Jetter and M. Reimer (eds.)
Mathematical Research, Vol. 101, pp. 45–62, ISBN 3–05–501770–6
Ⓒ Akademie–Verlag, Berlin 1997

Gauthier, Goldstein and Ow [21, 22] and Gauthier and Hengartner [23] in the early 1980's. An account of this subject may be found in the author's recent book [16]. The present paper will begin with some of the key theorems in [16], but will then proceed to describe several further results that have been obtained in the past two years.

Let Ω be an open set in Euclidean space $\mathbb{R}^n$ ($n \geq 2$) and let $E \subseteq \Omega$ be closed relative to Ω. Also, let $\mathcal{H}(\Omega)$ denote the collection of all functions which are harmonic on Ω. We are interested in uniformly approximating functions on E by members of $\mathcal{H}(\Omega)$. Since both continuity and harmonicity are preserved by uniform convergence, the functions to be approximated must at least belong to $C(E) \cap \mathcal{H}(E^\circ)$. Two fundamental questions are as follows:

(a) Which sets E have the property that every function in $C(E) \cap \mathcal{H}(E^\circ)$ can be uniformly approximated by functions in $\mathcal{H}(\Omega)$?

(b) Given a particular set E, which members of $C(E) \cap \mathcal{H}(E^\circ)$ can be uniformly approximated by functions in $\mathcal{H}(\Omega)$?

Solutions to both these questions will be described below. Further, we will discuss how the characterization in (a) changes when we restrict the class of functions to be approximated to:

(i) functions which are harmonic on some neighbourhood of E;

(ii) functions which are continuous on the closure of E in compactified space and are harmonic on E°;

(iii) bounded members of $C(E) \cap \mathcal{H}(E^\circ)$.

One of the most rewarding aspects of harmonic approximation theorems is the potential for applications, some of which are surprising. For example, despite the apparent lack of connection between the two topics, there is an application to the question of uniqueness for the Radon transform. To be more specific, suppose that a function $f : \mathbb{R}^n \to \mathbb{R}$ is integrable on each $(n-1)$–dimensional hyperplane P. The Radon transform $\hat{f}$ of f is defined on the collection of all such hyperplanes P by $\hat{f}(P) = \int_P f$, where the integral is with respect to $(n-1)$–dimensional measure. The uniqueness question mentioned above is this: is there a non–constant continuous function f on $\mathbb{R}^n$ such that $\hat{f} \equiv 0$? Remarkably, approximation theorems can be used to show that there exist non–constant harmonic functions f on $\mathrm{I\!R}^n$ with this property! (See [36] when $n = 2$ and [3] when $n \geq 3$; or see [16, Sect. 8.1].)

A deeper application concerns the Dirichlet problem on unbounded open sets Ω in $\mathrm{I\!R}^n$. Which open sets Ω have the property that, for each f in $C(\partial\Omega)$, there is a harmonic function h on Ω such that $h(x) \to f(y)$ as $x \to y$ for each (regular) boundary point y of Ω? The difficulty here is that, since $\partial\Omega$ may be non–compact, the function f may have wild behaviour near infinity. This version of the Dirichlet problem was first studied by R. Nevanlinna [28], who showed that it could be solved in the half–plane. Harmonic approximation results, of the type described below, have now been used to give a purely topological description of the open sets Ω with this property (see [14] or [16, Chapter 7]).

Approximation results can also be used to construct harmonic functions with pathological behaviour. Some examples of these can be found in [16, Chapter 8]. In Section 6 of this paper we will outline a further construction which sheds light on the boundary cluster sets of harmonic functions.

2   Holomorphic Approximation

Before stating the main results concerning harmonic approximation, we recall some of the background in the theory of holomorphic approximation. If $A \subseteq \mathbb{C}$, then let $\mathrm{Hol}\,(A)$ denote the collection of all functions which are holomorphic on some open set which contains A. Also, if Ω is an open set in $\mathbb{C}$ (or, later, in $\mathrm{I\!R}^n$), let $\Omega^* = \Omega \cup \{\mathcal{A}\}$ denote the Alexandroff (or one–point) compactification of Ω. The following classical result [32] marks the starting point in the history of holomorphic approximation.

Runge's Theorem. *Let $\Omega \subseteq \mathbb{C}$ be open and $K \subset \Omega$ be compact. The following are equivalent:*

(a) *for every f in $Hol(K)$ and every positive number ϵ, there exists g in $Hol(\Omega)$ such that $|g - f| < \epsilon$ on K;*

(b) $\Omega^* \backslash K$ *is connected.*

We remark that condition (b) is equivalent to saying that "holes" in K are matched by "holes" in Ω: or more precisely, each bounded component of $\mathbb{C} \backslash K$ intersects $\mathbb{C} \backslash \Omega$. The situation concerning approximation on non–compact sets is completely described in the next result [1, 2].

Arakelyan's Theorem. *Let $\Omega \subseteq \mathbb{C}$ be open and $E \subseteq \Omega$ be closed relative to Ω. The following are equivalent :*

   (a) *for every f in $C(E) \cap \text{Hol}(E^{\circ})$ and every positive number ϵ, there exists g in $\text{Hol}(\Omega)$ such that $|g - f| < \epsilon$ on E;*

   (b) *$\Omega^* \backslash E$ is connected and locally connected.*

Condition (b) above requires some explanation. Recall that a topological space is called *locally connected* if, for each point x and each neighbourhood N of x, there is a connected neighbourhood N' of x such that $N' \subseteq N$. In the context of Arakelyan's Theorem, the only point x where this condition can fail is the point $x = \mathcal{A}$. For example, let $\Omega = \mathbb{C}$ and $E = \partial S$, where

$$S = \left\{ x + iy : 0 < x < 1 \text{ and } y < \left(\frac{1}{x}\right) \sin^2 \left(\frac{1}{x}\right) \right\}.$$

Then $\mathbb{C}^* \backslash E$ is connected, but not locally connected: the set $\mathbb{C}^* \backslash (E \cup [0,1])$ is a neighbourhood of $\mathcal{A}$ (the point at infinity) which does not contain any connected neighbourhood of $\mathcal{A}$.

3   Runge and Arakelyan Pairs

If $A \subseteq \mathbb{R}^n$, then let $\mathcal{H}(A)$ denote the collection of all functions which are harmonic on some open set which contains A. Also, let $\Omega \subseteq \mathbb{R}^n$ be open and $E \subseteq \Omega$ be closed relative to Ω. Motivated by Section 2, we will call (Ω, E) a *Runge pair (for harmonic functions)* if, for each $u \in \mathcal{H}(E)$ and each positive number ϵ, there exists v in $\mathcal{H}(\Omega)$ such that $|v - u| < \epsilon$ on E.

An early result on harmonic approximation is due to J. L. Walsh [35], who modified Runge's arguments to show that, if K is compact set in $\mathbb{R}^n$ such that $\mathbb{R}^n \backslash K$ is connected, then $(\mathbb{R}^n, K)$ is a Runge pair. More recently, Gauthier, Goldstein and Ow [21, 22] showed that, if $\Omega^* \backslash E$ is connected and locally connected, then (Ω, E) is a Runge pair. However, unlike the holomorphic case, the converses of these results do not hold. It turns out that the characterization of Runge pairs for harmonic functions is more delicate, and involves both topological and potential theoretic (thinness) conditions.

One way of describing thinness is in terms of capacity. Thus, if $n \geq 3$, then a set A in $\mathbb{R}^n$ is said to be *thin* at a point y if the Wiener criterion holds:

$$\sum_{k=1}^{\infty} 2^{k(n-2)} C^* \left(\left\{ x \in A : 2^{-k-1} \leq |x - y| \leq 2^{-k} \right\} \right) < +\infty,$$

where C^* denotes outer Newtonian capacity. (A similar criterion involving logarithmic capacity applies when $n = 2$.) An example of a set A which satisfies this condition at 0 is the classical Lebesgue spine defined by

$$A = \{(x_1, x_2, x_3) \in \mathbb{R}^3 : x_1 > 0 \text{ and } x_2^2 + x_3^2 < \exp(-1/x_1)\}. \tag{3.1}$$

An equivalent way of defining thinness is in terms of potentials. Suppose again, for the sake of convenience, that $n \geq 3$, and given a measure μ on $\mathbb{R}^n$, let

$$G\mu(x) = \int |x - y|^{2-n} d\mu(y) \quad (x \in \mathbb{R}^n).$$

(We consider only those measures μ for which $G\mu \not\equiv +\infty$.) Then, for any point z,

$$\liminf_{x \to z} \frac{G\mu(x)}{|x - z|^{2-n}} = \mu(\{z\}).$$

It can be shown that a set A is thin at z if and only if there is a measure μ such that

$$\liminf_{x \to z, x \in A} \frac{G\mu(x)}{|x - z|^{2-n}} > \mu(\{z\}).$$

(For these and other facts concerning thin sets, see Doob [10, 1.XI].)

If V is a connected component of $\Omega \backslash E$ such that $\overline{V}$ is a compact subset of Ω, then we call V an Ω–*hole* of E. Let $\hat{E}$ denote the union of E with all the Ω–holes of E. We can now state the complete characterization of Runge pairs (see [15] or [16, Theorem 3.15]).

Theorem 3.1. *Let $\Omega \subseteq \mathbb{R}^n$ be open and $E \subseteq \Omega$ be closed relative to Ω. Then (Ω, E) is a Runge pair if and only if :*

(i) *$\Omega \backslash \hat{E}$ and $\Omega \backslash E$ are thin at the same points of E, and*

(ii) *for each compact set $K \subset \Omega$ there is a compact set $L \subset \Omega$ which contains all Ω–holes of $E \cup K$ whose closure intersects K.*

Conditions (i) and (ii) need some further comment. Firstly, condition (i) implies that $\partial\hat{E} = \partial E$, for otherwise it can be shown that there will be at least one point x of $\partial E \backslash \partial\hat{E}$ such that $\Omega \backslash E$ is non–thin at x whereas $\Omega \backslash \hat{E}$ is thin at x. In fact, when $n = 2$, condition (i) is equivalent to the equation $\partial\hat{E} = \partial E$. However, the picture is not so simple when $n \geq 3$. For example, let $\Omega = \mathbb{R}^3$ and $E = \partial(B \backslash A)$, where B is the closed unit ball and A is the Lebesgue spine defined in (3.1). Then $\partial\hat{E} = \partial E$, but condition (i) clearly fails. In general, condition (i) implies that any Ω–hole V of E must be regular; that is, $\mathbb{R}^n \backslash V$ must be non–thin at each point of ∂V. However, when $n \geq 3$, it is possible for condition (i) to fail even if $\partial\hat{E} = \partial E$ and each Ω–hole of E is regular, as the following example shows.

Example 3.2. Let $n \geq 4$ (slight modifications are necessary when $n = 3$), let $\Omega = \mathbb{R}^n$, let $(y^{(k)})$ be a dense sequence in $[0,1]^{n-1}$, and let

$$u(x',t) = \sum_k 2^{-k}|x' - y^{(k)}|^{3-n} \quad (x' \in \mathbb{R}^{n-1}, t \in \mathbb{R})$$

and

$$E = \partial([0,1]^{n-1} \times [-1,0]) \cup \{(x',t) \in [0,1]^{n-1} \times (0,1] : u(x',t) \leq t^{2-n}\}.$$

Then u is a potential on $\mathbb{R}^n$ with associated measure μ comprising some multiple of one–dimensional measure on each line $\{y^{(k)}\} \times \mathbb{R}$, and the lower semi-continuity of u ensures that E is closed and hence compact. Since $u = +\infty$ on each of these lines, it is easy to see that E has exactly one $\mathbb{R}^n$–hole, namely the cube $V = (0,1)^{n-1} \times (-1,0)$ which is regular, and that $\partial\hat{E} = \partial E$. However, if $z \in (0,1)^{n-1} \times \{0\}$, then

$$u(x',t) > t^{2-n} \geq |(x',t) - z|^{2-n} \quad ((x',t) \in (0,1)^n \backslash E),$$

so

$$\liminf_{x \to z, x \notin \hat{E}} \frac{u(x)}{|x - z|^{2-n}} \geq 1 > 0 = \mu(\{z\}).$$

Thus $\mathbb{R}^n \backslash \hat{E}$ is thin at each such point z, whereas $\mathbb{R}^n \backslash E$, which contains the cube V, is not.

We will refer to condition (ii) of Theorem 3.1 as the K, L–*condition*. This condition is equivalent to the combination of the following two conditions:

(a) $\Omega^* \backslash \hat{E}$ is locally connected, and

(b) for every compact set $K \subset \Omega$, the collection $\{V_\alpha\}$ of Ω–holes of E which intersect K are such that $\overline{\cup_\alpha V_\alpha}$ is a compact subset of Ω.

However, the K, L–condition does not imply that $\Omega^* \backslash E$ is locally connected. For example, if $E = \cup_k \partial B_k$, where (B_k) is a sequence of pairwise disjoint closed balls of radius 1 in $\mathbb{R}^n$, then $(\mathbb{R}^n, E)$ would satisfy the K, L–condition even though $(\mathbb{R}^n)^* \backslash E$ is not locally connected.

Again motivated by Section 2 we will call (Ω, E) an *Arakelyan pair (for harmonic functions)* if, for each $u \in C(E) \cap \mathcal{H}(E^\circ)$ and each positive number ϵ, there exists $v \in \mathcal{H}(\Omega)$ such that $|v - u| < \epsilon$ on E. Clearly an Arakelyan pair is a Runge pair which has the additional property that functions in $C(E) \cap \mathcal{H}(E^\circ)$ can be uniformly approximated by functions in $\mathcal{H}(E)$. We now recall the following result which is due to Keldyš [24] and Deny [8] in the case where E is compact, and was subsequently extended to non–compact sets by Labrèche [25].

Theorem A. *Let $\Omega \subseteq \mathbb{R}^n$ be open and $E \subseteq \Omega$ be closed relative to Ω. The following are equivalent :*

(a) *for each $u \in C(E) \cap \mathcal{H}(E^\circ)$ and each positive number ϵ, there exists $v \in \mathcal{H}(E)$ such that $|v - u| < \epsilon$ on E;*

(b) *$\Omega \backslash E$ and $\Omega \backslash E^\circ$ are thin at the same points of E.*

Combining Theorem A with Theorem 3.1, we immediately obtain the desired characterization of Arakelyan pairs.

Theorem 3.3. *Let $\Omega \subseteq \mathbb{R}^n$ be open and $E \subseteq \Omega$ be closed relative to Ω. Then (Ω, E) is an Arakelyan pair if and only if :*

(i) *$\Omega \backslash \hat{E}$ and $\Omega \backslash E^\circ$ are thin at the same points of E, and*

(ii) *the K, L–condition holds.*

If a set E in $\mathbb{R}^2$ is thin at a point y, then there are arbitrarily small circles centred at y which do not intersect E. This special feature of thinness in the plane yields a rather simple characterization of Runge and Arakelyan pairs.

Theorem 3.4. *Let $\Omega \subseteq \mathbb{R}^2$ be open and $E \subseteq \Omega$ be closed relative to Ω. The following are equivalent :*

(a) *(Ω, E) is a Runge pair;*

(b) *(Ω, E) is an Arakelyan pair;*

(c) *$\partial \hat{E} = \partial E$ and the K, L–condition holds.*

Before closing this section we remark (see [16, Theorem 3.15]) that every Runge pair (Ω, E) has the following apparently stronger property, which permits better–than–uniform approximation and is often useful in applications:

For each $u \in \mathcal{H}(E)$ and each positive superharmonic function w defined on some open set containing E, there exists v in $\mathcal{H}(\Omega)$ such that $|v - u| < w$ on E.

4   Accessibility and Approximation

In Section 3 we described the pairs (Ω, E) which have the property that all members of $\mathcal{H}(E)$ (respectively, of $C(E) \cap \mathcal{H}(E^\circ)$) can be uniformly approximated on E by members of $\mathcal{H}(\Omega)$. We will now deal with further classes of functions on E. Let $C_e(E)$ denote the collection of continuous functions on E which have a continuous extension to the closure of E in compactified space $(\mathbb{R}^n)^*$. We characterize below those pairs (Ω, E) which have the property that all members of $C_e(E) \cap \mathcal{H}(E^\circ)$ (respectively, all bounded members of $C(E) \cap \mathcal{H}(E^\circ)$) can be uniformly approximated by members of $\mathcal{H}(\Omega)$. To do this, we need a further topological notion.

Let ω be a connected open subset of Ω, and let $\mathcal{A}$ again be the Alexandroff point for Ω. Then $\mathcal{A}$ is said to be *accessible* from ω if there is a continuous function $f : [0, +\infty) \to \omega$ such that $f(t) \to \mathcal{A}$ as $t \to +\infty$. We denote by $\tilde{E}$ (or, sometimes, by $(E)^{\widetilde{}}$) the union of E with all the connected components of $\Omega \backslash E$ from which $\mathcal{A}$ is not accessible.

Clearly $\hat{E} \subseteq \tilde{E}$, but the inclusion may be strict. To see this, let $\Omega = (-2, 2)^2$ and $E = \Omega \cap \partial\omega$, where

$$\omega = \{(x_1, x_2) \in \mathbb{R}^2 : 0 < x_1 < 1 \text{ and } -1 < x_2 < (2 - x_1)\sin^2(1/x_1)\}.$$

Then $\hat{E} = E$, since $\overline{\omega}$ is not a subset of Ω. However, $\tilde{E} = E \cup \omega$.

One reason why the notion of accessibility has a rôle to play here lies in the following generalization of the maximum principle. This result is implied by work of Fuglede [12, Section 4]; an elementary proof of it may be found in [6].

Lemma A. *Let ω be a connected open subset of Ω from which $\mathcal{A}$ is not accessible. If s is a subharmonic function on ω and*

$$\limsup_{x \to y} s(x) \leq a \quad (y \in \partial\omega \cap \Omega),$$

then $s \leq a$ on ω.

Let $f|_A$ denote the restriction of a function f to a set A. The main results of this section are as follows.

Theorem 4.1. *Let $\Omega \subset \mathbb{R}^n$ be open and $E \subseteq \Omega$ be closed relative to Ω. The following are equivalent :*

(a) *for each $u \in C_e(E) \cap \mathcal{H}(E^\circ)$ and each positive number ϵ, there exists $v \in \mathcal{H}(\Omega)$ such that $|v - u| < \epsilon$ on E;*

(b) *for each $u \in C_e(E) \cap \mathcal{H}(E^\circ)$ and each positive number ϵ, there exists $v \in \mathcal{H}(\Omega)$ such that $v|_E \in C_e(E)$ and $|v - u| < \epsilon$ on E;*

(c) *$\Omega \setminus \tilde{E}$ and $\Omega \setminus E^\circ$ are thin at the same points of E.*

Theorem 4.2. *Let $\Omega \subseteq \mathbb{R}^n$ be open and $E \subseteq \Omega$ be closed relative to Ω. The following are equivalent :*

(a) *for each bounded function u in $C(E) \cap \mathcal{H}(E^\circ)$ and each positive number ϵ, there exists $v \in \mathcal{H}(\Omega)$ such that $|v - u| < \epsilon$ on E;*

(b)   (i) *$\Omega \setminus \tilde{E}$ and $\Omega \setminus E^\circ$ are thin at the same points of E, and*

     (ii) *for each compact set $K \subset \Omega$ there is a compact set $L \subset \Omega$ such that $\Omega \setminus \tilde{E}$ and $\Omega \setminus (\tilde{E} \cup K)\tilde{\ }$ are thin at the same points of $E \setminus L$.*

Proofs of Theorems 4.1 and 4.2 (and several variants of Theorem 4.2) may be found in [18] and [17] respectively. When $n = 2$, condition (c) of Theorem 4.1 (and condition (b) (i) of Theorem 4.2) simplifies to $\partial \tilde{E} = \partial E$, and condition (b) (ii) of Theorem 4.2 simplifies to:

(*) *for each compact set $K \subset \Omega$ there is a compact set $L \subset \Omega$ such that* $\partial \tilde{E} \setminus L = \partial((\tilde{E} \cup K)\tilde{\ }) \setminus L$.

This special case of Theorem 4.2 is essentially due to Nersesyan [27]. Condition (*) is illustrated by the following example.

Example 4.3. Let $\Omega = \mathbb{R}^2$, and for each m in $\mathbb{N}$, let $W_m = ((m+1)^{-1}, m^{-1}) \times (0, m)$. Further, let

$$F_1 = \partial\left(\left(\bigcup_{m=1}^{\infty} W_m\right) \cup (-\infty, 1)^2\right), \qquad F_2 = \partial\left(\left(\bigcup_{m=1}^{\infty} W_{2m}\right) \cup (-\infty, 1)^2\right),$$

$$F_3 = F_2 \cup \left\{\left(\frac{4m+1}{4m(2m+1)}, m\right) \in \mathbb{R}^2 : m \in \mathbb{N}\right\}.$$

Then condition $(*)$ is satisfied when $E = F_2$. However, it is not satisfied when $E = F_1$ or $E = F_3$, as can be seen by taking K to be $[0,1] \times \{1\}$.

In the case of holomorphic approximation, it was shown by Stray [33] that the pairs (Ω, E) for which every bounded member of $C(E) \cap \mathrm{Hol}(E^\circ)$ can be uniformly approximated by members of $\mathrm{Hol}(\Omega)$ are precisely the Arakelyan pairs (i.e. the pairs described in Arakelyan's Theorem above). It is interesting to note from Theorems 3.3 and 4.2 that the corresponding statement about harmonic approximation is false.

5 Approximable Functions

Suppose that (Ω, E) is not an Arakelyan pair. In that case not every function in $C(E) \cap \mathcal{H}(E^\circ)$ can be uniformly approximated by members of $\mathcal{H}(\Omega)$. Thus we are led to investigate which functions can be so approximated. (The corresponding problem for holomorphic approximation was posed by Arakelyan [2] and solved by Stray [34].)

Let $A(E)$ denote the collection of real–valued functions on E which are uniformly approximable by functions harmonic on Ω; that is, $u \in A(E)$ if and only if, for each $\epsilon > 0$, there exists $v_\epsilon \in \mathcal{H}(\Omega)$ such that $|v_\epsilon - u| < \epsilon$ on E. If, in addition, the approximating functions can be chosen so that the restrictions $v_\epsilon|_E$ belong to $C_e(E)$, then we write $u \in A_e(E)$. Clearly $A_e(E) \subseteq C_e(E)$.

From Theorem 4.1 (and the remarks following Theorem 4.2) it is clear that, when $n = 2$, any function in $C_e(\tilde{E}) \cap \mathcal{H}((\tilde{E})^\circ)$ is also in $A(E)$. This gives one of the implications in the following result (see [19]).

Theorem 5.1. *Let $\Omega \subseteq \mathrm{I\!R}^2$ be open, let $E \subseteq \Omega$ be closed relative to Ω, and let $u : E \to \mathrm{I\!R}$. Then $u \in A(E)$ if and only if $u = v + h$ on E for some $v \in C_e(\tilde{E}) \cap \mathcal{H}((\tilde{E})^\circ)$ and $h \in \mathcal{H}(\Omega)$.*

However, it is again the case that more subtlety is required in higher dimensions. Recall that superharmonic functions are lower semi–continuous, but not necessarily continuous. The fine topology on $\mathrm{I\!R}^n$ is defined to be the coarsest topology with respect to which all superharmonic functions on $\mathrm{I\!R}^n$ are continuous. It is strictly finer than the Euclidean topology. Finely open sets (that is, sets which are open in the fine topology on $\mathrm{I\!R}^n$) can be shown to relate to the notion of thinness in the following way: a set ω is finely open if and only if $\mathrm{I\!R}^n \backslash \omega$ is thin at every point of ω. Such sets may have empty Euclidean interior. Finely harmonic functions are a natural generalization of harmonic functions to functions defined on finely open sets. We refer to Doob [10, 1.XI] for an account of the fine topology, and to Fuglede [11] for an account of finely harmonic functions. The general result concerning approximable functions is as follows (see [19]).

Theorem 5.2. *Let $\Omega \subseteq \mathrm{IR}^n$ be open and $E \subseteq \Omega$ be closed relative to Ω, and let $u : E \to \mathrm{IR}$. The following are equivalent :*

(i) $u \in A(E)$;

(ii) $u = v + h$ on E for some $v \in A_e(\tilde{E})$ and $h \in \mathcal{H}(\Omega)$;

(iii) $u = v + h$ on E for some $v \in C_e(\tilde{E})$ which is finely harmonic on the fine interior of $\tilde{E}$ and some $h \in \mathcal{H}(\Omega)$.

Corollary. *If $u \in A(E)$, then for each $\epsilon > 0$ there exists w in $\mathcal{H}(\Omega)$ such that $(w - u)|_E \in C_e(E)$ and $|w - u| < \epsilon$ on E.*

To see why the Corollary holds, let v and h be as in condition (ii) of Theorem 5.2, choose $v_\epsilon \in \mathcal{H}(\Omega)$ such that $v_\epsilon|_{\tilde{E}} \in C_e(\tilde{E})$ and $|v_\epsilon - v| < \epsilon$ on $\tilde{E}$, and define $w = v_\epsilon + h$ (whence $w - u = v_\epsilon - v$ on E).

There are two main ingredients in the proof of Theorem 5.2. To describe the first, we recall that Debiard and Gaveau [7] have shown that, if K is a compact set in IR^n, then the functions on K which can be uniformly approximated by members of $\mathcal{H}(K)$ are precisely those which are continuous on K and finely harmonic on the fine interior of K. This result is generalized in part (a) of the theorem below. Part (b) shows that (iii) $\Rightarrow$ (i) in Theorem 5.2.

Theorem 5.3. *Let $\Omega \subseteq \mathrm{IR}^n$ be open and $E \subseteq \Omega$ be closed relative to Ω. Suppose that $u \in C_e(E)$ and that u is finely harmonic on the fine interior of E. Then :*

(a) *for each $\epsilon > 0$, there exist an open set U in $(\mathrm{IR}^n)^*$ containing the closure of E in $(\mathrm{IR}^n)^*$ and a function w in $C((\mathrm{IR}^n)^*) \cap \mathcal{H}(U \cap \Omega)$ such that $|w - u| < \epsilon$ on E;*

(b) *under the additional assumption that $\tilde{E} - E$, for each $\epsilon > 0$ there exists a function w in $\mathcal{H}(\Omega)$ such that $w|_E \in C_e(E)$ and $|w - u| < \epsilon$ on E.*

It is easily seen that any member of $A_e(\tilde{E})$ is finely harmonic on the fine interior of $\tilde{E}$, so (ii) $\Rightarrow$ (iii) in Theorem 5.2. For the final implication, some careful estimation is involved in first establishing the following lemma.

Lemma 5.4. *Let K be a compact subset of Ω, let $a > 3$, let $w \in \mathcal{H}(\Omega)$ and suppose that $|w| < 1$ on E. Then there is a function s in $\mathcal{H}(\Omega)$ such that $|s - w| < a$ on $K \cup \tilde{E}$ and $s|_{\tilde{E}} \in C_e(\tilde{E})$.*

Given Lemma 5.4, we argue as follows to show that (i) $\Rightarrow$ (ii) in Theorem 5.2. Let $u \in A(E)$ and let (K_m) be a sequence of compact sets such that $K_m \subset K^{\circ}_{m+1}$ for each m and $\cup_m K_m = \Omega$. Since $u \in A(E)$, there exists u_0 in $\mathcal{H}(\Omega)$ such that $|u_0 - u| < 1$ on E. Then $u_0 - u \in A(E)$ so there exists u_1 in $\mathcal{H}(\Omega)$ such that $|u_1 - (u - u_0)| < 2^{-1}$ on E. We proceed inductively. Given $u_0, \dots, u_{m-1}$, we choose u_m in $\mathcal{H}(\Omega)$ such that

$$|u_m - (u - u_0 - u_1 - \cdots - u_{m-1})| < 2^{-m} \text{ on } E. \tag{5.1}$$

We thus have $|u_m| < 3/2^m$ on E, and hence on $\tilde{E}$ by Lemma A.

For each m in $\mathbb{N}$ we apply Lemma 5.4 with $w = 3^{-1}2^m u_m$ and $K = K_m$ and $a = 10/3$ to obtain s_m in $\mathcal{H}(\Omega)$ such that $|s_m - u_m| < 10/2^m$ on $K_m \cup \tilde{E}$ and $s_m|_{\tilde{E}} \in C_e(\tilde{E})$. Thus the series

$$u_0 + \sum_{m=1}^{\infty} (u_m - s_m)$$

converges locally uniformly on Ω to a harmonic function h. Further, on $\tilde{E}$ we have

$$|s_m| \le |s_m - u_m| + |u_m| < 13/2^m,$$

so $v \in A(\tilde{E})$, where

$$v = \sum_{m=1}^{\infty} s_m.$$

From (5.1) we see that $u = u_0 + u_1 + \ldots = v + h$ on E, and so (ii) holds.

6   Boundary Behaviour

Finally, we present a sample application to illustrate how some of the above approximation results can be used to construct harmonic functions with specified boundary behaviour.

A classical theorem of Fatou asserts that a bounded harmonic function on the unit disc has a finite non–tangential limit at almost every point of the unit circle. More generally, by Plessner's Theorem (see [29, Chapter 6]), if h is any harmonic function on the unit disc, then at almost every boundary point ξ, either:

(i) h has a finite non–tangential limit at ξ, or

(ii) the image under h of any Stolz angle with vertex at ξ is $\mathbb{R}$.

However, the boundary behaviour of harmonic functions can also be examined using the notion of minimal fine limits, which we discuss below. There are many advantages in this approach: minimal fine limits make sense in any domain in $\mathbb{R}^n$, are conformally invariant when $n = 2$, and are suitable also for describing the boundary behaviour of superharmonic functions.

It will be more convenient to work in the context of the half–space $D = \mathbb{R}^{n-1} \times (0, +\infty)$. The Poisson kernel for D with pole at $z \in \partial D$ is given by

$$p_z(x) = x_n |x - z|^{-n} \quad (x \in D),$$

where $x = (x', x_n) \in \mathbb{R}^{n-1} \times \mathbb{R}$. Minimal thinness is a notion of thinness at boundary points which is analogous to the classical thinness described in Section 3. A set $A \subseteq D$ is said to be *minimally thin at $z \in \partial D$ (with respect to D)* if there is a positive superharmonic function s on D such that

$$\liminf_{x \to z, x \in A} \frac{s(x)}{p_z(x)} > \liminf_{x \to z, x \in D} \frac{s(x)}{p_z(x)}.$$

For example, if $n = 2$ and $\phi : \mathbb{R} \to [0, +\infty)$ is a Lipschitz function such that

$$\int_{-1}^{1} \frac{\phi(t)}{t^2}\, dt < +\infty,$$

then it is not hard to check that the set $\{(x_1, x_2) \in \mathbb{R}^2 : 0 < x_2 \leq \phi(x_1)\}$ is minimally thin at 0.

Now let $f : D \to [-\infty, +\infty]$ and let $z \in \partial D$. We write B_z for the non–tangential cluster set of f at z; that is, $l \in B_z$ if and only if there is a sequence $(x^{(m)})$ of points in D which approaches z non–tangentially and satisfies $f(x^{(m)}) \to l$. Also, we write C_z for the minimal fine cluster set of f at z; that is, $l \in C_z$ if and only if there is a subset E of D which is not minimally thin at z and which satisfies $f(x) \to l$ as $x \to z$ along E. If $C_z = \{l\}$, then f is said to have minimal fine limit l at z. (This is equivalent to the existence of a set $A \subseteq D$ which is minimally thin at z such that $f(x) \to l$ as $x \to z$ along $D \backslash A$.) Let λ denote $(n-1)$–dimensional measure on ∂D.

Doob [9] proved the following.

Theorem B. *If f is a harmonic function on D, then one of the following situations holds at λ-almost every point z in ∂D:*

(a) $B_z = C_z = [-\infty, +\infty]$;

(b) $B_z = [-\infty, +\infty]$, *but f has a finite minimal fine limit at z;*

(c) *f has a finite non–tangential limit and an equal minimal fine limit at z.*

Further, Doob observed that when $n = 2$ situation (b) can be omitted from this result, and asked whether this was also true when $n \geq 3$. Burkholder and Gundy [4] used probability and approximation methods to show that condition (b) can occur almost everywhere. Below we give a slightly stronger result (from [20]) and indicate how it can be obtained from the approximation results described in Section 3.

Theorem 6.1. *If $n \geq 3$, then there is a harmonic function f on D such that $B_z = [-\infty, +\infty]$ and $C_z = \{0\}$ for every z in ∂D.*

We indicate the proof of Theorem 6.1 when $n \geq 4$, slight modifications being necessary when $n = 3$. Let $S_m = \{2^{-m}k : k \in \mathbf{Z}\}$ for each m in $\mathbb{N}$, let

$$\bigcup_{m=1}^{\infty} S_m^{n-1} = \{y^{(k)} : k \in \mathbb{N}\},$$

$$s(x', x_n) = \sum_{k=1}^{\infty} 2^{-k}|x' - y^{(k)}|^{3-n} \quad ((x', x_n) \in \mathbb{R}^{n-1} \times \mathbb{R}),$$

$$J = \{(x', x_n) \in \mathbb{R}^{n-1} \times (0, 1] : s(x', x_n) \leq x_n^{1-n}\},$$

and

$$F = \bigcup_{m=1}^{\infty} \left(S_m^{n-1} \times (0, 2^{-m}]\right).$$

Arguing as in Example 3.2, we see that J is a relatively closed subset of D.

If $z \in \partial D$, then

$$s(x', x_n) > x_n^{1-n} \geq p_z(x) \quad ((x', x_n) \in (\mathbb{R}^{n-1} \times (0, 1])\backslash J).$$

However,

$$\liminf_{x' \to y'} \frac{s(x', 0)}{|x' - y'|^{3-n}} < +\infty \quad (y' \in \mathbb{R}^{n-1}),$$

so

$$\liminf_{\substack{x \to z \\ x \in D}} \frac{s(x)}{p_z(x)} = 0 \quad (z \in \partial D).$$

Thus $D \backslash J$ is minimally thin at each point of ∂D.

It is easy to see that (D, F) satisfies conditions (i) and (ii) of Theorem 3.3, so the continuous function on D given by $(x', x_n) \mapsto x_n^{-1} \sin(\pi/x_n)$ can be uniformly approximated on F by a function u_1 harmonic on D:

$$|u_1(x', x_n) - x_n^{-1} \sin(\pi/x_n)| < 1 \qquad ((x', x_n) \in F). \tag{6.1}$$

Since $s = +\infty$ on F, we see that $J \cap F = \emptyset$. Hence we can choose disjoint open subsets U_1 and U_2 of D such that $F \subset U_1$ and $J \subset U_2$. We define $u = u_1$ on U_1 and $u = 0$ on U_2. The set E, defined by $E = J \cup F$, is relatively closed in D, and u is defined and harmonic on an open set which contains E. Further, conditions (i) and (ii) of Theorem 3.1 hold.

We now apply Theorem 3.1 (as extended by the closing remark of Section 3) with u and E as above, with $\Omega = D$ and $w(x', x_n) = x_n$, to obtain a harmonic function f on D such that $|f - u| < w$ on E. Since $D \backslash J$ is minimally thin at each point of ∂D, we have $C_z = \{0\}$ for every z in ∂D, in view of the facts that $u = 0$ on J and w has limit 0 at points of ∂D. Further, the rapid oscillation of u on F (see (6.1)) shows that $B_z = [-\infty, +\infty]$, as required.

References

[1]   N. U. Arakelyan: *Uniform and tangential approximations by analytic functions,* Izv. Akad. Nauk Armjan. SSR Ser. Mat. **3** (1968), 273–286 (Russian); Amer. Math. Soc. Transl. (2) **122** (1984), 85–97.

[2]   N. U. Arakelyan: *Approximation complexe et propriétés des fonctions analytiques,* Actes, Congrès Intern. Math., 1970, Tome **2**, 595–600.

[3]   D. H. Armitage, M. Goldstein: *Nonuniqueness for the Radon transform,* Proc. Amer. Math. Soc. **117** (1993), 175–178.

[4]   D. L. Burkholder, R. F. Gundy: *Boundary behaviour of harmonic functions in a half-space and Brownian motion,* Ann. Inst. Fourier (Grenoble) **23**, 4 (1973), 195–212.

[5]   T. Carleman: *Sur un théorème de Weierstrass*, Ark. Mat. Astronom. Fys. **20B** (1927), 1–5.

[6]   Chen Huaihui, P. M. Gauthier: *A maximum principle for subharmonic functions and plurisubharmonic functions*, Canad. Math. Bull. **35** (1992), 34–39.

[7]   A. Debiard, B. Gaveau: *Potentiel fin et algèbres de fonctions analytiques I*, J. Funct. Anal. **16** (1974), 289–304.

[8]   J. Deny: *Systèmes totaux de fonctions harmoniques*, Ann. Inst. Fourier (Grenoble) **1** (1949), 103–113.

[9]   J. L. Doob: *Some classical function theory theorems and their modern versions*, Ann. Inst. Fourier (Grenoble) **15** (1965), 113–136.

[10]   J. L. Doob: *Classical Potential Theory and its Probabilistic Counterpart*, Springer, New York, 1983.

[11]   B. Fuglede: *Finely Harmonic Functions*, Lect. Notes Math. **289**, Springer, Berlin 1972.

[12]   B. Fuglede: *Asymptotic paths for subharmonic functions*, Math. Ann. **213** (1975), 261–274.

[13]   D. Gaier: *Lectures on Complex Approximation*, Birkhäuser, Boston 1987.

[14]   S. J. Gardiner: *The Dirichlet problem with non–compact boundary*, Math. Z. **213** (1993), 163–170.

[15]   S. J. Gardiner: *Superharmonic extension and harmonic approximation*, Ann. Inst. Fourier (Grenoble) **44** (1994), 65–91.

[16]   S. J. Gardiner: *Harmonic Approximation*, London Math. Soc. Lect. Notes **221**, Cambridge Univ. Press 1995.

[17]   S. J. Gardiner: *Uniform harmonic approximation of bounded functions*, Trans. Amer. Math. Soc. **348** (1996), 251–265.

[18]   S. J. Gardiner: *Uniform harmonic approximation with continuous extension to the boundary*, J. Analyse Math. **68** (1996), 95–106.

[19] S. J. Gardiner: *Decomposition of approximable harmonic functions,* Math. Ann., in press (1997).

[20] S. J. Gardiner: *Cluster sets of harmonic functions at the boundary of a half–space,* Bull. London Math. Soc., in press (1997).

[21] P. M. Gauthier, M. Goldstein, W. H. Ow: *Uniform approximation on unbounded sets by harmonic functions with logarithmic singularities,* Trans. Amer. Math. Soc. **261** (1980), 169–183.

[22] P. M. Gauthier, M. Goldstein, W. H. Ow: *Uniform approximation on closed sets by harmonic functions with Newtonian singularities,* J. London Math. Soc. (2) **28** (1983), 71–82.

[23] P. M. Gauthier, W. Hengartner: *Approximation uniforme qualitative sur des ensembles non bornées,* (NATO ASI) Press. Univ. Montréal 1982.

[24] M. V. Keldyš: *On the solvability and stability of the Dirichlet problem,* Uspehi Mat. Nauk **8** (1941), 171–231 (Russian); Amer. Math. Soc. Transl. **51** (1966), 1–73.

[25] M. Labrèche: *De l'Approximation Harmonique Uniforme,* Doctoral thesis, Université de Montréal 1982.

[26] A. A. Nersesyan: *Carleman sets,* Izv. Akad. Nauk Armjan. SSR Ser. Mat. **6** (1971), 465–471 (Russian); Amer. Math. Soc. Transl. (2) **122** (1984), 99–104.

[27] A. A. Nersesyan: *Approximation by harmonic functions on closed subsets of planar domains,* Soviet J. Contemp. Math. Anal. **25** (1990), 58–66.

[28] R. Nevanlinna: *Über eine Erweiterung des Poissonschen Integrals,* Ann. Acad. Sci. Fenn. Ser. A **24**, No. 4 (1925), 1–15.

[29] C. Pommerenke: *Boundary Behaviour of Conformal Maps,* Springer, Berlin 1992.

[30] A. Roth: *Approximationseigenschaften und Strahlengrenzwerte meromorpher und ganzer Funktionen,* Comment. Math. Helv. **11** (1938), 77–125.

[31]   A. Roth: *Uniform and tangential approximations by meromorphic functions on closed sets*, Can. J. Math. **28** (1976), 104–111.

[32]   C. Runge: *Zur Theorie der eindeutigen analytischen Funktionen*, Acta Math. **6** (1885), 228–244.

[33]   A. Stray: *On uniform and asymptotic approximation*, Math. Ann. **234** (1978), 61–68.

[34]   A. Stray: *Decomposition of approximable functions*, Ann. Math. **120** (1984), 225–235.

[35]   J. L. Walsh: *The approximation of harmonic functions by harmonic polynomials and by harmonic rational functions*, Bull. Amer. Math. Soc. (2) **35** (1929), 499–544.

[36]   L. Zalcman: *Uniqueness and non–uniqueness for the Radon transform*, Bull. London Math. Soc. **14** (1982), 241–245.

Address:

STEPHEN J. GARDINER
Department of Mathematics
University College Dublin
Dublin 4
Ireland

Properties of Bivariate Refinable Spline Pairs

T. N. T. Goodman

Abstract

We study the pairs of refinable bivariate spline functions ψ_n of degree n introduced in [4]. To motivate the definition we show how these spline functions are analogues of basic pairs of univariate spline functions with uniform double knots in the way that box splines are analogues of B–splines with uniform simple knots. We then study infinite linear combinations of integer translates of ψ_n and show, in particular, that any polynomial of degree n can be expressed in such a form. We also show that when the integer translates of ψ_n form a Riesz basis then, for $n \geq 1$, finite linear combinations of integer translates of ψ_n give all C^{n-1} spline functions of degree n with compact support on the given mesh and, provided that the mesh includes certain directions, any C^{n-1} spline function of degree n on the mesh can be expressed as an infinite linear combination of integer translates of ψ_n.

Introduction

For $r \geq 1$, $k \geq 1$, we shall say that a set of integrable functions $\{\phi_1, \ldots, \phi_r\}$ of compact support in $\mathbb{R}^k$ are *refinable* if for $\phi = (\phi_1, \ldots, \phi_r)^T$,

$$\phi(x) = \sum_{j \in \mathbb{Z}^k} A_j \phi(2x - j), \quad x \in \mathbb{R}^k, \tag{0.1}$$

for some $r \times r$ matrices A_j which vanish for all but a finite set of j in $\mathbb{Z}^k$.

We define the Fourier transform of a function ϕ on $\mathbb{R}^k$, $k \geq 1$, as

$$\widehat{\phi}(u) = \int_{\mathbb{R}^k} \phi(x) e^{-ixu}\, dx, \quad u \in \mathbb{R}^k.$$

Taking Fourier transforms of (0.1) we see that ϕ is refinable if and only if

Multivariate Approximation: Recent Trends and Results; W. Haußmann, K. Jetter and M. Reimer (eds.)
Mathematical Research, Vol. 101, pp. 63–82, ISBN 3-05-501770-6
© Akademie–Verlag, Berlin 1997

$$\widehat{\phi}(2u) = A(z)\widehat{\phi}(u), \quad z = e^{-iu}, \quad u \in \mathbb{R}^k \tag{0.2}$$

for some $r \times r$ matrix A of Laurent polynomials.

In [2] we characterised pairs of univariate refinable spline functions with knots at the integers, extending work in [6]. In [3] we investigated necessary conditions for the form of pairs of refinable bivariate spline functions, while such pairs were constructed in [4]. In both [2] and [4] the refinable pairs formed basic sets in the following sense.

Let S be a space of integrable functions with compact support in $\mathbb{R}^k$. Then a set $\{\psi_1, \ldots, \psi_r\}$ is called a *basic set* for S provided that a function ϕ lies in S if and only it is a finite linear combination of integer translates of $\psi_1, \ldots, \psi_r$, i.e.

$$\phi = \sum_{j=1}^{r} \sum_{l \in \mathbb{Z}^k} a_{j,l} \psi_j(. - l), \tag{0.3}$$

where only a finite number of the coefficients $a_{j,l}$ are non–zero. This can be expressed more neatly by putting $\psi = (\psi_1, \ldots, \psi_r)^T$, where we shall also refer to ψ as a basic set. Then (0.3) becomes

$$\phi = \sum_{l \in \mathbb{Z}^k} A_l \psi(. - l), \tag{0.4}$$

where each A_l is a $1 \times r$ matrix. Taking Fourier transforms, (0.4) is equivalent to

$$\widehat{\phi}(u) = M(z)\widehat{\psi}(u), \quad z = e^{-iu}, \quad u \in \mathbb{R}^k, \tag{0.5}$$

where M is a $1 \times r$ matrix of Laurent polynomials. The connection with refinability is given by the following.

Theorem 0.1. *Suppose that for any ϕ in S, $\phi(./2)$ also lies in S. Then any basic set for S is refinable.*

Proof: Let ψ be a basic set for S. Since $\psi(./2)$ is in S, (0.5) gives

$$\widehat{\psi}(2u) = M(z)\widehat{\psi}(u), \quad z = e^{-iu}, \quad u \in \mathbb{R}^k,$$

for some $r \times r$ matrix M of Laurent polynomials. By (0.2), ψ is refinable.

In Section 1 we consider basic sets for the space $S_{n,r}$ of univariate spline functions of degree n with knots of multiplicity r at Z. This is based on a simple characterisation of elements of $S_{n,r}$ in terms of their Fourier transforms. For the case $r = 1$ we consider an analogous characterisation of a space of bivariate functions for which a basic set comprises a box spline. In looking for an analogous space of functions for $r = 2$ we are naturally led to the basic pairs of functions ψ_n introduced in [4], which are C^{n-1} piecewise polynomials of degree n on a mesh of lines in $n + 3$ directions. These are defined in Section 2, where we recall the results of [4]. In particular we examine when the integer translates of ψ_n form a Riesz basis.

We recall that for a set of integrable functions $\{\phi_1, \ldots, \phi_r\}$ of compact support in $\mathbb{R}^k$, with $\phi = (\phi_1, \ldots, \phi_r)^T$, the integer translates of ϕ are said to form a Riesz basis if there are constants $A > 0$, $B > 0$, such that for any $a = (a_j)_{j \in \mathsf{Z}^k}$, $a_j \in \mathbb{R}^r$,

$$A \sum_{j \in \mathsf{Z}^k} |a_j|^2 \leq \int_{\mathbb{R}^k} |\sum_{j \in \mathsf{Z}^k} a_j \phi(. - j)|^2 \leq B \sum_{j \in \mathsf{Z}^k} |a_j|^2.$$

The following characterisation of Riesz bases follows from Theorem 5.1 of [5].

Theorem 0.2. ([5]) *If $\phi = (\phi_1, \ldots, \phi_r)^T$ comprises integrable functions with compact support in $\mathbb{R}^k$, then the integer translates of ϕ form a Riesz basis if and only if for any u in $\mathbb{R}^k$, there are integers $k_1, \ldots, k_r$ such that the matrix $[\widehat{\phi}(u + 2\pi k_1) \ldots \widehat{\phi}(u + 2\pi k_r)]$ is non-singular.*

We shall see that whereas the integer translates of a box spline can only form a Riesz basis on a mesh of at most 3 directions, the integer translates of ψ_n can form a Riesz basis for $n \leq 3$, thus allowing C^2 cubic splines on a 6 direction mesh. We also show that if $n \geq 1$ and the integer translates of ψ_n form a Riesz basis, then *any* C^{n-1} piecewise polynomial of degree n with compact support on the mesh can be expressed as a finite linear combination of integer translates of ψ_n.

In Section 3 we consider infinite linear combinations of integer translates of ψ_n and show, in particular, that any bivariate polynomial of degree n can be expressed in such a form. It is further shown that if the integer translates of ψ_n form a Riesz basis and the mesh includes lines in the directions of $(1,0)$, $(0,1)$ and $(1,1)$, then *any* C^{n-1} piecewise polynomial of degree n on the mesh can be expressed as an infinite linear combination of integer translates of ψ_n.

1   B–Splines and Box Splines

For $n \geq 0$, $1 \leq r \leq n+1$, we shall denote by $S_{n,r}$ the space of all univariate spline functions of degree n *with compact support* with knots of multiplicity r at Z, i.e. all C^{n-r} functions with compact support which are polynomials of degree n on $[j, j+1)$, $j \in \mathsf{Z}$. (If $r = n+1$, we ignore the values of the functions on Z.) It can be shown by integration by parts (see [2]) that a function ϕ lies in $S_{n,r}$ if and only if it has a Fourier transform of the form

$$\widehat{\phi}(u) = \sum_{j=0}^{r-1} P_j(z) u^{j-n-1}, \quad z = e^{-iu}, \quad u \in \mathbb{R}, \tag{1.1}$$

where $P_0, \ldots, P_{r-1}$ are Laurent polynomials.

Now suppose $r = 1$ and take ϕ in $S_{n,1}$. By (1.1), $\widehat{\phi}(u) = P(z) u^{-n-1}$ for some Laurent polynomial P, and since this is continuous when $u = 0$, $P(z)$ must be divisible by $(z-1)^{n+1}$. Thus for some Laurent polynomial Q,

$$\widehat{\phi}(u) = Q(z)(z-1)^{n+1} u^{-n-1} = Q(z)\widehat{B}_n(u),$$

where

$$\widehat{B}_n(u) := \left(\frac{z-1}{u} \right)^{n+1}. \tag{1.2}$$

Hence $\{B_n\}$ forms a basic set for $S_{n,1}$. Indeed B_n is, up to normalisation, the well known B–spline with knots at $0, 1, \ldots, n+1$.

We shall now extend this idea to construct a basic set for $S_{n,r}$. For $1 \leq k \leq n$, let $B_{n,k}$ denote the function defined by

$$\widehat{B}_{n,k}(u) = \frac{(-iu)^k - P_{n,k}(z)}{u^{n+1}}, \tag{1.3}$$

where $P_{n,k}(z)$ is the Taylor polynomial of degree n at $z = 1$ for $(\log z)^k$. These functions were introduced in [2].

Theorem 1.1. *For $n \geq 0$, $1 \leq r \leq n+1$, $\{B_n, B_{n,1}, \ldots, B_{n,r-1}\}$ is a basic set for $S_{n,r}$.*

Proof: Let ϕ be any function in $S_{n,r}$. Then by (1.1) we may write

$$\widehat{\phi}(u) = \sum_{j=0}^{r-1} P_j(z)(-iu)^{j-n-1}$$

where, since $\widehat{\phi}$ is continuous, $\sum_{j=0}^{r-1} P_j(z)(-iu)^j$ has a zero of order $n+1$ at $u = 0$. Defining $P_{n,k}$, $1 \le k \le n$, as in (1.3), this is equivalent to $P_0(z) + \sum_{j=1}^{r-1} P_j(z)P_{n,j}(z)$ having a zero of order $n+1$ at $z = 1$, i.e.

$$P_0(z) + \sum_{j=1}^{r-1} P_j(z)P_{n,j}(z) = Q(z)(z-1)^{n+1}$$

for some Laurent polynomial Q. Thus

$$\widehat{\phi}(u) = (-iu)^{-n-1}\{Q(z)(z-1)^{n+1} - \sum_{j=1}^{r-1} P_j(z)P_{n,j}(z)\} + \sum_{j=1}^{r-1} P_j(z)(-iu)^{j-n-1}$$

$$= (-iu)^{-n-1}\{Q(z)(z-1)^{n+1} + \sum_{j=1}^{r-1} P_j(z)[(-iu)^j - P_{n,j}(z)]\}$$

$$= (-i)^{-n-1}\{Q(z)\widehat{B}_n(z) + \sum_{j=1}^{r-1} P_j(z)\widehat{B}_{n,j}(z)\},$$

by (1.2) and (1.3). So ϕ is a finite linear combination of integer translates of $B_n, B_{n,1}, \ldots, B_{n,r-1}$ and so $\{B_n, B_{n,1}, \ldots, B_{n,r-1}\}$ is a basic set for $S_{n,r}$.

We note that for $r = 2$, $n \ge 1$, the basic set is given by

$$\begin{bmatrix} \widehat{B}_n(u) \\ \widehat{B}_{n,1}(u) \end{bmatrix} = \begin{bmatrix} (z-1)^{n+1} & 0 \\ \sum_{j=1}^n \frac{(1-z)^j}{j} & 1 \end{bmatrix} \begin{bmatrix} u^{-n-1} \\ -iu^{-n} \end{bmatrix}$$

$$= \begin{bmatrix} z-1 & 0 \\ \frac{(-1)^n}{n} & 1 \end{bmatrix} \begin{bmatrix} z-1 & 0 \\ \frac{(-1)^{n-1}}{n-1} & 1 \end{bmatrix} \cdots \begin{bmatrix} z-1 & 0 \\ -1 & 1 \end{bmatrix} \begin{bmatrix} z-1 & 0 \\ 0 & 1 \end{bmatrix} \begin{bmatrix} u^{-n-1} \\ -iu^{-n} \end{bmatrix},$$

where, as before, $z = e^{-iu}$. For general $n \ge 0$, $1 \le r \le n+1$, we have

$$(\widehat{B}_n(u), \widehat{B}_{n,1}(u), \ldots, \widehat{B}_{n,r-1}(u))^T = R(z)u^{-n-1}(1, (-iu), \ldots, (-iu)^{r-1})^T,$$
$$(1.4)$$

where R is a $r \times r$ matrix of Laurent polynomials with $\det R(z) = (z-1)^{n+1}$.

Another choice for a basic set for $S_{n,r}$ is the set of B–splines $\{M_0, \ldots M_{r-1}\}$, where M_j has knots at $t_j, \ldots, t_{j+n+1}$, where we define

$$t_{kr+l} = k, \quad k \in \mathbf{Z}, \quad l = 0, \ldots, r-1.$$

A recurrence relation for $(\widehat{M}_0, \ldots, \widehat{M}_{r-1})$ is given in [7]. Our next result characterises all basic sets for $S_{n,r}$.

Theorem 1.2. *For $n \geq 0$, $1 \leq r \leq n+1$, let $\psi = (B_n, B_{n,1}, \ldots, B_{n,r-1})^T$. Then η is a basic set for $S_{n,r}$ if and only if*

$$\widehat{\eta}(u) = N(z)\widehat{\psi}(u), \quad u \in \mathbb{R}, \quad z = e^{-iu}, \tag{1.5}$$

for some $r \times r$ matrix N of Laurent polynomials with $\det N(z) = cz^l$ *for* $c \neq 0$, *$l \in \mathbb{Z}$.*

Proof : Suppose that η satisfies (1.5). Since ψ is a basic set for $S_{n,r}$, there is a matrix M of Laurent polynomials with

$$\widehat{\phi}(u) = M(z)\widehat{\psi}(u) = M(z)N^{-1}(z)\widehat{\eta}(u).$$

Since N^{-1} is a matrix of Laurent polynomials, it follows that η is a basic set for $S_{n,r}$.

Conversely, suppose that η is a basic set for $S_{n,r}$. Then for $r \times r$ matrices M, N of Laurent polynomials,

$$\begin{aligned}
\widehat{\eta}(u) &= N(z)\widehat{\psi}(u), \\
\widehat{\psi}(u) &= M(z)\widehat{\eta}(u) = M(z)N(z)\widehat{\psi}(u).
\end{aligned}$$

We note from (1.4) that for $z \neq 0, 1$, the $r \times r$ matrix

$$\Psi_r := [\widehat{\psi}(u) \quad \widehat{\psi}(u+2\pi) \quad \ldots \quad \widehat{\psi}(u+2\pi(r-1))]$$

is non–singular. Since $\Psi_r = M(z)N(z)\Psi_r$, it follows that $\det M(z)N(z) \neq 0$. Thus $\det N(z) \neq 0$ for all $z \neq 0$ and so $\det N(z) = cz^l$ for $c \neq 0$, $l \in \mathbb{Z}$.

Note that $\{\phi\}$ forms a basic set for $S_{n,1}$ if and only if $\widehat{\phi}(u) = cz^l\widehat{B}(u)$, i.e. $\phi = cB(.-l)$, for $c \neq 0$, $l \in \mathbb{Z}$.

It is well–known that the integer translates of $(M_0, \ldots, M_{r-1})$ form a Riesz basis. It then follows from Theorem 0.2 and Theorem 1.2 that the integer translates of any basic set for $S_{n,r}$ form a Riesz basis.

We now turn our attention to box splines. For simplicity we restrict our attention to two dimensions. Take $n \geq 0$ and vectors $v_1, \ldots, v_{n+2}$ in $\mathbb{Z}^2$ which span $\mathbb{R}^2$, where we assume that for $1 \leq j \leq n+2$, the components of v_j are coprime. Let $S_1(v_1, \ldots, v_{n+2})$ denote the space of all functions ϕ on $\mathbb{R}^2$ with continuous Fourier transforms of the form

$$\widehat{\phi}(u) = \frac{P(z)}{(v_1 u) \ldots (v_{n+2} u)}, \quad u \in \mathbb{R}^2, \quad z = e^{-iu} = (e^{-iu_1}, e^{-iu_2}), \tag{1.6}$$

(where $v_j u$ denotes the scalar product of v_j and u). Since $\widehat{\phi}$ is continuous, $P(z)$ must be divisible by $(z^{v_1} - 1)\ldots(z^{v_{n+2}} - 1)$ and so for some Laurent polynomial Q,

$$
\begin{aligned}
\widehat{\phi}(u) &= \frac{Q(z)(z^{v_1} - 1)\ldots(z^{v_{n+2}} - 1)}{(v_1 u)\ldots(v_{n+2} u)} \\
&= Q(z)\widehat{B}(.|v_1,\ldots,v_{n+2}),
\end{aligned}
$$

where

$$
\widehat{B}(u|v_1,\ldots,v_{n+2}) := \prod_{j=1}^{n+2} \frac{z^{v_j} - 1}{v_j u}. \tag{1.7}
$$

Thus every function ϕ in $S_1(v_1,\ldots,v_{n+2})$ is a finite linear combination of integer translates of $B(.|v_1,\ldots,v_{n+2})$, i.e. $\{B(.|v_1,\ldots,v_{n+2})\}$ is a basic set for $S_1(v_1,\ldots,v_{n+2})$. The function $B(.|v_1,\ldots,v_{n+2})$ is, up to normalisation, the well known box–spline with directions $v_1,\ldots,v_{n+2}$, see [1].

Each function in $S_1(v_1,\ldots,v_{n+2})$ is a piecewise polynomial of degree n with compact support on the mesh $M(v_1,\ldots,v_{n+2})$ formed by lines through points of Z^2 parallel to $v_1,\ldots,v_{n+2}$. For $1 \leq j \leq n+2$, if the vectors $v_j, -v_j$ occur r_j times in $(v_1,\ldots,v_{n+2})$, then the elements of $S_1(v_1,\ldots,v_{n+2})$ are C^{n-r_j} across mesh lines parallel to v_j. Thus if $r = \max\{r_1,\ldots,r_{n+2}\}$, then the functions in $S_1(v_1,\ldots,v_{n+2})$ are C^{n-r}.

For $r > 1$, $S_1(v_1,\ldots,v_{n+2})$ does not contain all piecewise polynomials of degree n with compact support and the above smoothness. Indeed in any region not intersected by mesh lines, functions in $S_1(v_1,\ldots,v_{n+2})$ coincide with a space $P(v_1,\ldots,v_{n+2})$ of polynomials which contains Π_{n+1-r} but not Π_{n+2-r}. (Here we use the notation that for $d \geq 0$, $k \geq 1$, $\Pi_d(\mathbb{R}^k)$ denotes all polynomials of degree d on $\mathbb{R}^k$.) Moreover for any p in $\Pi_n(\mathbb{R}^2)$, p lies in $P(v_1,\ldots,v_{n+2})$ if and only if

$$
p = \sum_{j\in Z^2} a_j B(. - j|v_1,\ldots,v_{n+2})
$$

for some coefficients (a_j).

Finally we recall that the integer translates of $B(.|v_1,\ldots,v_{n+2})$ form a Riesz basis if and only if no two mesh lines intersect other than in points of Z^2.

2   Bivariate Refinable Pairs

The space $S_1(v_1, \ldots, v_{n+2})$ of the previous section can be considered as a 2–dimensional analogue of the space $S_{n,1}$. A natural question is then to find a corresponding 2–dimensional analogue of $S_{n,r}$ for $r \geq 2$. For $r = 2$ this was partially answered in [4], as we now proceed to describe.

Take $n \geq 0$ and let $v_1, \ldots, v_{n+3}$ be pairwise linearly independent vectors in Z^2, where for $1 \leq j \leq n + 3$, the components of v_j are coprime. We denote by $S_2(v_1, \ldots, v_{n+3})$ the space of all functions ϕ in $\mathbb{R}^2$ with continuous Fourier transforms of the form

$$\widehat{\phi}(u) = \frac{P_1(z)u_1 + P_2(z)u_2}{(v_1 u) \ldots (v_{n+3} u)}, \quad u \in \mathbb{R}^2, \quad z = e^{-iu}, \tag{2.1}$$

for some Laurent polynomials P_1, P_2. We recall that $M(v_1, \ldots, v_{n+3})$ denotes the mesh formed by lines through Z^2 parallel to $v_1, \ldots, v_{n+3}$.

Theorem 2.1. ([4]) *A function ϕ lies in $S_2(v_1, \ldots, v_{n+3})$ if and only if it is a C^{n-1} piecewise polynomial of degree n with compact support on the mesh $M(v_1, \ldots, v_{n+3})$ and the jumps in the nth order derivatives of ϕ across mesh lines remain constant except across points of Z^2.*

Following [4] we now give a construction of a basic pair for $S_2(v_1, \ldots, v_{n+3})$, $n \geq 0$, i.e. a basic set of two functions

$$\psi(\cdot \, | v_1, \ldots, v_{n+3}) = \psi_n = (\psi_{n,1}, \psi_{n,2})^T.$$

We define ψ_n by

$$\widehat{\psi}_n(u) = \frac{M_n(z)u}{(v_1 u) \ldots (v_{n+3} u)}, \quad u \in \mathbb{R}^2, \quad z = e^{-iu}, \tag{2.2}$$

where M_n is a 2×2 matrix of Laurent polynomials which we shall define recursively on n. Let

$$M_{-1}(z) = \left[\begin{array}{c} (z^{v_2} - 1)v_1 \\ (z^{v_1} - 1)v_2 \end{array} \right], \tag{2.3}$$

where v_1, v_2 are here regarded as row vectors. Now suppose that we have defined M_{n-1} for some $n \geq 0$. Write $v = v_{n+3}$ and let $v^\perp$, $\tilde{v}$ be in Z^2 with coprime components such that $vv^\perp = 0$, $v^\perp \tilde{v} = 1$. Let p and q be univariate

Laurent polynomials with no common factors such that for some univariate Laurent polynomial r,

$$M_{n-1}(w^{v^\perp})v^\perp = \begin{bmatrix} p(w)r(w) \\ q(w)r(w) \end{bmatrix}, \quad w \in \mathbb{C}, \tag{2.4}$$

where $w^{v^\perp} = (w^{v_1^\perp}, w^{v_2^\perp})$.

Choose univariate Laurent polynomials α, β so that

$$\alpha(w)p(w) + \beta(w)q(w) = -1, \quad w \in \mathbb{C}.$$

Then we define for z in $\mathbb{C}^2$,

$$M_n(z) = \begin{bmatrix} 1 & 0 \\ 0 & z^v - 1 \end{bmatrix} \begin{bmatrix} -q(z^{\tilde{v}}) & p(z^{\tilde{v}}) \\ \alpha(z^{\tilde{v}}) & \beta(z^{\tilde{v}}) \end{bmatrix} M_{n-1}(z). \tag{2.5}$$

Note that

$$\det M_n(z) = (z^v - 1)\det M_{n-1}(z) = \det[v_1^T \ v_2^T]\prod_{j=1}^{n+3}(z^{v_j} - 1). \tag{2.6}$$

As an example, when $v_1 = (1,0)$, $v_2 = (0,1)$, $v_3 = (1,1)$, we can take $v^\perp = (1,-1)$, $\tilde{v} = (1,0)$, $p = 1$, $q(w) = w$, $\alpha = -1$, $\beta = 0$, to get

$$M_0(z) = \begin{bmatrix} z_1(1 - z_2) & z_1 - 1 \\ (z_1 z_2 - 1)(1 - z_2) & 0 \end{bmatrix}.$$

It is shown in [4] that for $n \geq 0$, ψ_n as defined above is indeed a basic pair for $S_2(v_1, \ldots, v_{n+3})$. The following is also proved, in a similar manner to the proof of Theorem 1.2.

Theorem 2.2. ([4]) *If ψ_n is defined as above for $n \geq 0$, then η is a basic pair for $S(v_1, \ldots, v_{n+3})$ if and only if*

$$\widehat{\eta}(u) = N(z)\widehat{\psi}_n(u), \quad u \in \mathbb{R}^2, \quad z = e^{-iu},$$

for some 2×2 matrix of Laurent polynomials with $\det N(z) = cz^l$ for $c \neq 0$, $l \in \mathbb{Z}^2$.

We note that $\psi(.|v_1, \ldots, v_{n+3})$ depends on the order of the vectors $v_1, \ldots, v_{n+3}$. So permuting the vectors will give different basic pairs for $S_2(v_1, \ldots, v_{n+3})$.

Recalling (2.5) we define

$$\widehat{\eta}_{n-1}(u) = -i \begin{bmatrix} -q(z^{\tilde{v}}) & p(z^{\tilde{v}}) \\ \alpha(z^{\tilde{v}}) & \beta(z^{\tilde{v}}) \end{bmatrix} \widehat{\psi}_{n-1}(u), \quad u \in \mathrm{IR}^2.$$

Then by Theorem 2.2, η_{n-1} is a basic pair for $S_2(v_1,\ldots,v_{n+2})$. Now (2.2) and (2.5) give, for $v = v_{n+3}$,

$$\widehat{\psi}_n(u) = \frac{-1}{ivu} \begin{bmatrix} 1 & 0 \\ 0 & z^v - 1 \end{bmatrix} \widehat{\eta}_{n-1}(u), \quad u \in \mathrm{IR}^2.$$

Thus we have the derivative recurrence relation

$$D_v\psi_{n,1} = \eta_{n-1,1}, \quad D_v\psi_{n,2} = \eta_{n-1,2}(.-v) - \eta_{n-1,2}. \tag{2.7}$$

Our next result characterises when the integer translates of basic pairs form Riesz bases and is a natural analogue of the result for box splines in the final paragraph of Section 1.

Theorem 2.3. ([4]) *The following are equivalent for* $S(v_1,\ldots,v_{n+3})$.

(a) *The integer translates of some basic pair form a Riesz basis.*

(b) *The integer translates of any basic pair form a Riesz basis.*

(c) *No more than two mesh lines in* $M(v_1,\ldots,v_{n+3})$ *intersect other than in points of* Z^2.

(d) *For each* $z = (z_1,z_2) \in \mathbb{C}^2$ *with* $|z_1| = |z_2| = 1$, $z \neq (1,1)$, *there are at most two vectors* v_j, $1 \leq j \leq n+3$, *with* $z^{v_j} = 1$.

Condition (d) implies that $\{v_1,\ldots,v_{n+3}\}$ comprises at most two vectors of each of the parities (odd, odd), (odd, even) and (even, odd), and in particular, $n \leq 3$. It is shown in [4] that there are an infinite number of different choices for $\{v_1,\ldots,v_6\}$ which include $(1,0)$, $(0,1)$ and $(1,1)$ for which the conditions of Theorem 2.3 are satisfied. Under these conditions we may simplify, for $n \geq 1$, the characterisation of $S_2(v_1,\ldots,v_{n+3})$ given in Theorem 2.1, because of the following result.

Lemma 2.4. *Let* l_1 *and* l_2 *be straight lines in* IR^2 *which intersect in a single point. For* $n \geq 1$, *let* ϕ *be a* C^{n-1} *function on* IR^2 *which is a polynomial of degree* n *on each of the four regions bounded by* l_1 *and* l_2. *Then the jumps in the* nth *order derivatives of* ϕ *are constant across* l_1 *and across* l_2.

Proof : By making an affine transformation we may assume that l_1 is given by $x_2 = 0$ and l_2 is given by $x_1 = 0$. Suppose that for $x_1 > 0$, $x_2 > 0$, ϕ coincides with a polynomial p, while for $x_1 > 0$, $x_2 < 0$, ϕ coincides with a polynomial q. Since ϕ is continuous, for $x_1 > 0$ we have $p(x_1,0) = q(x_1,0)$ and so $D_1^n p(x_1,0) = D_1^n q(x_1,0)$. Since the nth order derivatives of p and q are constant, we see that $D_1^n \phi$ is constant for $x_1 > 0$. Similarly $D_1^n \phi$ is constant for $x_1 < 0$ and so the jump in $D_1^n \phi$ across l_2 is constant.

For $1 \le k \le n - 1$, we may apply the same argument to $D_2^k \phi$ to show that the jump in $D_1^{n-k} D_2^k \phi$ across l_2 is constant. But the jump in $D_2^n \phi$ across l_2 is zero for $x_2 > 0$ and also for $x_2 < 0$. Thus the jumps in all the nth order derivatives of ϕ across l_2 are constant. Similarly the jumps in all the nth order derivatives of ϕ across l_1 are constant.

From Lemma 2.4 and Theorem 2.1 we can immediately deduce the following result (which is *not* in [4]).

Theorem 2.5. *If $n \ge 1$ and the conditions of Theorem 2.3 are satisfied, then ϕ lies in $S_2(v_1,\ldots,v_{n+3})$ if and only if it is a C^{n-1} piecewise polynomial of degree n with compact support on the mesh $M(v_1,\ldots,v_{n+3})$.*

3   Infinite Combinations of Bivariate Refinable Pairs

Our space $S_2(v_1,\ldots,v_{n+3})$ comprises only finite linear combinations of integer translates of the basic pair $\psi_n = (\psi_{n,1}, \psi_{n,2})^T$. We now consider linear combinations of an infinite number of integer translates of ψ_n and show, in particular, that any polynomial in $\Pi_n(\mathbb{R}^2)$ can be expressed in such a form.

We first note the behaviour of $\widehat{\psi}_n(u)$ when $z = 1$, i.e. $u = 2\pi k$, $k \in \mathbb{Z}^2$.

Lemma 3.1. *For $n \ge 0$, $\widehat{\psi}_{n,2}(0) \ne 0$, while for $k \in \mathbb{Z}^2$, $k \ne 0$, $\widehat{\psi}_{n,2}(2\pi k) = 0$, and we have $\widehat{\psi}_{n,1}(2\pi k) = 0$ if and only if $v_j k \ne 0$ for $j = 1,\ldots,n + 3$.*

Proof : Take $n \ge 0$ and $k \in \mathbb{Z}^2$. By (2.3) and (2.5) we have $M_n(1) = 0$. So from (2.2), if $v_j k \ne 0$, $j = 1,\ldots,n + 3$, then $\widehat{\psi}_n(2\pi k) = 0$. If $v_{n+3} k \ne 0$, then by (2.2) and (2.5) (recalling $v = v_{n+3}$) we have $\widehat{\psi}_{n,2}(2\pi k) = 0$. If $v_{n+3} k = 0$, $k \ne 0$, then $v_j k \ne 0$, $j = 1,\ldots,n + 2$, and by (2.2) and (2.5),

$$\widehat{\psi}_{n,2}(2\pi k) = \lim_{vu \to 0} \frac{z^v - 1}{vu}[\alpha(1)\ \beta(1)]\frac{M_{n-1}(1)}{(v_1 u)\ldots(v_{n+2} u)} = 0.$$

It is shown in part (4) of the proof of Theorem 7 in [4] that $\widehat{\psi}_{n,2}(0) \neq 0$. It is also shown there that if $v_1 k = 0$, $k \neq 0$, then $\widehat{\psi}_{n,1}(2\pi k) \neq 0$. By permuting the vectors it follows from Theorem 2.2 that if $k \neq 0$ and $v_j k = 0$ for some j, $1 \leq j \leq n+2$, then $\widehat{\psi}_{n,1}(2\pi k) \neq 0$.

Corollary 3.2. *For $n \geq 0$, $\sum_{j \in Z^2} \psi_{n,2}(. - j)$ is a non–zero constant.*

Proof: The Fourier coefficients of the $(1,1)$ periodic function $\sum_{j \in Z^2} \psi_{n,2}(.-j)$ are given by $\widehat{\psi}_{n,2}(2\pi k)$, $k \in Z^2$. Since $\widehat{\psi}_{n,2}(2\pi k) = 0$ for $k \in Z^2$, $k \neq 0$, and $\widehat{\psi}_{n,2}(0) \neq 0$, the result follows.

We shall extend Corollary 3.2 to the following result.

Theorem 3.3. *Let $w \in Z^2$, $w \neq 0$, satisfy $w v_{n+3} = 0$. Then for any $n \geq 0$ and any polynomial p in $\Pi_n(\mathbb{R})$ of exact degree k, $0 \leq k \leq n$, $\sum_{j \in Z^2} p(wj)\psi_{n,2}(.-j)$ is a polynomial in wx of exact degree k.*

First we need two lemmas.

Lemma 3.4. *Let $(u_1, \ldots, u_{n+3})$ be a permutation of $(v_1, \ldots, v_{n+3})$ and let $\phi_n = \psi(.|u_1, \ldots, u_{n+3})$, $\psi_n = \psi(.|v_1, \ldots, v_{n+3})$. If w is any non–zero vector in $\mathbb{R}^2$ and p in $\Pi_n(\mathbb{R})$ has exact degree k, $0 \leq k \leq n$, then*

$$\sum_{j \in Z^2} p(wj)\psi_{n,2}(. - j) = \sum_{j \in Z^2} q(wj)\phi_{n,1}(. - j) + \sum_{j \in Z^2} r(wj)\phi_{n,2}(. - j),$$

where r in $\Pi_k(\mathbb{R})$ has exact degree k and q is in $\Pi_{k-1}(\mathbb{R})$.

Proof: From Theorem 2.2 we may write

$$\widehat{\psi}_n(u) = M(z)\widehat{\phi}_n(u) \tag{3.1}$$

for a matrix M of Laurent polynomials with $\det M(z) = cz^l$, $c \neq 0$, $l \in Z^2$. Writing

$$M_{21}(z) = \sum_{j \in Z^2} a_j z^j, \quad M_{22}(z) = \sum_{j \in Z^2} b_j z^j,$$

we have

$$\psi_{n,2} = \sum_{j \in Z^2} a_j \phi_{n,1}(. - j) + \sum_{j \in Z^2} b_j \phi_{n,2}(. - j).$$

Then

$$\sum_{j\in Z^2} p(wj)\psi_{n,2}(.-j) = \sum_{j\in Z^2} p(wj)\sum_{l\in Z^2} a_l\phi_{n,1}(.-j-l) +$$

$$\sum_{j\in Z^2} p(wj)\sum_{l\in Z^2} b_l\phi_{n,2}(.-j-l)$$

$$= \sum_{j\in Z^2} q(wj)\phi_{n,1}(.-j) + \sum_{j\in Z^2} r(wj)\phi_{n,2}(.-j),$$

where

$$q = \sum_{l\in Z^2} a_l p(. - wl), \quad r = \sum_{l\in Z^2} b_l p(. - wl).$$

Thus $q, r \in \Pi_k(\mathbb{R})$. Now choose $k \in Z^2$, $k \neq 0$, with $kv_1 = 0$. ¿From Lemma 3.1 we have $\widehat{\psi}_{n,1}(2\pi k) \neq 0 \neq \widehat{\phi}_{n,1}(2\pi k)$, $\widehat{\psi}_{n,2}(2\pi k) = \widehat{\phi}_{n,2}(2\pi k) = 0$. So putting $u = 2\pi k$ in (3.1) gives $M_{21}(1,1) = 0$, i.e. $\sum_{j\in Z^2} a_j = 0$. Thus $q \in \Pi_{k-1}(\mathbb{R})$. Since $\det M(1,1) \neq 0$, we must have $M_{2,2}(1,1) \neq 0$, i.e. $\sum_{j\in Z^2} b_j \neq 0$, and so r has exact degree k.

Lemma 3.5. *Let p, q be any bivariate polynomials. Then for any basic pair ψ for $S_2(v_1,\ldots,v_{n+3})$,*

$$\sum_{j\in Z^2} p(j)\psi_1(.-j) = \sum_{j\in Z^2} q(j)\psi_2(.-j) = 0 \tag{3.2}$$

implies $p = q = 0$.

Proof: We shall prove this by induction on the degree of p and q. First suppose that p and q are constants. Choose $k \in Z^2$, $k \neq 0$, with $v_1 k = 0$. Then by Lemma 3.1, $\widehat{\psi}_n(0)$ and $\widehat{\psi}_n(2\pi k)$ are linearly independent. From Theorem 2.2 it follows that for any basic pair ψ, $\widehat{\psi}(0)$ and $\widehat{\psi}(2\pi k)$ are linearly independent. Now suppose that (3.2) holds. If $p \neq 0$, then $\sum_{j\in Z^2} \psi_1(.-j) = 0$ and so $\widehat{\psi}_1(2\pi j) = 0$ for all j in Z^2, which contradicts $\widehat{\psi}(0)$ and $\widehat{\psi}(2\pi k)$ being linearly independent. So $p = 0$ and similarly $q = 0$.

Now suppose that the result is true if p and q have degree $k \geq 0$. Suppose that (3.2) is satisfied, where p and q have degree $k + 1$. Then

$$\sum_{j\in Z^2} \{p(j) - p(j - (1,0))\}\psi_1(.-j) = \sum_{j\in Z^2} p(j)\psi_1(.-j) -$$

$$\sum_{j\in Z^2} p(j)\psi_1(. - (1,0) - j) = 0.$$

Since $p - p(. - (1,0))$ has degree k, our inductive hypothesis shows that $p - p(. - (1,0)) = 0$. Similarly $p - p(. - (0,1)) = 0$ and so p is constant. Similarly q is constant and it follows from the first part of the proof that $p = q = 0$. Thus the result holds for degree $k + 1$ and hence by induction for all degrees.

Proof of Theorem 3.3 : This is by induction on n. By Corollary 3.2 it is true for $n = 0$. Assume that it is true for $n = m \geq 0$. Then take $n = m + 1$, w and p as in the statement of the theorem, and let

$$f = \sum_{j \in Z^2} p(wj)\psi_{n,2}(. - j).$$

By the derivative recurrence relation (2.7) with $v = v_{n+3}$, we have

$$D_v f = \sum_{j \in Z^2} (p(wj + wv) - p(wj))\eta_{n,2}(. - j) = 0,$$

since $wv = 0$. So $f(x) = g(wx)$, $x \in R^2$, for a univariate function g.

We now apply Lemma 3.4 to write

$$f = \sum_{j \in Z^2} q(wj)\phi_{n,1}(. - j) + \sum_{j \in Z^2} r(wj)\phi_{n,2}(. - j),$$

for $q \in \Pi_{k-1}(\mathbb{R})$, $r \in \Pi_k(\mathbb{R})$, where $\phi_n := \psi(.|v_1, \ldots, v_{n+1}, v_{n+3}, v_{n+2})$. Letting $\phi_{n-1} = \psi(.|v_1, \ldots, v_{n+1}, v_{n+3})$,

$$D_{v_{n+2}} f = \sum_{j \in Z^2} s(wj)\phi_{n-1,1}(. - j) + \sum_{j \in Z^2} t(wj)\phi_{n-1,2}(. - j)$$

for s, t in $\Pi_{k-1}(\mathbb{R})$. Now $D_{v_{n+3}} D_{v_{n+2}} f = 0$. so applying the derivative recurrence relation (2.7) and Lemma 3.5 gives $s = 0$. Our inductive hypothesis then implies that $D_{v_{n+2}} f$ is a polynomial of degree at most $k - 1$. Since $D_{v_{n+2}} f(x) = (v_{n+2}w)g'(wx)$, we see that $g \in \Pi_k(\mathbb{R})$ and so f is a polynomial in wx of degree at most k.

It remains to show that f has exact degree k. This we do by induction on k. By Corollary 3.2 it is true for $k = 0$. Suppose that it is true for $k = l \geq 0$. For q in $\Pi_{k-1}(\mathbb{R})$, define Tq by

$$Tq(wx) = \sum_{j \in Z^2} q(wj)\psi_{n,2}(x - j).$$

Then T is a non–singular linear map from $\Pi_{k-1}(\mathbb{R})$ to itself. If the exact degree of $\sum_{j\in\mathbb{Z}^2} p(wj)\psi_{n,2}(.-j)$ does not equal k, then there is a non–zero q in $\Pi_k(\mathbb{R})$ with $\sum_{j\in\mathbb{Z}^2} \psi_{n,2}(.-j) = 0$, which contradicts Lemma 3.5.

Corollary 3.6. *For $0 \leq k \leq n$ and any p in $\Pi_k(\mathbb{R}^2)$ there are unique q in $\Pi_{k-1}(\mathbb{R}^2)$ and r in $\Pi_k(\mathbb{R}^2)$ with*

$$p = \sum_{j\in\mathbb{Z}^2} q(j)\psi_{n,1}(.-j) + \sum_{j\in\mathbb{Z}^2} r(j)\psi_{n,2}(.-j). \tag{3.3}$$

Moreover the map $p \mapsto r$ is a non–singular linear map from $\Pi_k(\mathbb{R}^2)$ to itself.

Proof : Take $0 \leq k \leq n$. For $3 \leq l \leq n+3$, take $w_l \in \mathbb{Z}^2$, $w_l \neq 0$, with $w_l v_l = 0$ and let $\phi_n := \psi(.|v_1,\ldots,\widehat{v_l},\ldots,v_{n+3},v_l)$, where $\widehat{v_l}$ means that v_l is omitted. Then Theorem 3.3 shows that for any p in $\Pi_k(\mathbb{R})$ of exact degree k, there exists q in $\Pi_k(\mathbb{R})$ of exact degree k with

$$
\begin{aligned}
p(w_l x) &= \sum_{j\in\mathbb{Z}^2} q(w_l j)\phi_{n,2}(x-j) \\
&= \sum_{j\in\mathbb{Z}^2} r(w_l j)\psi_{n,1}(x-j) + \sum_{j\in\mathbb{Z}^2} s(w_l j)\psi_{n,2}(x-j),
\end{aligned}
$$

for $r \in \Pi_{k-1}(\mathbb{R})$, $s \in \Pi_k(\mathbb{R})$ of exact degree k, by Lemma 3.4. Moreover the map $p \mapsto s$ is a non–singular linear map from $\Pi_k(\mathbb{R})$ to itself. Finally we note that for $j = 1,\ldots,n$, the bivariate polynomials $(w_3 x)^j,\ldots,(w_{j+3})^j$ are linearly independent and so $\{p(w_3 x),\ldots,p(w_{k+3}x) : p \in \Pi_k(\mathbb{R})\}$ spans $\Pi_k(\mathbb{R}^2)$, and so (3.3) holds for all p in $\Pi_k(\mathbb{R}^2)$. That q is unique follows from Lemma 3.5.

We have shown that any polynomial p in $\Pi_n(\mathbb{R}^2)$ can be written as a linear combination of integer translates of ψ_n. We shall now consider expressing more general spline functions ϕ in this form, i.e.

$$\phi = \sum_{j\in\mathbb{Z}^2} a_j \psi_{n,1}(.-j) + \sum_{j\in\mathbb{Z}^2} b_j \psi_{n,2}(.-j). \tag{3.4}$$

for constants a_j, b_j, $j \in \mathbb{Z}^2$. We first note that if (3.4) holds, then we can express ϕ as such a linear combination of any basic pair η for $S_2(v_1,\ldots,v_{n+3})$. For by Theorem 2.2, we can write

$$\psi_n = \sum_{k\in S} A_k \eta(.-k),$$

where S is a finite subset of Z^2 and for $k \in S$, A_k is a 2×2 constant matrix. Substituting into (3.4) gives

$$
\begin{aligned}
\phi &= \sum_{j \in Z^2} [a_j \ b_j] \psi_n(. - j) \\
&= \sum_{j \in Z^2} \sum_{k \in S} [a_j \ b_j] A_k \eta(. - j - k) \\
&= \sum_{l \in Z^2} \sum_{k \in S} [a_{l-k} \ b_{l-k}] A_k \eta(. - l).
\end{aligned}
$$

We shall assume the conditions of Theorem 2.3 and write $v = v_{n+3}$ and define $v^\perp$, $\tilde{v}$ as in the definition of ψ_n in (2.2) - (2.5). We note that any x in $\mathbb{R}^2$ can be uniquely written $x = s\tilde{v} + tv$, where $v^\perp x = sv^\perp \tilde{v} = s$. Moreover $x \in Z^2$ if and only if $s, t \in Z$.

Our next lemma can be regarded as an extension of Corollary 3.2. We recall that B_n is defined by (1.2) and is, up to normalisation, the B–spline with knots $0, 1, \ldots, n + 1$.

Lemma 3.7. *If the conditions of Theorem 2.3 hold, then for some non–zero constant C,*

$$
\sum_{j \in Z} \psi_{n,2}(x - jv) = C B_n(v^\perp x), \quad x \in \mathbb{R}^2.
$$

Proof : By (2.7), $D_v \sum_{j \in Z} \psi_{n,2}(. - jv) = 0$ and so for some function f,

$$
\sum_{j \in Z} \psi_{n,2}(x - jv) = f(v^\perp x), \quad x \in \mathbb{R}^2.
$$

Then for each $x \in \mathbb{R}^2$,

$$
\sum_{j \in Z} \psi_{n,2}(x - jv) = \int_{-\infty}^{\infty} \psi_{n,2}(x + tv)\, dt = \int_{-\infty}^{\infty} \psi_{n,2}((v^\perp x)\tilde{v} + tv)\, dt
$$

and so

$$
f(s) = \int_{-\infty}^{\infty} \psi_{n,2}(s\tilde{v} + tv)\, dt, \quad s \in \mathbb{R}.
$$

Then for $y \in \mathbb{R}$,

$$
\begin{aligned}
\hat{f}(y) &= \int_{-\infty}^{\infty} e^{-isy} f(s)\, ds \\
&= \int_{-\infty}^{\infty} \int_{-\infty}^{\infty} e^{-isy} \psi_{n,2}(s\tilde{v} + tv)\, dt\, ds
\end{aligned}
$$

$$= c \int_{\mathbb{R}^2} e^{-i(v^\perp x)y} \psi_{n,2}(x)\, dx$$

$$= c\widehat{\psi}_{n,2}(yv^\perp),$$

for a non-zero constant c. By (2.2) and (2.5), with $w = e^{-iy}$,

$$\widehat{\psi}_{n,2}(yv^\perp) = \frac{M_n(w^{v^\perp})yv^\perp}{(v_1 yv^\perp)\dots(v_{n+3} yv^\perp)}$$

$$= \lim_{t\to\infty} \frac{e^{-it}-1}{t}\, \frac{[\alpha(w)\ \beta(w)]M_{n-1}(w^{v^\perp})v^\perp}{y^{n+1}(v_1 v^\perp)\dots(v_{n+2} v^\perp)}.$$

By (2.4),

$$[\alpha(w)\ \beta(w)]M_{n-1}(w^{v^\perp})v^\perp = [\alpha(w)\ \beta(w)]\begin{bmatrix} p(w)r(w) \\ q(w)r(w) \end{bmatrix} = -r(w).$$

Thus for a non-zero constant d,

$$\widehat{f}(y) = \frac{dr(w)}{y^{n+1}}, \qquad w = e^{-iy}, \qquad y \in \mathbb{R}. \tag{3.5}$$

Now suppose that $r(w) = 0$ for some $w \neq 1$. Then by (2.4), $M_{n-1}(w^{v^\perp})v^\perp = 0$. Then $M_{n-1}(w^{v^\perp})$ is singular. By (2.6) there is some j, $1 \leq j \leq n+2$, with $w^{v^\perp v_j} = 1$. So $|w| = 1$ and we can write $w^{v^\perp} = e^{-iu}$, $u \in \mathbb{R}^2$, where $uv_j = 0$, $u \neq 0$. Since $v^\perp$ and u are linearly independent and $M_{n-1}(e^{-iu})v^\perp = 0$, we see that $M_{n-1}(e^{-iu}) = 0$.

Now suppose $e^{-iuv_l} \neq 1$ for $l = 1,\dots,n+2$, $l \neq j$. Then for $r \in \mathbb{Z}^2$, $\widehat{\psi}_{n-1}(u + 2\pi r) = 0$ unless $v_j(u + 2\pi r) = 0$, in which case $u + 2\pi r = \lambda u$, for some $\lambda \neq 0$, and so $\widehat{\psi}_{n-1}(u + 2\pi r)$ is a scalar multiple of $\widehat{\psi}_{n-1}(u)$. But this contradicts the criteria of Theorem 0.2 for the integer translates of ψ_{n-1} to form a Riesz basis, and hence condition (b) of Theorem 2.3. Thus there is some l, $1 \leq l \leq n+2$, $l \neq j$, with $e^{-iuv_l} = 1$. Then $e^{-iuv_j} = e^{-iuv_l} = 1$ and also $e^{-iuv} = w^{v^\perp v} = 1$, which contradicts (d) of Theorem 2.3.

We have thus shown that $r(w)$ is a constant multiple of $(w\ \ 1)^m$ for some $m \geq 0$. If $m \leq n$, then (3.5) is not well–defined. However if $m \geq n+2$, then (3.5) gives $\widehat{f}(0) = 0$ and so $\widehat{\psi}_{n,2}(0) = 0$ which contradicts Lemma 3.1. So $m = n+1$ and comparing (3.5) and (1.2) shows that f is a constant multiple of B_n.

Theorem 3.8. *Assume that the conditions of Theorem 2.3 hold and let ϕ be any univariate spline function of degree n with simple knots at the integers. Then there is a sequence (c_j), $j \in \mathsf{Z}$, of constants such that*

$$\phi(v^{\perp}x) = \sum_{j \in \mathsf{Z}^2} c_{v^{\perp}j}\psi_{n,2}(x - j), \quad x \in \mathrm{IR}^2.$$

Proof: We may write $\phi = \sum_{k \in \mathsf{Z}} a_k B_n(. - k)$ for some sequence (a_k). Then for x in IR^2,

$$
\begin{aligned}
\phi(v^{\perp}x) &= \sum_{k \in \mathsf{Z}} a_k B_n(v^{\perp}x - k) \\
&= \sum_{k \in \mathsf{Z}} a_k B_n(v^{\perp}(x - \tilde{v}k)) \\
&= C^{-1} \sum_{k \in \mathsf{Z}} \sum_{j \in \mathsf{Z}} a_k \psi_{n,2}(x - k\tilde{v} - jv) \\
&= C^{-1} \sum_{l \in \mathsf{Z}^2} a_{v^{\perp}l}\psi_{n,2}(x - l).
\end{aligned}
$$

Theorem 3.9. *Assume that the conditions of Theorem 2.3 hold and let ϕ be any C^{n-1} piecewise polynomial of degree n on the mesh $M(v_1, \ldots, v_{n+3})$, where $\{v_1, \ldots, v_{n+3}\}$ contains $(1,0)$, $(0,1)$ and $(1,1)$. Then if $\psi = (\psi_1, \psi_2)$ is any basic pair for $S_2(v_1, \ldots, v_{n+3})$, there are constants a_j, b_j, $j \in \mathsf{Z}^2$, so that*

$$\phi = \sum_{j \in \mathsf{Z}^2} a_j \psi_1(. - j) + \sum_{j \in \mathsf{Z}^2} b_j \psi_2(. - j). \tag{3.6}$$

Proof: By the remarks before Lemma 3.7 it is sufficient to verify (3.6) for *some* basic pair ψ. Without loss of generality we may assume $v_1 = (1,0)$, $v_2 = (0,1)$, $v_3 = (1,1)$. We shall prove the result by induction on n. When $n = 0$ a basic pair is given by the characteristic functions of the triangles with vertices $(0,0)$, $(1,0)$, $(1,1)$ and with vertices $(0,0)$, $(0,1)$, $(1,1)$. With this choice of ψ it is clear that any piecewise constant function ϕ on $M(v_1, v_2, v_3)$ can be written in the form (3.6) and so the result is true for $n = 0$.

Assume that the result is true for $n = m - 1 \geq 0$. Let ϕ be a C^{m-1} piecewise polynomial of degree m on $M(v_1, \ldots, v_{m+3})$. Letting $v = v_{m+3}$, $D_v\phi$ is a piecewise polynomial of degree $m - 1$ on $M(v_1, \ldots, v_{m+2})$. We define a basic pair η_{m-1} for $S_2(v_1, \ldots, v_{m+2})$ as in (2.7). Our induction hypothesis ensures that there are constants a_j, b_j, $j \in \mathsf{Z}^2$, with

$$D_v\phi = \sum_{j \in \mathsf{Z}^2} a_j \eta_{m-1,1}(. - j) + \sum_{j \in \mathsf{Z}^2} b_j \eta_{m-1,2}(. - j).$$

The recurrence relation (2.7) then gives constants c_j, d_j so that

$$f = \sum_{j \in \mathsf{Z}^2} c_j \psi_{m,1}(. - j) + \sum_{j \in \mathsf{Z}^2} d_j \psi_{m,2}(. - j) \qquad (3.7)$$

satisfies $D_v f = D_v \phi$ and hence

$$\phi(x) = f(x) + g(v^\perp x), \quad x \in \mathbb{R}^2, \qquad (3.8)$$

for some function g. Now $g(v^\perp.)$ is a C^{m-1} piecewise polynomial of degree m on $M(v_1, \ldots, v_{m+3})$. Thus g is a C^{m-1} piecewise polynomial of degree m which can have a knot at t only if the line $v^\perp x = t$ is a mesh line, i.e. $t \in \mathsf{Z}$. By Theorem 3.8, there is a sequence (e_j), $j \in \mathsf{Z}$, with

$$g(v^\perp x) = \sum_{j \in \mathsf{Z}^2} e_{v^\perp j} \psi_{m,2}(x - j), \quad x \in \mathbb{R}^2. \qquad (3.9)$$

Combining (3.7), (3.8) and (3.9) gives the result.

We remark that the vectors (1,0), (0,1) and (1,1) in the statement of Theorem 3.9 could be replaced by any vectors v_1, v_2, v_3 for which the result holds when $n = 0$. We do not know how to classify such vectors.

References

[1]   C. deBoor, K. Höllig, S. Riemenschneider: *Box Splines.* Springer, Berlin–Heidelberg–New York 1993.

[2]   T. N. T. Goodman: *Characterising pairs of refinable splines,* J. Approx. Theory, to appear.

[3]   T. N. T. Goodman: *Pairs of refinable bivariate splines,* in: *Advanced Topics in Multivariate Approximation,* F. Fontanella, K. Jetter, P.–J. Laurent (eds.), 125–138, World Scientific, Singapore 1996.

[4]   T. N. T. Goodman: *Constructing pairs of refinable bivariate spline functions,* University of Dundee report AA967, 1996.

[5]   R.–Q. Jia, C. A. Micchelli: *On linear independence for integer translates of a finite number of functions,* Proc. Edinburgh Math. Soc. **36** (1992), 69–85.

[6]   W. Lawton, S. L. Lee, Z. Shen: *Complete characterization of re-
          finable splines,* Adv. Comput. Math. **3** (1995), 137–145.

[7]   G. Plonka: *Two–scale symbol and autocorrelation symbol for B–
          splines with multiple knots,* Adv. Comput. Math. **3** (1995),
          1–22.

Address:

Tim N. T. Goodman
Department of Mathematics
Dundee University
Dundee DD1 4HN
Scotland

Polar Cone Splines

R. Gormaz

Abstract

Polyhedral splines have been defined analytically as some kind of distributions, or geometrically, as measures of slices of a polyhedron. H. P. Seidel and A. H. Vermeulen propose to define the polar form of a simplex spline using a variation of the known recurrence formula which relates simplex splines of different degrees. Here we proceed on a similar idea, but we apply it to a cone spline which is a simpler object. We propose new and simple proofs for its basic properties.

1   Introduction

Simplex splines are a natural generalization of B–splines to the multivariate case. However, its definition is not well suited for computations. Given $n+1$ knots $t_0, t_1, ..., t_n \in \mathbb{R}^s$, the associated simplex spline can be defined as the distribution:

$$\varphi \longrightarrow \; < M, \varphi > \; := n! \int_{\Sigma_n} \varphi\Big(t_0 + \sum_{i=1}^{n} \lambda_i\,(t_i - t_0)\Big) d\lambda_1 \ldots d\lambda_n \;, \qquad (1.1)$$

where $\Sigma_n = \{\lambda \in \mathbb{R}^n \mid \lambda_i \geq 0 \; i = 1, \ldots, n, \; \sum_{i=1}^{n} \lambda_i \leq 1\}$. From this short definition, some basic properties of simplex splines can be derived with little effort. In particular, it is easy to prove that, in the case that the knots are in general position, the distribution M is in fact an $L^2(\mathbb{R}^s)$ function characterized by

$$M(u) := Vol_k(\Delta \cap P^{-1}(\{u\})), \qquad (1.2)$$

where Δ is any simplex in $\mathbb{R}^n$ such that its vertices $x_0, x_1, ..., x_n$ verify $P(x_i) = t_i$. Here P is the standard projection $\mathbb{R}^n \to \mathbb{R}^s$, keeping the s first components of a point in R^n. In this characterization, $\Delta \cap P^{-1}(\{u\})$ corresponds to the k–dimensional slice of Δ above $u \in \mathbb{R}^s$. For a good overview of the simplex splines theory and other "polyhedral" splines see [4].

Multivariate Approximation: Recent Trends and Results; W. Haußmann, K. Jetter and M. Reimer (eds.)
Mathematical Research, Vol. 101, pp. 83–94, ISBN 3-05-501770-6
© Akademie–Verlag, Berlin 1997

This geometric characterization, assuming the general position hypothesis, provides an equivalent definition which gives us some intuition about simplex splines. For example, we can feel that they are piecewise polynomials whose support is $[t_0, ..., t_n]$, the convex hull of the knots.

However, both formulae (1.1) and (1.2) have proved to be useless to provide practical evaluation algorithms for simplex splines. The key property for evaluating a simplex spline is the recurrence formula, proved independently by W. Dahmen [2, 3] and C. Micchelli [6, 7]. This formula can be stated for $u = \sum_{i=0}^{n} \lambda_i\, t_i$, with $\sum_{i=0}^{n} \lambda_i = 1$, as

$$M(u|t_0, ..., t_n) = \sum_{i=0}^{n} \lambda_i\, M(u|t_0, ..., t_{i-1}, t_{i+1}, ..., t_n). \qquad (1.3)$$

From the distributional definition of M, it is clear that

$$M(u|t_0, ..., t_s) = \frac{1}{Vol_s([t_0, ..., t_s])}$$

if u is in $[t_0, ..., t_s]$, and $M(u|t_0, ..., t_s) = 0$ otherwise. With this recurrence formula, the evaluation of a simplex spline is reduced to the computation of some determinants.

The idea developed by H. P. Seidel and A. H. Vermeulen [9] was to modify slightly formula (1.3), in order to define a (piecewise) k–affine function m, the polar form of M. They merged formula (1.3) with polar form or blossoming. The polar form or blossoming of a polynomial was defined and studied by L. Ramshaw [8]. Given a degree k polynomial P, its polar form is the unique k–affine symmetric function $p : \mathbb{R}^s \times ... \times \mathbb{R}^s \to \mathbb{R}$ such that $P(u) = p(u, u, ..., u)$ for all $u \in \mathbb{R}^s$. In the piecewise polynomial case, we have a polar form for each region. Given a point v, we denote by $m_v(u_1, ..., u_k)$ the polar form of M corresponding to the region to which v belongs.

W. Dahmen introduces the cone splines [2], also called truncated power functions, and writes simplex splines as sum of cone splines. Cohen, Lyche and Riesenfeld showed [1] that there are even stronger connections between cone splines and simplex splines. For example, a simplex spline can be seen as a sum of cone splines along hyperplanes. In this work, we will see that cone splines are homogenized versions of simplex splines.

The outline of this paper is as follows. In Section 2 we introduce the notation and present a technical result (Lemma 2.1). In Section 3 we define the polar cone spline and derive its properties. In Section 4 we study the cone spline (whose polar form is the polar cone spline) and obtain trivially all its properties. Finally, we apply it to the computation of the jump of the kth derivative of a cone spline and show a relation between cone splines and simplex splines.

Acknowledgment. This work was partially supported by the Facultad de Ciencias Fisicas y Matemáticas, Universidad de Chile, under a 1996 research grant.

2   Some Preliminary Results.

Given a sequence $Y = y^0, y^1, y^2, ...$ in $\mathbb{R}^{s+1}$, we denote by

$$Y^n := (y^0, y^1, ..., y^n)$$

the corresponding *ordered list* or *matrix* of size $n + 1$. Moreover,

$$Y_i^n := (y^0, ..., y^{i-1}, y^{i+1}, ..., y^n)$$

will be the matrix Y^n without the ith column, and more generally, $Y_{i_1,i_2,...,i_j}^n$ denotes Y^n with columns $i_1, i_2, ..., i_j$ omitted. We assume from here on that Y is in general position, that is, all its subsets of $s + 1$ vectors are linearly independent.

The convex hull of Y^n is denoted by $[Y^n]$ and its cone convex hull by $cc(Y^n)$. An s–face of Y^n is any set of the form $S = cc\{y^{i_1}, ..., y^{i_s}\}$ $(1 \leq i_j \leq n)$. The characteristic function χ_A of a set A is the function that takes the value 1 on A and vanishes outside A.

The following lemma is a technical result on which the theory rests.

Lemma 2.1.   *Suppose that $0 \notin [Y^{s+1}]$ and that v is not on an s–face of $cc(Y^{s+1})$. Then, for any reals $\mu_0, ..., \mu_{s+1}$ such that $\mu_0 y^0 + \cdots + \mu_{s+1} y^{s+1} = 0$ we have that*

$$\sum_{i=0}^{s+1} \mu_i \, c_v(Y_i^{s+1}) = 0, \tag{2.1}$$

where $c_v(Z)$ is $|\det Z|^{-1}$ if v is in $cc(Z)$ and is 0 otherwise.

We can get the geometric meaning of this result by transforming this statement into a statement about convex cones in R^{s+1} as follows. Because of the general position assumption for Y, $\{y^0, ..., y^s\}$ is a basis of $\mathbb{R}^{s+1}$ and there exist $s + 1$ linear functionals (the coordinates functionals) $\ell_0, ..., \ell_s$ such that

$$v = \sum_{i=0}^{s} \ell_i(v)\, y^i \quad \forall v \in \mathbb{R}^{s+1}.$$

With these functionals, writing the condition over the reals μ_i as $\mu_0\, y^0 + \cdots + \mu_s\, y^s = -\mu_{s+1}\, y^{s+1}$, and writing t instead of $-\mu_{s+1}$, we have that $\mu_i = t\, \ell_i(y^{s+1})$, $i = 0, ..., s$. The lemma is then clearly equivalent to the relation

$$\sum_{i=0}^{s} t\, \ell_i(y^{s+1})\, c_v(Y_i^{s+1}) - t\, c_v(Y^s) = 0 \quad \text{for all } t \in \mathbb{R},$$

from where, taking $t = 1$, we obtain

$$c_v(Y^s) = \sum_{i=0}^{s} \ell_i(y^{s+1})\, c_v(Y_i^{s+1}).$$

We can obtain an explicit expression for ℓ_i applying Cramer's rule to solve the system $\ell_i(y^j) = \delta_{ij}$, $j = 0, ..., s$. This leads us to

$$\frac{det\, Y^s}{|det\, Y^s|}\chi_{cc(Y^s)}(v) = \sum_{i=0}^{s} \frac{det\, [y^0, ..., y^{s+1}, ..., y^s]}{|det\, [y^0, ..., y^i, ..., y^s]|}\, \chi_{cc(Y_i^{s+1})}(v)$$

where y^{s+1} is on the ith column of the right hand side determinants. Let us suppose that the basis $\{y^0, ..., y^s\}$ is positively oriented, that is, $det\, Y^s > 0$, and let us denote by ε_i the sign of the orientation of the basis $\{y^0, ..., y^{i-1}, y^{s+1}, y^{i+1}, ..., y^s\}$. From this last relation we show finally that formula (2.1) is equivalent to

$$\chi_{cc(Y^s)}(v) = \sum_{i=0}^{s} \varepsilon_i \chi_{cc(Y_i^{s+1})}(v) \tag{2.2}$$

for all v not in an s–face of Y^{s+1}. This geometric relation can be easily checked in $\mathbb{R}^2$ or even in $\mathbb{R}^3$. A general proof of (2.2) is beyond the scope of this paper and will be published elsewhere.

From Lemma 2.1, a first simple result can be obtained.

Proposition 2.2. *Suppose that* $0 \notin [Y^{s+1}]$, *then*

$$c_v(u \,|Y^{s+1}) := \sum_{i=0}^{s+1} \mu_i\, c_v(Y_i^{s+1}) \tag{2.3}$$

is a well defined linear functional with respect to $u = \sum_{i=0}^{s+1} \mu_i\, y^i$.

Proof : The functional c_v is well defined because if there were two expressions for u, for example, $u = \sum \mu_i\, y^i = \sum \mu_i'\, y^i$, then

$$\sum_{i=0}^{s+1} (\mu_i - \mu_i')\, y^i = 0$$

which implies, using Lemma 2.1, that

$$\sum_{i=0}^{s+1} (\mu_i - \mu_i')\, c_v(Y_i^{s+1}) = 0.$$

The linearity follows immediately by taking a particular choice of μ_i, for example, $\mu_{s+1} = 0$ and $\mu_i = \ell_i(u)$, $i = 0, ..., s$. $\qquad\square$

The following result is a direct consequence of Proposition 2.2. It follows from the fact that two points not separated by an s–face of Y^{s+1}, necessarily belong to the same convex cones $cc(Y_i^{s+1})$.

Proposition 2.3. *Under the same assumptions, if v^1 and v^2 are not separated by an s–face of Y^{s+1}, then*

$$c_{v^1}(\ \cdot\ |Y^{s+1}) = c_{v^2}(\ \cdot\ |Y^{s+1}).$$

3 The Polar Cone Spline.

Definition 3.1. *A polar cone spline is defined recursively by*

$$c_v(u^1, ..., u^k|Y^{s+k}) := \sum_{i=0}^{s+k} \mu_i\, c_v(u^1, ..., u^{k-1}|Y_i^{s+k}) \qquad (3.1)$$

where $u^k = \sum_i \mu_i\, y^i$. For $k = 0$, we define

$$c_v(Y^s) := \frac{\chi_{cc(Y^s)}(v)}{|det\, Y^s|}. \qquad (3.2)$$

The next result justifies the use of the adjective "polar" for c_v's name.

Theorem 3.2. *The polar cone spline $c_v(u^1, ..., u^k|Y^{s+k})$ is a well defined k–linear symmetric function of its arguments $u^1, ..., u^k$.*

Proof : We proceed by induction. The case $k = 1$ is given by Proposition 2.2. Suppose that the theorem is true for $1, 2, ..., k - 1$ arguments.

To prove that $c_v(u^1, ..., u^k | Y^{s+k})$ is well defined (that is, it doesn't depend on the way we write u^k as a linear combination of the vectors y^i) it is enough to prove that

$$\text{if } \sum_{i=0}^{s+k} \mu_i \, y^i = 0, \quad \text{then } \sum_{i=0}^{s+k} \mu_i \, c_v(u^1, ..., u^{k-1} | Y_i^{s+k}) = 0. \tag{3.3}$$

Let us note first that by writing $y^i = 1 \cdot y^i$, and using our induction hypothesis, we have that, for any $j \neq i$,

$$c_v(u^1, ..., u^{k-2}, y^i | Y_j^{s+k}) = c_v(u^1, ..., u^{k-2} | Y_{i,j}^{s+k}),$$

where we recall that $Y_{i,j}^{s+k}$ is Y^{s+k} without its ith and j^{th} columns.

The same reasoning produces also the formula

$$c_v(u^1, ..., u^{k-2}, y^j | Y_i^{s+k}) = c_v(u^1, ..., u^{k-2} | Y_{i,j}^{s+k}).$$

Combining the last two formulae we find that

$$c_v(u^1, ..., u^{k-2}, y^i | Y_j^{s+k}) = c_v(u^1, ..., u^{k-2}, y^j | Y_i^{s+k}). \tag{3.4}$$

The proof of (3.3) continues as follows: We write $u^{k-1} = \sum_j \alpha_j \, y^j$ for some reals α_j. By induction hypothesis, the c_v functions are linear in all its $k-1$ arguments, hence

$$
\begin{aligned}
&\sum_i \mu_i \, c_v(u^1, ..., u^{k-2}, u^{k-1} | Y_i^{s+k}) \\
&= \sum_i \mu_i \, c_v(u^1, ..., u^{k-2}, \sum_j \alpha_j \, y^j | Y_i^{s+k}) \\
&= \sum_i \sum_j \mu_i \, \alpha_j \, c_v(u^1, ..., u^{k-2}, y^j | Y_i^{s+k}) \\
&= \sum_j \alpha_j \sum_i \mu_i \, c_v(u^1, ..., u^{k-2}, y^i | Y_j^{s+k}) \\
&= \sum_j \alpha_j \, c_v(u^1, ..., u^{k-2}, 0 | Y_j^{s+k}) \\
&= 0 \, .
\end{aligned}
$$

Next, consider the linearity of c_v. By induction, c_v is already linear with respect to $u^1, ..., u^{k-1}$. To prove the linearity with respect to u^k, we just choose $\mu_i = \ell_i(u)$, $i = 0, ..., s$.

The symmetry of c_v comes from the symmetry of $Y_{i,j}^{s+k}$ with respect to i and j. In fact

$$c_v(u^1, ..., u^{k-2}, y^i, y^j | Y^{s+k})$$
$$= c_v(u^1, ..., u^{k-2} | Y_{i,j}^{s+k})$$
$$= c_v(u^1, ..., u^{k-2}, y^j, y^i | Y^{s+k}).$$

$\square$

Theorem 3.3. *For polar cone splines the following properties hold :*

i) $c_v(u^1, ..., u^k, w | Y^{s+k}w) = c_v(u^1, ..., u^k | Y^{s+k})$ *for all* $w \in \mathbb{R}^{s+1}$, *where* $Y^{s+k}w$ *denotes the matrix* Y^{s+k} *augmented with the column* w.

ii) $c_v(u^1, ..., u^k | Y^{s+k}) = \sum_i \mu_i c_v(u^1, ..., u^k | Y_i^{s+k}y)$, *where* $y = \sum_i \mu_i y^i$.

iii) $c_v(u^1, ..., u^k | Y^{s+k})$ *is symmetric with respect to the ordering of* Y^{s+k}.

iv) *If* v^1 *and* v^2 *are not separated by an* s–*face of* Y^{s+k}, *then*

$$c_{v^1}(\cdot | Y^{s+k}) = c_{v^2}(\cdot | Y^{s+k}).$$

Proof : Property i) comes from writing $w = 1 \cdot w$.

Next, applying it with $w = y$, we prove ii) as follows:

$$c_v(u^1, ..., u^k | Y^{s+k}) = c_v(u^1, ..., u^k, y | Y^{s+k}y)$$
$$= \sum_{i=0}^{s+k} \mu_i\, c_v(u^1, ..., u^k, y^i | Y^{s+k}y)$$
$$= \sum_{i=0}^{s+k} \mu_i\, c_v(u^1, ..., u^k | Y_i^{s+k}y).$$

The iii) property is obtained by choosing $y = y^i$ in ii) and observing that

$$c_v(u^1, ..., u^k | Y^{s+k}) = c_v(u^1, ..., u^k | Y_j^{s+k}y^j).$$

Finally, iv) is obtained from Proposition 2.3. $\square$

4   The Cone Spline and its Properties.

Definition 4.1. *The cone spline corresponding to Y^n, with $n = s + k$, is given by*

$$C(u|Y^n) := c_u(\underbrace{u, ..., u}_{k \ times}|Y^n).$$

Let us show how to obtain the main properties of a cone spline.

- $C(\ \cdot\ |Y^n)$ is a piecewise homogeneous polynomial of degree k, because c_u is k–linear. Theorem 3.2 tell us that c_{u_0}, for a fixed $u_0 \in \mathbb{R}^{s+1}$, is k–linear. On the other side, the s–faces of Y^{s+k} induce a partition of $\mathbb{R}^{s+k}$. Thus, as long as u belongs to the same region of the partition as u_0, Theorem 3.3 (iv) says that $c_u = c_{u_0}$, which is k–linear all over this region.

- Recurrence relation: For $u = \sum_i \mu_i\, y^i$, the k–linearity of c_u and property i) of Theorem 3.3 produces:

$$
\begin{aligned}
C(u|Y^n) &= c_u(u, ..., u, \sum_i \mu_i\, y^i|Y^n) \\[2mm]
&= \sum_i \mu_i\, c_u(u, ..., u, y^i|Y^n) \\[2mm]
&= \sum_i \mu_i\, c_u(u, ..., u|Y_i^n) \\[2mm]
&= \sum_i \mu_i\, C(u|Y_i^n).
\end{aligned}
$$

- Derivative formula. This is obtained from the derivative of a k–linear symmetric functional. More precisely

$$D_h C(u|Y^n) = k\, c_u(u, ..., u, h|Y^n) \quad \forall h \in \mathbb{R}^{s+1}.$$

From here, we obtain the following particular cases:

- $h = u$

$$D_u C(u\,|Y^n) = k\, C(u\,|Y^n),$$

- $h = \sum_i \alpha_i\, y^i$

$$D_h C(u|Y^n) = k \sum_i \alpha_i\, C(u|Y_i^n),$$

- $h = y^j$

$$D_{y^j} C(u|Y^n) = k C(u|Y_j^n).$$

Applying the derivative formula ν times, we obtain the following result:

Proposition 4.2. *For any point v not separated from u by an s-face,*

$$D_{h^1 \dots h^\nu} C(u|Y^n) = \frac{k!}{(k-\nu)!} c_v(u, \dots, u, h^1, \dots, h^\nu|Y^n).$$

The following theorem is a fundamental classical result.

Theorem 4.3. *The cone spline $C(\,\cdot\,|Y^n)$ is $\mathcal{C}^{k-1}$ where $k = n - s$.*

Proof: Let us recall that we are always assuming that Y^n is in general position. The s-faces of Y^n produce a partition of the domain of the cone spline. Two adjacent regions of the resulting partition are separated by exactly one s-face. If a and b are points in two adjacent regions, they are separated by an s-face which we can assume is of the form $S = cc\{y^{j_1}, \dots, y^{j_s}\}$. But a and b are no more separated by an s-face of $Y_{j_i}^n$, for $i = 1, \dots, s$. From this simple observation, if $a \to \bar{u} \in S$, the $k-1$ derivative tends to $k!\, c_a(\bar{u}, h^1, \dots, h^{k-1}|Y^n)$. From the fact that $\bar{u} \in S$, there exist $\mu_1, \dots, \mu_s$ such that $\bar{u} = \sum_{i=1}^{s} \mu_i\, y^{j_i}$. With this, we finally obtain

$$k!c_a(\bar{u}, h^1, \dots, h^{k-1}|Y^n)$$

$$= k! \sum_{i=1}^{s} \mu_i\, c_a(y^{j_i}, h^1, \dots, h^{k-1}|Y^n)$$

$$= k! \sum_{i=1}^{s} \mu_i\, c_a(h^1, \dots, h^{k-1}|Y_{j_i}^n)$$

$$= k! \sum_{i=1}^{s} \mu_i\, c_b(h^1, \dots, h^{k-1}|Y_{j_i}^n).$$

In the last expression, we have replaced c_a by c_b. This is possible because, as we noticed before, a and b are not separated by an s-face of $Y_{j_i}^n$.

We conclude that the limit of the $k-1$ derivative of $C(\,\cdot\,|Y^n)$ at $\bar{u}$ is the same from both sides of S. $\qquad\qquad\square$

The next and last theorem is the cone spline version of a result due to II. Hakopian giving a formula for the size of the jump of the kth derivative of a simplex spline. The proof proposed here is based mainly on the algebraic properties of a cone spline and is much easier to follow than the original one [5].

Theorem 4.4. *Let $\bar{u} \in S = cc\{y^{j_1}, ..., y^{j_s}\}$ and $h \in \mathbb{R}^{s+1}$, a direction not in S. The jump of $D^k_{h...h} C(u|Y^n)$ at $\bar{u} \in S$, in the direction h, is given by*

$$\Delta^k_{S,h} = k!\, sgn(det[hy^{j_1}...y^{j_s}]) \frac{(det[h\, y^{j_1}...y^{j_s}])^k}{\Pi_{i \notin \{j_1,...,j_s\}} det[y^i y^{j_1}...y^{j_s}]}.$$

Proof: It is clear that the kth derivative is constant on each region not separated by an s–face. More precisely, if a is a point of the region, the kth derivative is $k!c_a(h, ..., h|Y^n)$. Assuming that h points towards a region where b is located, and a is a point on the adjacent region, the jump can be written as $\Delta^k_{S,h} = k!c_b(h, ..., h|Y^n) - k!c_a(h, ..., h|Y^n)$. For each i not in $\{j_1, ..., j_s\}$ we can consider the basis $\{y^i, y^{j_1}, ..., y^{j_s}\}$ and h can be written as the linear combination $h = \alpha_0\, y^i + \alpha_1\, y^{j_1} + ... + \alpha_s\, y^{j_s}$. From Cramer's rule, α_0 can be computed explicitly

$$\alpha_0 = \frac{det[h\, y^{j_1}...y^{j_s}]}{det[y^i y^{j_1}...y^{j_s}]}.$$

Using this expression, the jump can be computed as

$$\begin{aligned}
\Delta^k_{S,h} &= k!\alpha_0 \Big(c_b(h, ..., h, y^i|Y^n) - c_a(h, ..., h, y^i|Y^n)\Big) \\
&= k!\frac{det[h\, y^{j_1}...y^{j_s}]}{det[y^i y^{j_1}...y^{j_s}]} \Big(c_b(h, ..., h|Y^n_i) - c_a(h, ..., h|Y^n_i)\Big).
\end{aligned}$$

After k applications of the same idea, each time with a different vector $y^i \notin \{y^{j_1}, ..., y^{j_s}\}$ there will remain exactly one more y^i vector not used in this process. If we denote $\bar{Y} = (y^i y^{j_1}...y^{j_s})$, the formula obtained at this stage is as follows:

$$\Delta^k_{S,h} = k!\frac{\Big(det[h\, y^{j_1}...y^{j_s}]\Big)^k}{\Pi_{\ell \notin \{i,j_1,...,j_s\}} det[y^\ell y^{j_1}...y^{j_s}]} \Big(c_b(\bar{Y}) - c_a(\bar{Y})\Big).$$

To complete the proof, it is necessary to consider the two possibilities for y^i, that is, whether it points towards the a region or towards the b region. In each case, the difference $c_b(\bar{Y}) - c_a(\bar{Y})$ is reduced to one term and is evaluated as

$$\pm \frac{1}{det[y^i y^{j_1}...y^{j_s}]}.$$

The sign depends upon whether y^i and h point both towards the same region or not. It is clear that in all cases the sign must be $sgn(det[h\, y^{j_1}...y^{j_s}])$. $\qquad \square$

Consider the affine hyperplane H^s of $\mathbb{R}^{s+1}$ consisting of all points whose last component is 1, that is, points of the form $t^* = (t, 1)$ for each $t \in \mathbb{R}^s$. We say that $n+1$ points $\{t^0, ..., t^n\}$ in $\mathbb{R}^s$ are in general position if no subset of size $s+1$ is included in an affine hyperplane of $\mathbb{R}^s$. This is equivalent to the assertion that the vectors $\{t^{0*}, ..., t^{n*}\}$ are in general position in $\mathbb{R}^{s+1}$ according to the definition given in Section 2. We can define

$$N(u|t^0, ..., t^n) := C(u^*|t^{0*}, ..., t^{n*})$$

which corresponds to the restriction of C to the hyperplane H^s. It is possible to check that the respective polar forms share the same relation, that is,

$$n_v(u^1, ..., u^k|t^0, ..., t^n) = c_v(u^{1*}, ..., u^{k*}|t^{0*}, ..., t^{n*}),$$

and Definition 3.1 can be used also to compute n_v. We observe that the recurrence relation obtained in this particular case is exactly the definition given by H. P. Seidel and A. H. Vermeulen [9] of what they called the *polar B–spline* which correspond to a simplex spline up to a multiplicative factor. This relation between cone splines and simplex splines is the same existing between a degree k polynomial in s variables and its degree k homogeneous version in $s + 1$ variables obtained by adding an extra variable to a convenient power on each monomial term. For example, the degree 3 polynomial in two variable $2 - x + xy + y^3$ has the homogenized version $2w^3 - xw^2 + xyw + y^3$.

Starting from this observation, it is possible to obtain the properties for simplex splines rewriting the corresponding results for cone splines.

Some future work will correspond to the relaxation of the general position hypothesis, and then to a study of the degenerate cases. In particular, it is possible to obtain formulae for the jump of lower order derivatives.

As it is possible to write a box spline as a sum of cone splines, it is possible to derive formulae for the polar form of a box spline. From here we think that it will be possible to present the subdivision properties of box splines in terms of its polar form.

References

[1] E. Cohen, T. Lyche and R. F. Riesenfeld: *Cones and recurrence relations for simplex splines*, SIAM J. Numer. Anal. **17** (1980), 179–191.

[2] W. Dahmen: *Multivariate B–splines – recurrence relations and linear combinations of truncated powers*, in: *Multivariate Approximation Theory*, W. Schempp and K. Zeller (eds.), 64–82, Birkhäuser, Basel–Boston 1979.

[3] W. Dahmen: *On Multivariate B–splines*, SIAM J. Numer. Anal. **17** (1980), 179–191.

[4] T. N. T. Goodman: *Polyhedral splines*, in: *Computation of Curves and Surfaces*, W. Dahmen et al. (eds.), 347–382, Kluwer, Dordrecht 1990.

[5] H. A. Hakopian: *Multivariate divided differences and multivariate interpolation of Lagrange and Hermite type*, J. Approx. Theory **34** (1982), 286–305.

[6] C. A. Micchelli: *On a numerically efficient method for computing B–splines*, in: *Multivariate Approximation Theory*, W. Schempp and K. Zeller (eds.), 211–248, Birkhäuser, Basel–Boston 1979.

[7] C. A. Micchelli: *A constructive approach to Kergin interpolation in $\mathbb{R}^k$: Multivariate B–splines and Lagrange interpolation*, Rocky Mountain J. Math. **10** (1980), 485–497.

[8] L. Ramshaw: *Blossoming: A connect–the–dots approach to splines*, DEC System Research Center Report **19**, June 1987.

[9] H. P. Seidel, A. H. Vermeulen: *Simplex splines support surprisingly strong symmetric structures and subdivision*, in: *Curves and Surfaces in Geometric Design*, P. J. Laurent, A. Le Méhauté and L. L. Schumaker (eds.), 443–455, A K Peters, 1994.

Address:

RAÚL GORMAZ

Departamento de Ing. Matemática

Universidad de Chile

Casilla 170/3 Correo 3

Santiago, Chile

Discrepancies of Point Sequences on the Sphere and Numerical Integration

P. J. Grabner, B. Klinger and R. F. Tichy

Abstract

Several concepts of discrepancy for point sequences on the d–dimensional unit sphere S^d are studied. The different discrepancies are compared to each other and applied to error estimates in numerical integration.

Introduction

Numerical integration of continuous functions on the d–dimensional unit sphere S^d is an important application of Quasi–Monte Carlo methods. In principal, one has to generate a sequence of points $\mathbf{x}_n$, $n = 1, \ldots, N$ on the sphere in order to approximate the integral

$$I(f) = \int_{S^d} f(\mathbf{x}) \, d\sigma(\mathbf{x})$$

by functionals

$$I_N(f) = \frac{1}{N} \sum_{n=1}^{N} f(\mathbf{x}_n),$$

where σ denotes the normalized surface measure on S^d and f is a continuous real valued function. As a general reference on Quasi–Monte Carlo methods we mention Niederreiter [22]. The problem of distributing points on the sphere is also related to constructive multivariate approximation, see Reimer [24]. For the recent literature on spherical problems concerned with approximation and numerical integration we refer to the forthcoming book [7].

For applications in numerical integration it is necessary to have suitably smoothly distributed points on the sphere. There are several quantities measuring the distribution of point sets $\mathbf{x}_n$, $n = 1, \ldots, N$. The geometrically most natural concept is the spherical cap discrepancy

Multivariate Approximation: Recent Trends and Results; W. Haußmann, K. Jetter and M. Reimer (eds.)
Mathematical Research, Vol. 101, pp. 95–112, ISBN 3–05–501770–6
© Akademie–Verlag, Berlin 1997

$$D_N^C(\mathbf{x}_n) := \sup_C \left| \frac{1}{N} \sum_{n=1}^{N} \chi_C(\mathbf{x}_n) - \sigma(C) \right|, \tag{0.1}$$

where $C := \{\mathbf{x} : \langle \mathbf{x}, \mathbf{y} \rangle > \cos\phi\}$ is a spherical cap with center $\mathbf{y}$ and radius ϕ which is the intersection of the sphere and a half space; χ_C denotes the characteristic function of C. (Here and in the following we consider S^d to be embedded in $\mathbb{R}^{d+1}$ and $\langle \cdot, \cdot \rangle$ denotes the standard inner product on $\mathbb{R}^{d+1}$.) Roughly speaking, this discrepancy measures the maximal deviation between the empirical distribution of the points and uniform distribution.

The error $E_N(f) = |I_N(f) - I(f)|$ in numerical integration of continuous functions on S^d satisfying a Lipschitz condition $|f(\mathbf{x}) - f(\mathbf{y})| \leq C_f \arccos(\langle \mathbf{x}, \mathbf{y} \rangle)$ can be estimated by

$$E_N(f) \leq C_f \left(\frac{6d}{M} + \pi \sum_{m=1}^{2M} \sum_{\ell=1}^{Z(d,m)} \left| \frac{1}{N} \sum_{n=1}^{N} S_{m,\ell}(\mathbf{x}_n) \right| \right),$$

where M is an arbitrary positive integer and $S_{m,\ell}$, $\ell = 1, \ldots, Z(d,m)$ denotes an orthonormal basis of the spherical harmonics of order m (cf. [20]). The proof is given in [10] and makes use of an approximation kernel due to Newman and Shapiro [21].

Extending the well known Erdős–Turán inequality (cf. [17, 14]), Grabner [9] established the following bound for the cap discrepancy: For any positive integer M and constants $c_i(d)$ only depending on the dimension d the inequality

$$D_N^C(\mathbf{x}_n) \leq \frac{c_1(d)}{M+1} + \sum_{m=1}^{M} \left(\frac{c_2(d)}{m} + \frac{c_3(d)}{M+1} \right) \sum_{j=1}^{Z(d,m)} \left| \frac{1}{N} \sum_{n=1}^{N} S_{m,j}(\mathbf{x}_n) \right| \tag{0.2}$$

holds. The proof of this result mainly depends on Vaaler's approximation kernel [28], which is very suitable to approximate step functions by trigonometric polynomials.

The above upper bound for the cap discrepancy in terms of the spherical harmonics suggests that a notion of discrepancy based on spherical harmonics might be fruitful. In analogy to the so called polynomial discrepancy for sequences in $[0,1)^d$ (cf. [12, 27]) we introduce a spherical polynomial discrepancy defined by

$$D_N^S(\mathbf{x}_n) := \sup_{m \geq 1} \frac{1}{m^d} \max_{1 \leq j \leq Z(d,m)} \left| \frac{1}{N} \sum_{n=1}^{N} S_{m,j}(\mathbf{x}_n) \right|. \qquad (0.3)$$

It is a natural question to ask whether uniform distribution of infinite point sequences can be defined via cap discrepancy as well as via the polynomial discrepancy. Let us recall the definition of uniform distribution of a sequence $(\mathbf{x}_n)_{n=1}^{\infty}$: $(\mathbf{x}_n)$ is said to be uniformly distributed if

$$\lim_{N \to \infty} I_N(f) = I(f)$$

for all continuous functions f. By well known arguments (cf. [17]) this is equivalent to

$$\lim_{N \to \infty} D_N^C(\mathbf{x}_n) = 0 \quad \text{or} \quad \lim_{N \to \infty} D_N^S(\mathbf{x}_n) = 0.$$

A quantitative relation between these two concepts of discrepancy is due to Klinger and Tichy [15]

$$c_4(d) D_N^S(\mathbf{x}_n) \leq D_N^C(\mathbf{x}_n) \leq c_5(d) \left(D_N^S(\mathbf{x}_n) \right)^{\frac{1}{2d}} \qquad (0.4)$$

with constants only depending on the dimension.

An upper bound for the cap discrepancy in terms of Legendre polynomials P_m^d was proved by Grabner and Tichy [10]:

Theorem 0.1.

$$D_N^C(\mathbf{x}_n) \leq \frac{c_6(d)}{M+1} + c_7(d) \sum_{m=1}^{M} m^{\frac{d-3}{2}} \frac{1}{N} \sqrt{\sum_{i=1}^{N} \sum_{\ell=1}^{N} P_m^d(\langle \mathbf{x}_i, \mathbf{x}_\ell \rangle)}.$$

Remark 1 *In the following we make frequently use of the addition theorem for spherical harmonics*

$$\sum_{j=1}^{Z(d,n)} S_{n,j}(\mathbf{x}) S_{n,j}(\mathbf{y}) = Z(d,n) P_n^d(\langle \mathbf{x}, \mathbf{y} \rangle),$$

where the Legendre polynomials are normalized such that $|P_n^d(x)| \leq P_n^d(1) = 1$. Furthermore, they are given by the generating function (cf. [19])

$$\sum_{n=0}^{\infty} (n+1)(n+2)\cdots(n+d-2)P_n^d(t)z^n = \frac{1}{(1-2tz+z^2)^{\frac{d-1}{2}}}. \qquad (0.5)$$

The numerical value of the dimension $Z(d,n)$ *is known to be* $\frac{2n+d-1}{n+d-1}\binom{n+d-1}{d-1}$.

This suggests to define a discrepancy based on Legendre polynomials:

$$D_N^P(\mathbf{x}_n) := \sup_{m\geq 1} \frac{1}{m^d}\frac{1}{N}\sqrt{\sum_{i=1}^{N}\sum_{\ell=1}^{N} P_m^d(\langle \mathbf{x}_i, \mathbf{x}_\ell\rangle)}. \qquad (0.6)$$

Again this discrepancy is compatible with the above mentioned concepts:

$$c_8(d)D_N^P(\mathbf{x}_n) \leq D_N^C(\mathbf{x}_n) \leq c_9(d)\left(D_N^P\right)^{\frac{2}{3d+1}},$$

where the constants depend only on the dimension (cf. [15]).

For the case $d = 2$ Freeden [8] developed a discrepancy based on the Green function $G(x,y)$ of the Beltrami operator. The Fourier series of this function is given by

$$G(\mathbf{x},\mathbf{y}) = \sum_{n=1}^{\infty} \frac{\varphi_n(\mathbf{x})\varphi_n(\mathbf{y})}{\lambda_n},$$

where $\varphi_n(\mathbf{x})$ and λ_n are the eigenfunctions and eigenvalues of the operator. In order to make the series convergent, iterations of this kernel are considered:

$$G^{(r)}(\mathbf{x},\mathbf{y}) = \sum_{n=1}^{\infty} \frac{\varphi_n(\mathbf{x})\varphi_n(\mathbf{y})}{\lambda_n^r}.$$

This sum is uniformly convergent for $r \geq \lceil\frac{d+3}{2}\rceil$. The Green discrepancy (of order r) is defined by

$$D_N^{(G,r)}(\mathbf{x}_n) := \sup_{\mathbf{y}\in S^d}\left|\frac{4\pi}{N}\sum_{n=1}^{N} G^{(r)}(\mathbf{x}_n,\mathbf{y})\right| \qquad (0.7)$$

and might be a suitable measure for the quality of the distribution. In fact, Hlawka [13] considered an extension of this kind of discrepancy to Riemannian manifolds and proved that the cap discrepancy is compatible with the Green function discrepancy. Hlawka [13] established the following bounds

$$c_{10}(d)\left(D_N^{(G,r)}\right)^{\frac{d+1}{2r-d-1}} \le D_N^C \le c_{11}(d)\left(D_N^{(G,r)}\right)^{\frac{1}{4r+2d+3}}\left|\log D_N^{(G,r)}\right|. \qquad (0.8)$$

Note, that we suppose $r \ge [\frac{d+3}{2}]$. For $r > d+1$ the left hand of (0.8) can be improved to $c_{12}(d)D_N^{(G,r)}$ and for $r = d+1$ to $c_{13}(d)D_N^{(G,r)}|\log D_N^{(G,r)}|^{-1}$. The Green function discrepancy is extremely useful for estimating the approximation error $E_N(f)$ via a Koksma–Hlawka type inequality

$$E_N(f) \le V^{(r)}(f) \cdot D_N^{(G,r)}, \qquad (0.9)$$

where $V^{(r)}(f) = \int_{S^d} |\Delta^{(r)} f(\mathbf{x})|\, d\sigma(\mathbf{x})$ with $\Delta^{(r)}$ the r–times iterated Laplace–Beltrami operator, cf. [13].

A different approach was made by Cui and Freeden [5], who used elements of the theory of weighted Sobolev spaces to obtain an estimate for the approximation error in terms of the following modified polynomial discrepancy

$$\begin{aligned}
D_N^{P^*}(\mathbf{x}_n) &:= \frac{1}{N}\left[\sum_{i=1}^{N}\sum_{\ell=1}^{N}\sum_{m=1}^{\infty}\frac{1}{m(m+1)}P_m^d(\langle \mathbf{x}_i, \mathbf{x}_\ell\rangle)\right]^{\frac{1}{2}} \\
&= \frac{1}{N}\left[\sum_{i=1}^{N}\sum_{\ell=1}^{N}\left(1 - 2\ln\left(1 + \sqrt{\frac{1 - \langle \mathbf{x}_i, \mathbf{x}_\ell\rangle}{2}}\right)\right)\right]^{\frac{1}{2}}
\end{aligned}$$

for the case $d = 2$. The concept, however, is not restricted to the case $d = 2$ and in Section 1 we will give a generalization to aribitrary dimensions. We will prove that

$$D_N^P(\mathbf{x}_n) \le D_N^{P^*}(\mathbf{x}_n) \le c_{14}(d)\left(D_N^P\right)^{\frac{1}{2(d+1)}}. \qquad (0.10)$$

The appearance of Legendre polynomials and spherical harmonics in the expressions for discrepancies is not artificial but makes sense for another reason, namely the construction of optimally chosen integration points, so called spherical designs.

A point set $\mathbf{x}_1, \ldots, \mathbf{x}_N \in S^d$ is called a spherical t–design if

$$\frac{1}{N} \sum_{n=1}^{N} p(\mathbf{x}_n) = \int_{S^d} p(\mathbf{x}) \, d\sigma(\mathbf{x})$$

for all polynomials (in $d+1$ variables restricted to S^d) of degree not greater than t. Now, we would expect point sets which are spherical t–designs for large t to have small discrepancy. This is in fact the case; we may easily deduce from the definitions and a standard upper bound for the Legendre polynomials (cf. [19]) that for any spherical t–design $\mathbf{x}_1, \ldots, \mathbf{x}_N$ we have

$$D_N^P(\mathbf{x}_n) \ll \frac{1}{t^2}. \tag{0.11}$$

A similar bound is true for $D_N^{P^*}$, cf. [5]. For the cap discrepancy in [10] $D_N^C \ll \frac{1}{t}$ is established. Estimates concerning the number of points of a spherical t–design are due to Wagner [29] and Kuijlaars [16], where t–designs with $O(t^{\frac{d(d+1)}{2}})$ points are constructed. The construction makes use of Chebyshev quadrature formulae for ultraspherical weight functions. Delsarte, Goethals and Seidel [6] have proved that the number points of a t–design is bounded from below by

$$\binom{d+n-1}{d-1} + \binom{d+n-2}{d-1} \quad \text{for} \quad t = 2n,$$
$$2\binom{d+n-1}{d-1} \qquad \text{for} \quad t = 2n+1.$$

Bannai and Dammerell [1, 2] have shown that there exist no t–designs with equality in the above estimates.

Studying distribution properties of point sequences other quantities different from discrepancies are well known. For instance, the dispersion measures the denseness of infinite sequences. The dispersion

$$\Theta_N = \sup_{\mathbf{x} \in S^d} \min_{j \neq k} \delta(\mathbf{x}_k, \mathbf{x}) \tag{0.12}$$

(with respect to the geodesic metric δ on S^d) is just the radius of the largest spherical cap not containing one of the points $\mathbf{x}_1, \ldots, \mathbf{x}_N$. Note that sequences with dispersion tending to 0 need not be uniformly distributed. This spherical cap dispersion is of great interest, because it measures the approximation error of Quasi–Monte Carlo methods for global optimization; see Niederreiter [22] and the more recent contribution [4]. A related concept is the dispersion with respect to spherical slices (i. e. intersections of two half–spheres). It is defined as the angle of the largest slice not containing one of the points $\mathbf{x}_1, \ldots, \mathbf{x}_N$. This kind of dispersion is important for applications in computational geometry, cf. [25]. In [26] the following relation between dispersion and discrepancy is established:

$$\Theta_N(\mathbf{x}_n) \leq c_{15}(d) \left(D_N^C \right)^{\frac{1}{d}} . \tag{0.13}$$

Furthermore, we mention another quantity for measuring the distribution behaviour of a point set $\mathbf{x}_1, \ldots, \mathbf{x}_N$. For sequences in the unit cube Beardwood, Halton and Hammersley [3] studied the length $\mu_N(\mathbf{x}_n)$ of the shortest closed polygonal path joining the given points. For independently chosen random points these authors proved that

$$\mu_N \sim c_{16}(d) N^{\frac{d-1}{d}}, \quad \text{almost surely.} \tag{0.14}$$

In Section 2 of the present article we estimate $\mu_N(\mathbf{x}_n)$ from below by the spherical cap discrepancy obtaining

$$\mu_N(\mathbf{x}_n) \geq c_{17}(d) \left(D_N^C(\mathbf{x}_n) \right)^{-\frac{d-1}{d}} . \tag{0.15}$$

Finally, we remark that in principle there are several possibilities for constructing uniformly distributed sequences on the sphere. First, one can take a low–discrepancy point sequence in the unit cube and transform it via polar coordinates to the sphere. Secondly, there is a classical construction given by Pommerenke [23] using the representation of integers as sums of squares. Thirdly, Lubotzky, Phillips and Sarnak [18] established a construction using a free subgroup of the group of rotations. This methods uses a deep machinery, mainly modular forms and Hecke operators and is restricted to S^2.

1   Discrepancies Involving Pseudo–Differential Operators

In this section we want to describe a notion of discrepancy based on a certain class of pseudo–differential operators. As usual, we define the Fourier coefficient $f_{m,j}$ in the spherical harmonics expansion by

$$f_{m,j} := \int_{S^d} f(\mathbf{x}) S_{m,j}(\mathbf{x}) \, d\sigma(\mathbf{x}),$$

where σ is the normalized surface measure on S^d.

We define for $s \in \mathbb{R}^+$ the *weighted Sobolev space* $H^s(\Omega)$ by

$$H^s(S^d) := \left\{ f \mid f : S^d \to \mathbb{R}, \text{ such that } \sum_{m=0}^{\infty} \sum_{j=1}^{Z(d,m)} f_{m,j}^2 \, \widehat{m}^{2s} < \infty \right\}, \quad (1.16)$$

where

$$\widehat{m} = \begin{cases} 1 & \text{if } m = 0 \\ m & \text{otherwise} \end{cases}.$$

On this space we introduce an inner product and the corresponding norm in the obvious way

$$\langle f, g \rangle_{H^s} := \sum_{m=0}^{\infty} \sum_{j=1}^{Z(d,m)} f_{m,j} g_{m,j} \, \widehat{m}^{2s} \quad \text{and} \quad \|f\|_{H^s} := \sqrt{ \sum_{m=0}^{\infty} \sum_{j=1}^{Z(d,m)} f_{m,j}^2 \, \widehat{m}^{2s} }.$$

We proceed with an embedding theorem.

Proposition 1.1. *For $s > d/2$, $H^s(S^d)$ is a subspace of $C(S^d)$, the space of all continuous functions.*

Proof : Using the Cauchy–Schwarz inequality and the addition theorem for spherical harmonics we obtain

$$\left(\sum_{m=0}^{\infty} \sum_{j=1}^{Z(d,m)} |f_{m,j} S_{m,j}(\mathbf{x})| \right)^2 \leq$$

$$\left(\sum_{m=0}^{\infty} \sum_{j=1}^{Z(d,m)} f_{m,j}^2 \, \widehat{m}^{2s} \right) \left(\sum_{m=0}^{\infty} \sum_{j=1}^{Z(d,m)} S_{m,j}^2(\mathbf{x}) \widehat{m}^{-2s} \right) = \|f\|_{H^s} \sum_{m=0}^{\infty} Z(d,m) m^{-2s}.$$

$$(1.17)$$

Furthermore, $Z(d, m)$ satisfies

$$Z(d, m) \leq e^d m^{d-1},$$

and therefore the series (1.17) converges for $d - 1 - 2s < -1 \iff s > d/2.$ $\square$

The key concept for our further investigations is that of a pseudo–differential operator. We call $\mathbf{T}$ a *pseudo–differential operator* on S^d with symbol $\{T_{m,j},\ m = 0, 1, \ldots,\ 1 \leq j \leq Z(d, m)\}$ if

$$\mathbf{T} S_{m,j} = T_{m,j}\, S_{m,j}$$

holds for all spherical harmonics $S_{m,j}$. Furthermore we assume that there exist constants C_1, C_2, and s such that

$$C_1 m^s \leq T_{m,j} \leq C_2 m^s,$$

and call s the *order* of the operator $\mathbf{T}$. If $T_{m,j}$ does not depend on j, we set $T_m := T_{m,j}$.

Example 1 *Since the spherical harmonics are the eigenfunctions of the Laplace–Beltrami operator Δ corresponding to the eigenvalues $-m(m + d - 1)$, we conclude that Δ is a pseudo–differential operator of order 2.*

Let $\mathbf{T}$ be any pseudo–differential operator of order s. Then we get a different representation of the space H^s by

$$H^s(S^d) = \left\{ f \mid f : S^d \to \mathbb{R}, \mathbf{T} f \in \mathcal{L}_2(S^d) \right\}. \tag{1.18}$$

Now we are ready to prove an estimate for the approximation error.

Theorem 1.2. *Let $\mathbf{T}$ be a pseudo–differential operator of order s and symbol $\{T_m\}$, with $s > d/2$. Then for any function $f \in H^s$ we have*

$$E_N(f) \leq \frac{1}{N} \sqrt{\sum_{i=0}^{N} \sum_{\ell=0}^{N} \sum_{m=1}^{\infty} \frac{Z(d, m)}{T_m^2} P_m^d(\langle \mathbf{x}_i, \mathbf{x}_\ell \rangle)}\ \|\mathbf{T} f\|_{H^s}.$$

Proof: It is well known that the spherical harmonic expansion

$$f(\mathbf{x}) = \sum_{m=0}^{\infty} \sum_{j=1}^{Z(d,m)} f_{m,j} S_{m,j}(\mathbf{x})$$

converges uniformly since we have demanded $s > d/2$. Now we apply $\mathbf{T}$ to the above series, multiply by $S_{m,j}$ and integrate over S^d to obtain

$$f_{m,j} = \int_{S^d} \frac{\mathbf{T}f(\mathbf{y})S_{m,j}(\mathbf{y})}{T_m}\, d\sigma(\mathbf{y}), \text{ for } m > 1.$$

This means we can rewrite the spherical harmonic expansion

$$f(\mathbf{x}) = \int_{S^d} f(\mathbf{y})\, d\sigma(\mathbf{y}) + \sum_{m=1}^{\infty} \sum_{j=1}^{Z(d,m)} \int_{S^d} \frac{\mathbf{T}f(\mathbf{y})S_{m,j}(\mathbf{y})}{T_m}\, d\sigma(\mathbf{y})\, S_{m,j}(\mathbf{x}).$$

If we put $\mathbf{x} = \mathbf{x}_n$ and sum over n, we get

$$\frac{1}{N} \sum_{n=1}^{N} f(\mathbf{x}_n) = \int_{S^d} f(\mathbf{x})\, d\sigma(\mathbf{x})$$

$$+ \sum_{m=1}^{\infty} \sum_{j=1}^{Z(d,m)} \int_{S^d} \frac{\mathbf{T}f(\mathbf{y})S_{m,j}(\mathbf{y})}{T_m}\, d\sigma(\mathbf{y})\, \frac{1}{N}\left(\sum_{n=1}^{N} S_{m,j}(\mathbf{x}_n)\right). \quad (1.19)$$

Clearly, the approximation error can be estimated by the absolute value of the last term in (1.19):

$$E_N(f) \leq \frac{1}{N}\left| \int_{S^d} \mathbf{T}f(\mathbf{y}) \sum_{m=1}^{\infty} \sum_{j=1}^{Z(d,m)} \sum_{n=1}^{N} \frac{S_{m,j}(\mathbf{x}_n)S_{m,j}(\mathbf{y})}{T_m}\, d\sigma(\mathbf{y})\right|$$

$$\leq \frac{1}{N}\left(\int_{S^d} (\mathbf{T}f(\mathbf{y}))^2\, d\sigma(\mathbf{y})\right)^{\frac{1}{2}} \times$$

$$\times \left[\int_{S^d} \left(\sum_{m=1}^{\infty} \sum_{j=1}^{Z(d,m)} \sum_{n=1}^{N} \frac{S_{m,j}(\mathbf{x}_n)S_{m,j}(\mathbf{y})}{T_m}\right)^2 d\sigma(\mathbf{y})\right]^{\frac{1}{2}}$$

$$\leq \frac{1}{N}\|\mathbf{T}f\|_2 \left[\sum_{m=1}^{\infty} \sum_{j=1}^{Z(d,m)} \frac{1}{T_{m,j}^2}\left(\sum_{n=1}^{N} S_{m,j}(\mathbf{x}_n)\right)^2\right]^{\frac{1}{2}}$$

$$\leq \frac{1}{N}\|\mathbf{T}f\|_2 \left[\sum_{i=1}^{N} \sum_{\ell=1}^{N} \sum_{m=1}^{\infty} \sum_{j=1}^{Z(d,m)} \frac{1}{T_{m,j}^2} S_{m,j}(\mathbf{x}_i)\, S_{m,j}(\mathbf{x}_\ell)\right]^{\frac{1}{2}}$$

$$\leq \frac{1}{N}\|\mathbf{T}f\|_2 \left[\sum_{i=1}^{N} \sum_{\ell=1}^{N} \sum_{m=1}^{\infty} \frac{Z(d,m)}{T_m^2} P_m^d(\langle \mathbf{x}_i, \mathbf{x}_\ell\rangle)\right]^{\frac{1}{2}},$$

where we have used the Cauchy–Schwarz inequality, the orthonormality of $S_{m,j}$ and the addition theorem. Since by (1.18) $\|\mathbf{T}\cdot\|_2 = \|\cdot\|_{H^s}$ we have completed the proof. $\qquad\square$

The first factor in the estimate for the error is independent of the function and depends only on the point set. Therefore we can introduce a discrepancy by

$$D_N(\mathbf{x}_n;\mathbf{T}) := \frac{1}{N}\left[\sum_{i=1}^{N}\sum_{\ell=1}^{N}\sum_{m=1}^{\infty}\frac{Z(d,m)}{T_m^2}P_m^d(\langle\mathbf{x}_i,\mathbf{x}_\ell\rangle)\right]^{\frac{1}{2}}. \qquad (1.20)$$

Since $Z(d,m) \ll m^{d-1}$ and $T_m^2 \gg m^{2s}$ with $s > \frac{d}{2}$, this quantity is bounded for $N \to \infty$.

Theorem 1.3. *Let $f \in H^s$, where $s > d + \frac{1}{2}$. Then*

$$E_N(f) \leq c_s \|f\|_{H^s} D_N^C(\mathbf{x}_n).$$

Proof : Let $\mathbf{T}$ be a pseudo–differential operator of order s. By the result of Theorem 1.2 it suffices to show that

$$D_N(\mathbf{x}_n;\mathbf{T}) \ll D_N^C(\mathbf{x}_n).$$

Now

$$D_N(\mathbf{x}_n;\mathbf{T}) \leq \sum_{m=1}^{\infty}\frac{Z(d,m)\,m^d}{T_m^2}\,D_N^P(\mathbf{x}_n). \qquad (1.21)$$

Since $D_N^P(\mathbf{x}_n) \ll D_N^C(\mathbf{x}_n)$, we are finished once we have shown that the series (1.21) converges. But as $\mathbf{T}$ is of order s we have

$$T_m \ll m^s,$$

and the series (1.21) is bounded by

$$c_{18}(d)D_N^P(\mathbf{x}_n)\sum_{m=1}^{\infty}m^{2d}\,m^{-2s},$$

which is finite if $2d - 2s < -1 \iff s > d + \frac{1}{2}$. $\qquad\square$

A specialisation of the pseudo–differential operator leads to the discrepancy $D_N^{P^*}$, which was defined in the introduction for $d = 2$. Consider the operator $\mathbf{A}$ for S^2 with symbol $A_m^2 = (2m + 1)m(m + 1)$. Now the identity

$$\sum_{n=1}^{\infty} \frac{P_m^2(t)}{m(m + 1)} = 1 - 2\ln\left(1 + \sqrt{\frac{1 - t}{2}}\right)$$

together with formula (1.20) yields immediately

$$D_N(\mathbf{x}_n; \mathbf{A}) = D_N^{P^*}(\mathbf{x}_n).$$

We will now generalize this idea to arbitrary dimensions. To evaluate the sum in (1.20) we will use the generating functions of the Legendre polynomials (0.5). Consider the operator $\mathbf{A}$ for S^d with the symbol

$$A_m^2 = Z(d, m)m(m + d - 1),$$

which together with formula (1.20) suggests the following definition

$$D_N^{P^*}(\mathbf{x}_n) := D_N(\mathbf{x}_n; \mathbf{A}) = \frac{1}{N}\left[\sum_{i=1}^{N}\sum_{\ell=1}^{N}\sum_{m=1}^{\infty} \frac{1}{m(m + d - 1)}P_m^d(\langle \mathbf{x}_i, \mathbf{x}_\ell\rangle)\right]^{\frac{1}{2}}.$$

$$(1.22)$$

Theorem 1.4.

$$D_N^P(\mathbf{x}_n) \le D_N^{P^*}(\mathbf{x}_n) \le c_{14}(d)\left(D_N^P\right)^{\frac{1}{2(d+1)}}.$$

Proof : The left hand side inequality is obvious. For the proof of the right hand side choose $K = [(D_N^P)^{-\frac{1}{2(d+1)}}]$ and estimate

$$\left(D_N^{P^*}\right)^2 = \sum_{m=1}^{\infty} \frac{m^{2d}}{m(m + d - 1)}\frac{1}{m^{2d}}\frac{1}{N^2}\sum_{i=1}^{N}\sum_{\ell=1}^{N} P_m^d(\langle \mathbf{x}_i, \mathbf{x}_\ell\rangle)$$

$$\le \sum_{m=1}^{K} \frac{m^{2d-1}}{m + d - 1}\left(D_N^P\right)^2 + \sum_{m=K+1}^{\infty} \frac{1}{m(m + d - 1)}$$

$$\ll K^{2d-1}\left(D_N^P\right)^2 + \frac{1}{K} \ll \left(D_N^P\right)^{\frac{1}{d+1}},$$

which proves the theorem. $\qquad\qquad\square$

The following proposition shows how we can use generating functions to evaluate the series (1.22). We omit a detailed proof which runs by induction.

Proposition 1.5. *Let $\ell \geq 0$, $k \geq 0$ and*

$$g(\omega) = \sum_{m=1}^{\infty} c_m \omega^m.$$

Then

$$\sum_{m=1}^{\infty} \frac{c_m \omega^m}{(m+\ell)(m+\ell+1)\cdots(m+\ell+k)} = \frac{1}{k!} \sum_{i=0}^{k} (-1)^i \binom{k}{i} \frac{1}{\omega^{\ell+i}} \int_0^{\omega} g(y) y^{\ell+i-1} dy.$$

$$(1.23)$$

Corollary 1.6.

$$\sum_{m=1}^{\infty} \frac{P_m^d(x)}{m(m+d-1)} = \int_0^1 \left((y^2 - 2yx + 1)^{-\frac{d-1}{2}} - 1 \right) \frac{(1-y)^{d-1}}{y} \, dy.$$

Proof : We again use generating functions. Set $\ell = 0$, $k = d-1$,

$$c_m = (m+1)(m+2)\cdots(m+d-2)P_m^d(x),$$

and

$$g(\omega) = \frac{1}{(1 - 2x\omega + \omega^2)^{(d-1)/2}} - 1$$

in the proposition and the result follows setting $\omega = 1$. $\qquad\qquad\square$

The integrals, which appear in (1.23), can be evaluated using the formula

$$\int \frac{x^m}{(ax^2 + c)^{k+1/2}} \, dx = \frac{1}{(2k-1)c} \frac{x^{m+1}}{(ax^2 + c)^{k-1/2}}$$
$$+ \frac{2k-m-2}{(2k-1)c} \int \frac{x^m}{(ax^2 + c)^{k-1/2}} \, dx,$$

cf. [11], page 43. It is therefore clear that the expression on the right hand side in (1.23) can be expressed by elementary functions. We will use the program MAPLE to compute this expression for the case $d = 3$ and obtain

$$\frac{3}{2} - 2\sqrt{\frac{1-x}{1+x}} \arctan \sqrt{\frac{1+x}{1-x}} =: \psi(x)$$

for the right hand side in Corollary 1.6.

Thus we have for $d = 3$

$$D_N^{P^*}(\mathbf{x}_n) = \frac{1}{N} \left(\sum_{i=1}^{N} \sum_{\ell=1}^{N} \psi(\langle \mathbf{x}_i, \mathbf{x}_\ell \rangle) \right)^{\frac{1}{2}}.$$

2   Shortest Paths on the Sphere

In this section we will estimate the length of the shortest path through a point set on the sphere S^d in terms of the spherical cap discrepancy.

The following result is due to Wyner [30] (p. 1090).

Proposition 2.7. *The maximal number $M(\rho)$ of non–intersecting spherical caps with radius ρ satisfies*

$$M(\rho) \geq \frac{\omega_d}{A(2\rho)},$$

where $A(\rho)$ is the surface area of a cap with radius ρ and ω_d denotes the surface area of S^d.

Proof: Let $\{C(\mathbf{x}_1, \rho), \ldots, C(\mathbf{x}_{M(\rho)}, \rho)\}$ be a maximal set of non–intersecting caps. Then the set $\{C(\mathbf{x}_1, 2\rho), \ldots, C(\mathbf{x}_{M(\rho)}, 2\rho)\}$ covers the whole sphere, since

$$\mathbf{y} \notin \bigcup_{n=1}^{M(\rho)} C(\mathbf{x}_n, 2\rho)$$

implies that $C(\mathbf{y}, \rho)$ would not intersect any of the $C(\mathbf{x}_n, \rho)$. Therefore

$$M(\rho)\, A(2\rho) \geq \omega_d,$$

which proves the proposition. $\square$

Theorem 2.8. *Let $\mathbf{x}_1, \mathbf{x}_2, \ldots$ be a point sequence on S^d. Then for sufficiently large N*

$$\mu(N) \geq c_{17}(d) \left(D_N^C(\mathbf{x}_n) \right)^{-(d-1)/d},$$

where $c_{17}(d)$ only depends on the dimension d.

Proof : From [30] (p. 1090) we know

$$A(\rho) = \frac{d\,\pi^{d/2}}{\Gamma(\frac{d}{2}+1)} \int_0^\rho (\sin\phi)^{d-1}\,d\phi =: \kappa_d \int_0^\rho (\sin\phi)^{d-1}\,d\phi. \tag{2.24}$$

Now choose ρ such that

$$D_N^C(\mathbf{x}_n) = A(\alpha\rho), \tag{2.25}$$

where α is some for the moment arbitrary constant in the interval $(0,1)$. This is possible for large enough N since $A(\rho)$ is a continuous strictly increasing function. From Proposition 2.7 it follows that we can place $M(\rho) \geq \frac{\omega_d}{A(2\rho)}$ non–intersecting caps with angle ρ on S^d. Now we conclude to obtain

$$\mu_N(\mathbf{x}_n) \geq (1-\alpha)2\rho M(\rho) \geq (1-\alpha)2\rho\frac{\omega_d}{A(2\rho)}.$$

From (2.24) and $\sin x \leq x$ we get

$$\mu(N) \geq c_{19}(d)(1-\alpha)\rho^{1-d}.$$

On the other hand (2.25) and the fact that $\sin x \geq \frac{2}{\pi}x$ for $x \in [0, \frac{\pi}{2}]$ yield

$$D_N^C(\mathbf{x}_n) \leq c_{20}(d)\,\alpha^d\rho^d.$$

Combining these two results we obtain

$$\begin{aligned}
\mu_N(\mathbf{x}_n) \;&\geq\; c_{19}(d)(1-\alpha)\left[\left(\frac{D_N^C(\mathbf{x}_n)}{c_{20}(d)}\right)^{\frac{1}{d}}\alpha^{-1}\right]^{-(d-1)} \\
&=\; c_{19}(d)\,(c_{20}(d))^{\frac{d-1}{d}}\left(\alpha^{d-1}-\alpha^d\right)\left(D_N^C(\mathbf{x}_n)\right)^{-\frac{d-1}{d}},
\end{aligned}$$

where the optimal constant is attained for $\alpha = (d-1)/d$. $\qquad\square$

Remark 2 *We note that for a suitable infinite sequence of points $(\mathbf{x}_n)$ contained in one half sphere, $\mu_N(\mathbf{x}_n)$ may tend to infinity, whereas $D_N^C(\mathbf{x}_n) \geq \frac{1}{2}$ (for all N). This means that there are no upper bounds for $\mu_N(\mathbf{x}_n)$ of the same shape as the lower bound (0.15).*

References

[1] E. Bannai, R. M. Damerell: *Tight spherical designs I*, J. Math. Soc. Japan **31** (1979), 199–207.

[2] E. Bannai, R. M. Damerell: *Tight spherical designs II*, J. London Math. Soc. (2) **21** (1980), 13–30.

[3] J. Beardwood, J. H. Halton, J. M. Hammersley: *The shortest path through many points*, Proc. Cambridge Phil. Soc. **55** (1959), 299–327.

[4] C. Biester, P. J. Grabner, G. Larcher, R. F. Tichy: *Adaptive search in quasi–Monte–Carlo optimization*, Math. Comp. **64** (1995), 807–818.

[5] J. Cui, W. Freeden: *Equidistribution on the sphere*, to appear.

[6] P. Delsarte, J. M. Goethals, J. J. Seidel: *Spherical codes and designs*, Geom. Dedicata **6** (1977), 363–388.

[7] M. Drmota, R. F. Tichy: *Sequences, Discrepancies, and Applications*, Lecture Notes Math., Springer, Berlin–Heidelberg–New York 1996, to appear.

[8] W. Freeden: *On integral formulas of the (unit) sphere and their applications to numerical computations of integrals*, Computing **25** (1980), 131–146.

[9] P. J. Grabner: *Erdös–Turán type discrepancy bounds*, Monatsh. Math. **111** (1991), 127–135.

[10] P. J. Grabner, R. F. Tichy: *Spherical designs, discrepancy and numerical integration*, Math. Comp. **60** (1993), 327–336.

[11] W. Gröbner, N. Hofreiter: *Integraltafel, 1. Teil*, Springer, Wien 1949.

[12] E. Hlawka: *Zur quantitativen Theorie der Gleichverteilung*, Österr. Akad. Wiss. math.–nat. Kl. S–B. II **184** (1975), 335–365.

[13] E. Hlawka: *Gleichverteilung auf Produkten von Sphären*, J. Reine Angew. Math. **330** (1982), 1–43.

[14] E. Hlawka: *The Theory of Uniform Distribution*, Academic Publishers, Herts, UK, 1984.

[15] B. Klinger, R. F. Tichy: *Polynomial discrepancy of point sequences*, to appear, 1996.

[16] A. Kuijlaars: *The minimal number of nodes in Chebyshev quadrature formulas*, Indag. Math. **4** (1993), 339–362.

[17] L. Kuipers, H. Niederreiter: *Uniform Distribution of Sequences*, Wiley, New York 1974.

[18] A. Lubotzky, R. Phillips, P. Sarnak: *Hecke operators and distributing points on the sphere*, Comm. Pure Appl. Math. **39** (1986), S149–186.

[19] W. Magnus, F. Oberhettinger, R. P. Soni: *Formulas and Theorems for the Special Functions of Mathematical Physics*, Springer, Berlin 1966.

[20] C. Müller: *Spherical Harmonics*, Lecture Notes Math. 17, Springer, Berlin 1966.

[21] D. J. Newman, H. S. Shapiro: *Jackson's theorem in higher dimensions*, in: *Über Approximationstheorie*, P. L. Butzer and J. Korevaar (eds.), 209–219, Birkhäuser, Basel 1964.

[22] H. Niederreiter: *Random Number Generation and Quasi–Monte–Carlo Methods*, CBMS–NSF Regional Conf. Ser. Appl. Math. **63**, SIAM, 1992.

[23] C. Pommerenke: *Über die Gleichverteilung von Gitterpunkten auf m–dimensionalen Ellipsoiden*, Acta Arith. **5** (1959), 227–257.

[24] M. Reimer: *Constructive Theory of Multivariate Functions: With an Application to Tomography*, BI–Wissenschaftsverlag, Mannheim 1990.

[25] G. Rote, R. F. Tichy: *Spherical dispersion with an application to polygonal approximation of curves*, Österr. Akad. Wiss., Anzeiger II **132** (1995), 3–10.

[26]   G. Rote, R. F. Tichy: *Quasi–Monte–Carlo methods and the dispersion of point sequences,* Math. Comput. Modelling **23** (1996), 9–23.

[27]   R. F. Tichy: *Beiträge zur Polynomdiskrepanz,* Österr. Akad. Wiss., math.–nat. Kl. Abt. II **193** (1984), 513–519.

[28]   J. D. Vaaler: *Some extremal functions in Fourier analysis,* Bull. Amer. Math. Soc. **12** (1985), 183–216.

[29]   G. Wagner: *On averaging sets,* Monatsh. Math. **111** (1991), 69–78.

[30]   A. D. Wyner: *Capabilities of bounded discrepancy decoding,* Bell System Technical J. **44** (1965), 1061–1122.

Addresses:

PETER J. GRABNER, ROBERT F. TICHY
Department of Mathematics A
Technical University of Graz
Steyrergasse 30
A–8010 Graz
Austria

BERNHARD KLINGER
Department of Mathematics
Purdue University
West Lafayette, IN 47907–1395
USA

Regularity of Multivariate Hermite Interpolation

H. Hakopian

Abstract

In this paper a result due to H. Gevorgian, A. Sahakian and the author concerning the regularity of bivariate Hermite interpolation is generalized in two directions: in the bivariate case and for arbitrary dimensions. As a corollary a number inequality involving combinations is obtained.

1 Introduction and Main Results

In this paper Hermite interpolation by multivariate algebraic polynomials is studied. The interpolation parameters are the values of a function and its partial derivatives up to some order $n_\nu - 1$ at given points y_ν, where n_ν is the multiplicity of y_ν. The space of polynomials is $\pi_{n,k} := \{k\text{-variate polynomials of total degree} \leq n\}$.

We generalize a result from [1] (Theorem 3.3), which can be formulated in the following way:

Theorem 1.1. *Let $\mathcal{N} = \{n_i\}_{i=1}^s$ be a non-increasing sequence of multiplicities satisfying the conditions*

$$n_1 + n_2 + n_3 \leq n, \quad \sum_{i=1}^s n_i \leq 3n. \tag{1.1}$$

Then the interpolation parameters at the points of the set

$$\mathcal{Y} = \{y_i\}_{i=1}^s \subset \{(x, x^3) : x > 0\}$$

with the corresponding multiplicities from $\mathcal{N}$ are independent for $\pi_{n,2}$.

Note that this theorem, in its turn, is an extension of a result due to Nagata [2] where one has $s \leq 9$ instead of the second condition of (1.1).

Multivariate Approximation: Recent Trends and Results; W. Haußmann, K. Jetter and M. Reimer (eds.)
Mathematical Research, Vol. 101, pp. 113–121, ISBN 3-05-501770-6
© Akademie–Verlag, Berlin 1997

We generalize Theorem 1.1 in the mentioned bivariate case and for arbitrary dimensions. The result is a scale of type (1.1) conditions implying the above independence for $\pi_{n,k}$:

Theorem 1.2. *Let $\mathcal{N} = \{n_i\}_{i=1}^s$ be a non-increasing sequence of multiplicities satisfying for some q, $1 \leq q \leq s$, the conditions*

$$\sum_{i=1}^q n_i \leq A, \quad \sum_{i=1}^s n_i \leq q^{k-1} A, \tag{1.2}$$

where

$$A := A(n,q) := n + 1 - \frac{(q-1)(k-1)}{2}.$$

Then the interpolation parameters at the points of the set

$$\mathcal{Y} = \{y_i\}_{i=1}^s \subset \{(x, x^q, \ldots, x^{q^{k-1}}) : x > 0\}$$

with the corresponding multiplicities from $\mathcal{N}$ are independent for $\pi_{n,k}$.

Note that the case $q = 1$ of this result, in view of Remark 2.1 ii) below, is equivalent to a theorem due to Severi [3].

Now let us fix the set of points $\mathcal{Y}$ with

$$y_i = (x_{0i}, x_{0i}^q, \ldots, x_{0i}^{q^{k-1}}), \quad x_{0i} > 0, \quad i = 1, \ldots, s.$$

Denote $\mathcal{Y}^* := \{y_i^*\}_{i=1}^s$, where

$$y_i^* = (x_{0i}, x_{0i}^q, \ldots, x_{0i}^{q^{l-1}})$$

and l ($l = 1, \ldots, k - 1$) is the number of variables which will be clear from the context.

Also, for the multiindex $\alpha = (\alpha_o, \ldots, \alpha_{k-1}) \in Z_+^k$ we denote

$$D^\alpha P := \frac{\partial^{|\alpha|} P}{\partial x_0^{\alpha_0} \cdots \partial x_{k-1}^{\alpha_{k-1}}}, \quad |\alpha| := \alpha_0 + \cdots + \alpha_{k-1}.$$

Now note that, for proving the Theorem, it is enough to show that one can add

to $\mathcal{Y}$ a set of $m = m(k) \geq 0$ simple points $\mathcal{Z} = \{z_i\}_{i=1}^m$, so that the number of conditions imposed by the resulting set be equal to $dim\,\pi_{n,k}$ and the resulting Hermite interpolation problem be regular (cf. Remark 2.3 below).

The latter is equivalent to the following assertion: a polynomial from $\pi_{n,k}$, for which the above interpolation parameters are zero, is identically zero. Geometrically this means that there is no hypersurface of degree n passing through the points of the resulting set with corresponding multiplicities.

Actually we have that the interpolation parameters imposed by $\mathcal{N}$ are independent also at the points of the set $\mathcal{Y}^*$ for some spaces $\pi_{n,l,q}$ of l-variate polynomials (which we define later). Namely, we have

Proposition 1.3. *Let $1 \leq l \leq k$ and let $\{n_i\}_{i=1}^s$ be as in Theorem 1.2. Then there is a set of points $\mathcal{Z} = \mathcal{Z}(n, l, q) = \{z_i\}_{i=1}^{m(l)}$ with $m(l) = m(n, l, q) \geq 0$ determined by the equality*

$$\sum_{i=1}^{s} \binom{n_i + l - 1}{l} + m(l) = \dim \pi_{n,l,q}, \tag{1.3}$$

such that the conditions

$$D^\alpha p(y_i^*) = 0, \quad |\alpha| \leq n_i - 1, \quad i = 1, \ldots, s,$$

$$p(z_i) = 0, \quad i = 1, \ldots, m(l)$$

imply that the polynomial $p \in \pi_{n,l,q}$ vanishes identically.

As we will see later, the last of above polynomial spaces – $\pi_{n,k,q}$ coincides with $\pi_{n,k}$. Therefore in the case $l = k$ Theorem 1.2 coincides with Proposition 1.3. Hence we need to prove only this latter.

2 Sketch of Proof of Proposition 1.3

The proof is carried out in several steps.

STEP 0. Here we give a property of non-increasing sequences $\mathcal{N} = \{n_i\}_{i=1}^s$ satisfying (1.2). Namely, we prove that the sequences

$$\mathcal{N} - \nu_+ := \{(n_i - \nu)_+\}_{i=1}^s$$

satisfy the inequalities (1.2)

i) with $A(n - \nu q, q)$ for $\nu = 1, \ldots, n_q$,

and

ii) with $A(n - n_q q + n_q - \nu, 1)$ for $\nu = n_q + 1, \ldots, n - n_q q + n_q$.

STEP 1. In this step we prove Proposition 1.3 in the case $l = 1$. To this end we consider the following class of univariate polynomials:

$$\pi_{n,1,q} = \left\{ \sum_{|\alpha| \le n} a_\alpha x_0^{\alpha * q} : a_\alpha \text{ is real} \right\},$$

where

$$\alpha * q := \sum_{i=0}^{k-1} \alpha_i q^i \quad \text{for} \quad k - \text{multiindex} \quad \alpha = (\alpha_0, \dots, \alpha_{k-1}).$$

These classes are connected with the classes $\pi_{n,k}$ as follows:

$$\text{if} \quad p(x) = \sum_{|\alpha| \le n} a_\alpha x^\alpha \in \pi_{n,k} \quad \text{then} \quad p_0(x_0) := p(x_0, x_0^q, \dots, x_0^{q^{k-1}}) \in \pi_{n,1,q}.$$

Let us denote

$$M := M(n, 1, q) := \dim \pi_{n,1,q}.$$

Descartes sign change rule implies that a polynomial $p \in \pi_{n,1,q}$ identically equals to zero if the number of its positive roots equals to M (since $\pi_{n,1,1} = \pi_{n,1}$, in the case $q = 1$ we can omit here the word "positive"). Therefore, we need only to prove that

$$\sum_{i=1}^{s} n_i \le M. \tag{2.1}$$

First we obtain an explicit representation for M:

$$M(n, 1, q) = \#I(n, 1, q), \tag{2.2}$$

where

$$I(n, 1, q) := \{\alpha = (\alpha_0, \dots, \alpha_{k-1}) : |\alpha| \le n, \alpha_i \le q - 1, i = 0, \dots, k - 2\}.$$

Then we prove the inequality

$$M(n, 1, q) \ge q^{k-1} A(n, q),$$

implying (2.1) in view of (1.2).

In the next Steps 2 and 3 we prove Proposition 1.3 by the use of induction on l. For the case $l = 1$ this was proved in Step 1. Now we assume that it is true for $l - 1$ and prove for l.

STEP 2. Consider the following class of l-variate polynomials

$$\pi_{n,l,q} = \left\{ \sum_{|\alpha|+|\beta|\leq n} a_{\alpha,\beta}\bar{x}^{\alpha}x_{l-1}^{\beta*q} : a_{\alpha,\beta} \text{ is real} \right\},$$

where $\bar{x} = (x_0,\ldots,x_{l-2})$ and α, β are $l-1$ and $k-l+1$-multiindices respectively.

In this step we carry out a procedure described below.

First we note that

$$p(\bar{x}, x_{l-1}) \in \pi_{n,l,q} \quad \text{implies} \quad p(\bar{x}, x_{l-2}^q) \in \pi_{n,l-1,q}.$$

Then by imposing conditions at some simple points of hypersurface $x_{l-1} = x_{l-2}^q$ and using induction hypothesis we make the polynomial $p(\bar{x}, x_{l-1}) \in \pi_{n,l,q}$ vanish on this hypersurface. Now using Bézout's theorem we get the factorization:

$$p(\bar{x}, x_{l-1}) = (x_{l-1} - x_{l-2}^q)\, p_1(\bar{x}, x_{l-1}). \tag{2.3}$$

Then we observe that

$$p \in \pi_{n,l,q} \quad \text{and} \quad p(\bar{x}, x_{l-1}) = (x_{l-1} - x_{l-2}^q)\, p_1(\bar{x}, x_{l-1}) \quad \text{imply} \quad p_1 \in \pi_{n-q,l,q}.$$

On the other hand, in view of relation (2.3), the sequence of multiplicities of p_1 at the set $\mathcal{Y}^*$ coincides with $\mathcal{N}-1_+$. This sequence, according to Step 0, satisfies the inequalities (1.2) with $A(n-q, q)$, provided $n_q \neq 0$. Therefore, again using induction hypothesis and imposing conditions at some simple points we can make p_1 vanish on the hypersurface $x_{l-1} = x_{l-2}^q$ and factorize p_1 and the subsequent factors:

$$p_j(\bar{x}, x_{l-1}) = (x_{l-1} - x_{l-2}^q)\, p_{j+1}(\bar{x}, x_{l-1}), \tag{2.4}$$

where $p_i \in \pi_{n-iq,l,q}$. Note that we use

$$m(n - jq, l - 1, q) = M(n - jq, l - 1, q) - \sum_{i=1}^{s} \binom{(n_i-j)_+ + l - 2}{l-1}$$

simple points to get the factorizations (2.3) ($j = 0$) and (2.4).

In the case $q = 1$ we finish this process at the stage $j = n$. Otherwise, we end at the stage $j = n_q - 1$ and continue the consideration in the next step.

STEP 3. Here we show that slightly perturbing the simple points added in Step 2 we obtain a set $\mathcal{Z}$ satisfying Proposition 1.3.

We start with the following

Remark 2.1. i) *If the interpolation parameters at the points* $\{z_i\}_{i=1}^s$ *with some multiplicities are independent (for* $\pi_{n,k}$*), then the same is true for any set* $\{z'_i\}_{i=1}^s$*, where* z'_i *is sufficiently close to* z_i*. Moreover*

ii) *If the interpolation parameters at the points of any set* $\{z_i\}_{i=1}^s$*, from an* m*-dimensional* $(1 \leq m \leq k)$ *linear subspace, with some multiplicities, are independent (for* $\pi_{n,k}$*), then the same is true for arbitrary set* $\{z'_i\}_{i=1}^s$*.*

Next lemma is the basic tool we shall use in this step.

Lemma 2.2. *Let* $\mathcal{Z}(n - (j+1)q, l, q)$ *and* $\mathcal{Z}(n - jq, l - 1, q)$ *be some sets satisfying Proposition* 1.3*. Then there is a set* $\mathcal{Z}'(n - (j+1)q, l, q)$ *of points sufficiently close to those of the first set, such that*

$$\mathcal{Z}(n - jq, l, q) = \mathcal{Z}'(n - (j+1)q, l, q) \cup \mathcal{Z}(n - jq, l - 1, q),$$

i. e. the set in the right-hand side satisfies Proposition 1.3 *for the triplet of parameters* $(n - jq, l, q)$*.*

Sketch of proof of Lemma 2.2. First we slightly change positions of the points of the set $\mathcal{Z}(n-(j+1)q, l, q)$ in order to get a set $\mathcal{Z}'(n-(j+1)q, l, q)$ which has no points on the hypersurface $x_{l-1} = x_{l-2}^q$. Now, in view of the factorization (2.4) and Remark 2.1 i), we need only to check the equality

$$m(n - jq, l, q) = m(n - (j+1)q, l, q) + m(n - jq, l - 1, q),$$

which in view of (1.3) and the identity

$$\binom{\nu + 1}{l} = \binom{\nu}{l} + \binom{\nu}{l - 1}$$

reduces to the following equality:

$$M(n - jq, l, q) = M(n - (j + 1)q, l, q) + M(n - jq, l - 1, q).$$

To check this we use an explicit representation for $M(n, l, q)$ similar to one described in (2.2) in Step 1.

Now our aim is to accomplish the proof of Proposition 1.3. First let us check the case $q = 1$. In view of Lemma 2.2 we need only to show that there is a set $\mathcal{Z}(-1, l, 1)$. The latter is obvious since for the considered parameters (n, l, q) the sequence $\mathcal{N}$ satisfying (1.2) turns into a null sequence and the polynomial class consists only of zero.

Next, for proving the general case we need to show that there is a set $\mathcal{Z}(n - n_q q, l, q)$ satisfying the Proposition 1.3. To this end note that, according to the just proved case, for any set $\mathcal{Y}^*$ of points on the line $x_1 = \cdots = x_l$ and the space of polynomials

$$\pi_{n - n_q q, l} = \pi_{n - n_q q, l, 1}$$

there is a set $\mathcal{Z}(n - n_q q, l, 1)$. Now making use of Remark 2.1 ii) we can assume that the set of points coincides with $\mathcal{Y}^*$. To end the proof it remains only to note that

$$\pi_{n - n_q q, l, 1} \subset \pi_{n - n_q q, l, q}$$

and therefore the set $\mathcal{Z}(n - n_q q, l, 1)$ can be extended up to a set $\mathcal{Z}(n - n_q q, l, q)$ satisfying the Proposition 1.3.

At the end of this section we give the following remark concerning the addition of simple points to $\mathcal{Y}$, up to obtaining a regular Hermite interpolation problem.

Remark 2.3. *Let the Hermite interpolation parameters at the points of some set $\mathcal{Y}$ (with the corresponding multiplicities) be independent and let $\mathcal{L}$ be any set of points at which the Lagrange interpolation problem is regular (both problems considered for $\pi_{n,k}$). Then the process of successive addition to $\mathcal{Y}$ of simple points from $\mathcal{L}$, checking the independence of resulting conditions, leads to a regular Hermite interpolation problem.*

3   A Number Inequality

An immediate consequence of Theorem 1.2 is the following number inequality.

Corollary 3.1. *Let* $\mathcal{N} = \{n_i\}_{i=1}^{s}$ *be a non-increasing sequence of nonnegative integers satisfying for some* q, $1 \le q \le s$, *the conditions* (1.2), *then*

$$\sum_{i=1}^{s} \binom{n_i+k-1}{k} \le \binom{n+k}{k}. \tag{3.1}$$

This inequality can be proved also directly, and for any sequence of nonnegative numbers, by the use of the following theorem, which is a generalization of Lemma 2.1 from [1].

Theorem 3.2. *Let* $n_1 \ge n_2 \ge \cdots \ge n_s$ *and* $m_1 \ge m_2 \ge \cdots \ge m_s$ *be arbitrary collections of nonnegative numbers and* q, $1 \le q \le s$, *be an integer. Then*

$$\sum_{i=1}^{s} n_i m_i \le \frac{M}{q} \sum_{i=1}^{q} n_i,$$

where

$$M = \max(q m_1, \sum_{i=1}^{s} m_i).$$

Moreover we have equality in (1.1) *if and only if*

    i) $q m_1 \le \sum_{i=1}^{s} m_i$ *in the case of* $n_q \ne 0$,

and there is an integer j, $1 \le j \le q - 1$, *for which*

    ii) $m_1 = \cdots = m_j = \frac{M}{q}$ *in the case of* $j \ne 0$,

    iii) $n_1 = \cdots = n_l$ *with* $q \le l < s$ *and* $m_{l+1} = 0$.

Indeed, applying Theorem 3.2 for the sequences $\{n_i + 1\}_{i=1}^{s}$ and $\{n_i\}_{i=1}^{s}$, where the last one satisfies (1.2), we obtain

$$\sum_{i=1}^{s} n_i(n_i + 1) \le q^{k-2} A(A + q).$$

Then applying it again for the sequences $\{n_i + 2\}_{i=1}^{s}$ and $\{n_i(n_i + 1)\}_{i=1}^{s}$, we obtain

$$\sum_{i=1}^{s} n_i(n_i + 1)(n_i + 2) \le q^{k-3} A(A + q)(a + 2q).$$

Continuing in this way finally we get

$$\sum_{i=1}^{s} n_i \cdots (n_i + k - 1) \le A(A + q) \cdots (A + (k - 1)q).$$

In view of the inequality

$$A(A + q) \cdots (A + (k - 1)q) \le (n + 1) \cdots (n + k)$$

(which follows from the equality $A + (A + (k - 1)q) = (n + 1) + (n + k)$) we come to (3.1).

Let us mention that similarly one can show that inequalities (1.2) (for arbitrary A) imply

$$\sum_{i=1}^{s}(n_i)^j \le q^{k-j} A, \quad j = 1, \ldots, k.$$

References

[1]   H. Gevorgian, H. Hakopian, A. Sahakian: *On the bivariate Hermite interpolation problem,* Constr. Approx. **11** (1995), 23–36.

[2]   M. Nagata: *On rational surfaces II,* Mem. Coll. Sci. Kyoto **33** (1960), 271–293.

[3]   F. Severi, E. Löfler: *Vorlesungen über Algebraische Geometrie,* Teubner, Berlin 1921.

Address:

HAKOP HAKOPIAN
Department of Mathematics
The Yerevan State University
Alex Manoukian St. 1
Yerevan–375049
Armenia

Extremizers for the Multivariate Landau–Kolmogorov Inequality

O. Kounchev

Abstract

In the present paper we prove the following sharp inequality

$$\|\Delta f\|_\infty \le 2\sqrt{\frac{d}{d+2}}\sqrt{\|f\|_\infty\,\|\Delta^2 f\|_\infty}$$

in $\mathbb{R}^d$, $d \ge 2$, and we write the explicit solution (the extremizer) for the equality case. This result is an analogue to the famous inequality of Landau in $\mathbb{R}$: $\|f'\|_\infty \le \sqrt{2}\sqrt{\|f\|_\infty\,\|f''\|_\infty}$. In both cases $\|.\|_\infty$ means the L_∞ norm on $\mathbb{R}^d$, resp. $\mathbb{R}$.

1 The Classical Inequality of Landau–Kolmogorov

We will give a general setting for inequalities of the type of Landau–Kolmogorov. For a function $f \in C^{n-1}(\mathbb{R})$ which has a bounded n–th derivative a.e. in $\mathbb{R}$ we use $M_k(f) = \sup_{x\in\mathbb{R}} |f^{(k)}(x)|$ for $0 \le k \le n-1$ and

$$\| f^{(n)} \|_\infty \;=\; M_n(f) = \operatorname{ess\,sup}_{x\in\mathbb{R}} \left| f^{(n)}(x) \right|.$$

First we will recall the complete result of A. N. Kolmogorov, see [7], which includes the inequality of Landau as a special case.

Theorem 1.1. *For every bounded f as above we have*

$$M_k(f) \le C_{nk} M_0(f)^{1-\frac{k}{n}} M_n(f)^{\frac{k}{n}} \tag{1.1}$$

where $C_{nk} = K_{n-k}/K_n^{1-\frac{k}{n}}$ for $0 < k < n$ with

$$K_i = \begin{cases} \frac{4}{\pi}\left(1 - \frac{1}{3^{i+1}} + \frac{1}{5^{i+1}} - \frac{1}{7^{i+1}} + \ldots\right) & \text{for even } i, \\[2mm] \frac{4}{\pi}\left(1 + \frac{1}{3^{i+1}} + \frac{1}{5^{i+1}} - \frac{1}{7^{i+1}} + \ldots\right) & \text{for odd } i. \end{cases}$$

Multivariate Approximation: Recent Trends and Results; W. Haußmann, K. Jetter and M. Reimer (eds.)
Mathematical Research, Vol. 101, pp. 123–132, ISBN 3–05–501770–6
© Akademie–Verlag, Berlin 1997

The extremizing function for which (1.1) *becomes equality for all k is given by*

$$f_n(x) = \frac{4}{\pi} \sum_{m=0}^{\infty} \frac{\sin\left((2m+1)x - \frac{\pi}{2}n\right)}{(2m+1)^{n+1}}.$$

Remark 1.2. It is well known that the function f_n is a cardinal spline of degree n, i.e. $f_n(x)$ is a polynomial of degree n on the intervals $(k\pi, (k+1)\pi)$, $k \in \mathbb{Z}$, and $f_n(x) \in C^{(n-1)}(\mathbb{R})$. It is a *perfect spline* in the sense that $f_n^{(n)}(x) = \pm Const.$ a.e., cf. Schoenberg [10]. Another equivalent formulation of Kolmogorov's inequality using the extremizing function f_n is: Suppose that for some function f the following inequalities hold:

$$\|f\|_\infty \le 1, \quad \left\|f^{(n)}\right\|_\infty \le \left\|f_n^{(n)}\right\|_\infty.$$

Then it follows that

$$\left\|f^{(k)}\right\|_\infty \le \left\|f_n^{(k)}\right\|_\infty, \qquad \text{for } 1 \le k \le n.$$

That is the way of formulating the inequalities which appeared in the original paper of E. Landau, [8]. J. Hadamard was the first who formulated in [4] the inequality of Landau in the more compact form (1.1). Let us point to the paper of de Boor and Schoenberg [1] (see also [10]), where an interesting proof of (1.1) and the uniqueness of the extremizers is given.

2 General Framework for Inequalities of Landau–Kolmogorov Type

Here we provide a general setting for inequalities of the Landau–Kolmogorov type which will be very useful for generalizations.

Let D_1, D_2 be homogeneous differential operators (with constant or variable coefficients), of orders $d_1 = \deg(D_1) < d_2 = \deg(D_2)$. Now the generalized inequality of Landau–Kolmogorov type may be formulated in the following way:

$$\|D_1 f\| \le K \|f\|^{1 - \frac{d_1}{d_2}} \|D_2 f\|^{\frac{d_1}{d_2}}. \tag{2.1}$$

It is clear that in the case of inequality (1.1) we have $D_1 = \frac{d^k}{dx^k}$, $D_2 = \frac{d^n}{dx^n}$.

Equality is obtained for a function f_0 which is a piecewise solution to $D_2 f_0(x) = \pm Const.$ a.e. in $\mathbb{R}$. It is natural to call f_0 *perfect L–spline*, for the operator $L = D_2$, cf. Schumaker [11].

Let us remark that the powers $1 - \frac{d_1}{d_2}$ and $\frac{d_1}{d_2}$ in (2.1) come in a natural way from the requirement that inequality (2.1) should hold for all functions of the type

$$af(bx) \qquad a, b \in \mathbb{R}$$

with the same constant K !

3 The Multivariate Case

In the space $\mathbb{R}^d$, $d \geq 2$, we consider elliptic operators D_2, which is natural in view of a priori estimates (cf. Hörmander [6]). The simplest and best understood elliptic differential operator in $\mathbb{R}^d$ is the Laplace operator:

$$\Delta = \sum_{j=1}^{d} \frac{\partial^2}{\partial x_j^2}.$$

The first result in the multivariate case is due to Timofeev [12]. He proved for every bounded $f \in C(\mathbb{R}^d)$ the following inequality:

$$\left\| \frac{\partial f}{\partial \xi} \right\|_\infty \leq \sqrt{2} \sqrt{\|f\|_\infty \|\Delta f\|_\infty}.$$

Here $\frac{\partial f}{\partial \xi}$ is an arbitrary directional derivative. The expression Δf is taken in distributional (Sobolev) sense, and it is assumed that $\Delta f \in L_\infty\left(\mathbb{R}^d\right)$. In view of the general framework, we have in the Timofeev's inequality the operator $D_1 = \frac{\partial}{\partial \xi}, D_2 = \Delta$.

The result of Timofeev was later generalized by Kh. Boyadzhiev [2] and finally by Z. Ditzian [3]. The main result in [3, Theorem 6.1], is the following inequality:

$$\left\| \frac{\partial^k f}{\partial \xi_1 ... \partial \xi_k} \right\|_\infty \leq C(n, k) \|f\|_\infty^{1 - \frac{k}{2n}} \|\Delta^n f\|_\infty^{\frac{k}{2n}}, \qquad 0 < k < 2n, \tag{3.1}$$

which holds for functions $f \in C\left(\mathbb{R}^d\right)$, such that f is bounded and $\Delta^n f \in L_\infty\left(\mathbb{R}^d\right)$, where $\Delta^n f$ is taken in distributional (Sobolev) sense.

Clearly, from the point of view of the general framework in the above inequality we have $D_1 = \frac{\partial^k}{\partial \xi_1 ... \partial \xi_k}, D_2 = \Delta^n$.

Let us remark that no question of finding extremizers or proving sharpness of the constants $C(n, k)$ has been solved in Ditzian [3].

4   Extremizers in the Multivariate Case

The main point in the present paper is to prove the following sharp inequality for bounded $f \in C(\mathbb{R}^d)$ satisfying $\Delta f \in L_\infty(\mathbb{R}^d)$:

$$\|\Delta f\|_\infty \leq K_d \sqrt{\|f\|_\infty \, \|\Delta^2 f\|_\infty}.$$

In addition, we find the extremizers and the sharp constant K_d.

From the point of view of the general framework in that case we have $D_1 = \Delta$, $D_2 = \Delta^2$.

Let us remark that the existence of some constant $K_d > 0$ follows from (3.1). This will be proved independently in Theorem 4.1 below.

We will need some preliminary definitions in order to formulate and prove the main theorem of the paper.

Let us introduce the $2-$periodic function E_2 as follows:

$$E_2(x) \;=\; 1 - 4x^2 \qquad \text{in} \quad \left[-\frac{1}{2}, \frac{1}{2}\right],$$

$$E_2(x) \;=\; -E_2(x-1) \quad \text{in} \quad \left[\frac{1}{2}, \frac{3}{2}\right].$$

For $x \notin \left[-\frac{1}{2}, \frac{3}{2}\right]$, E_2 is defined by $2-$periodicty.

Following Schoenberg [10] we introduce the function:

$$g(x) = \begin{cases} 1 - 4x^2 & \text{if} \;\; -1/\sqrt{2} \leq x \leq 0, \\ E_2(x) & \text{if} \;\;\;\;\; 0 \leq x < \infty, \end{cases}$$

and we define

$$f_0(x) = g\left(x - 1/\sqrt{2}\right) \qquad \text{for } x \geq 0.$$

We will prove the following

Theorem 4.1.   *Let $d \geq 2$. For every bounded function $f \in C(\mathbb{R}^d)$ with $\Delta^2 f \in L_\infty(\mathbb{R}^d)$, taken in the distributional sense, we have $\Delta f \in L_\infty(\mathbb{R}^d)$, and the following sharp inequality holds:*

$$\|\Delta f\|_\infty \leq 2\sqrt{\frac{d}{d+2}} \sqrt{\|f\|_\infty \, \|\Delta^2 f\|_\infty}. \tag{4.1}$$

The function defined by

$$F_0(x) = f_0\left(|x|^2\right)$$

is extremizer.

We will need some preparatory work for the proof.

First we introduce the mean value operator: For a domain $D \subset \mathbb{R}^d$, $d \geq 2$, for $f \in C(D)$, $x \in D$ and $h > 0$ such that the open ball $B_h(x) := \{y \in \mathbb{R}^d : |x - y| < h\}$ is strictly contained in D, i. e. $\overline{B_h(x)} \subset D$, we denote by

$$\mu(f, x; h) := \frac{1}{\sigma_d} \int_{S_{d-1}} f(x + h\xi)d\sigma_\xi = \frac{1}{\sigma_d h^{d-1}} \int_{S_{d-1}(h)} f(x + y)d\sigma(y)$$

the surface mean of f over $\partial B_h(x)$. Here $d\sigma_\xi$ denotes the area element of the $(d-1)$–sphere S_{d-1} and $d\sigma(y)$ is the area element of the sphere $S_{d-1}(h)$ with radius h centred at 0. For $h = 0$ we have $\mu(f, x; 0) = f(x)$.

We introduce the linear integral operator J defined for a given $h > 0$ by

$$J(\phi; t) := \int_0^t \left(r - \frac{r^{d-1}}{t^{d-2}}\right) \phi(r)dr \qquad \text{for} \quad \phi \in C[0, h], \ 0 \leq t \leq h$$

if $d \geq 3$. In the case $d = 2$, J is defined by

$$J(\phi; t) := \int_0^t r \log\frac{t}{r}\phi(r)dr \qquad \text{for} \quad \phi \in C[0, h], \ 0 \leq t \leq h.$$

The powers of J are defined as usual by $J^1 := J$, and $J^{k+1} := J(J^k)$ for $k \geq 1$. When $\phi = 1$, then

$$J^p(1; t) := \begin{cases} a_{dp}(d-2)^p\, t^{2p} & \text{for } d \geq 3, \\ \frac{1}{4^p(p!)^2}t^{2p} & \text{for } d = 2, \end{cases}$$

where $p \geq 1$ and

$$a_{dp} := \frac{1}{2^p p! d(d+2)...(d+2p-2)} \qquad (d \geq 2).$$

Note that $a_{2p} = \frac{1}{4^p(p!)^2}$. We put $a_{d0} = 1$.

Now we state the formula of Pizzetti–Nicolescu (see [9] and [5]):

Proposition 4.2.   (i) *Suppose f is in the Sobolev space $H^{2p}(D)$. Then for any ball $B_h(x)$ strictly contained in D, the following equation holds:*

$$\mu(f, x; h) = f(x) + \sum_{j=1}^{p-1} a_{dj} h^{2j} \Delta^j f(x) + J^p(\mu(\Delta^p f, x; \cdot); h) \cdot c_{p,d},$$

where

$$c_{p,d} = \begin{cases} \frac{1}{(d-2)^p} & \text{for } d \geq 3, \\ 1 & \text{for } d = 2. \end{cases}$$

(ii) *The remainder in the Pizzetti–Nicolescu formula can be written as:*

$$\begin{aligned} J^p(\mu(\Delta^p f, x; \cdot); h) &= \mu(\Delta^p f, x; \vartheta_p h) \cdot J^p(1; h) = \\ &= \Delta^p f(\xi_{x,p}) \begin{cases} a_{dp} h^{2p} (d-2)^p & \text{for } d \geq 3 \\ a_{dp} h^{2p} & \text{for } d = 2 \end{cases} \end{aligned}$$

with suitable $\xi_{x,p} \in B_h(x)$.

Proof of Theorem 4.1 : We will prove that the function

$$F_0(x) = f_0\left(r^2\right), \quad r = |x| \tag{4.2}$$

is an extremizer.

Let us put $h = \left(\frac{1}{2}\right)^{\frac{1}{4}}$. Note that $F_0(0) = -1$, $F_0(h) = 1$. Applying the formula of Pizzetti–Nicolescu for $p = 2$ and $x = 0$ we obtain:

$$\mu(f, 0; h) = f(0) + a_{d1} h^2 \Delta f(0) + J^2(\mu(\Delta^2 f, 0; \cdot); h) \cdot c_{2,d} \tag{4.3}$$

From this equality it follows that Δf belongs to $L_\infty(\mathbb{R}^d)$ since it is seen from it that Δf can be expressed by a.e. bounded quantities.

On the other hand we have

$$a_{d1} = \frac{1}{2d}, \quad a_{d2} = \frac{1}{8d(d+2)} \quad \text{for} \quad d \geq 2$$

and

$$J^2(1; t) = \begin{cases} a_{d2} (d-2)^2 t^4 & \text{for } d \geq 3, \\ \frac{1}{64} t^4 & \text{for } d = 2. \end{cases}$$

Taking into account that the radial part of the Laplace operator is

$$\Delta^* = \frac{\partial^2}{\partial r^2} + \frac{d-1}{r}\frac{\partial}{\partial r}$$

and that by definition we have

$$F_0(x) = f_0\left(|x|^2\right) = 1 - 4\left(|x|^2 - \frac{1}{\sqrt{2}}\right)^2, \qquad \text{for } |x| = r \le h,$$

we obtain

$$\begin{aligned}
\Delta F_0(x) &= \Delta^* F_0(r) &= \Delta^* f_0\left(r^2\right)\\
&= \left(f_0\left(r^2\right)\right)'' + \frac{d-1}{r}\left(f_0\left(r^2\right)\right)'\\
&= \left(2r f_0'\left(r^2\right)\right)' + 2(d-1) f_0'\left(r^2\right)\\
&= 2f_0'\left(r^2\right) + 4r^2 f_0''\left(r^2\right) + 2(d-1) f_0'\left(r^2\right)\\
&= 4r^2 f_0''\left(r^2\right) + 2d f_0'\left(r^2\right)\\
&= 4r^2(-8) + 2d\left(-8\left(r^2 - \frac{1}{\sqrt{2}}\right)\right)\\
&= 16\left(-2r^2 - dr^2 + \frac{d}{\sqrt{2}}\right).
\end{aligned}$$

It follows that the maximum of $\left|\Delta^* F_0(r)\right|$ in the interval $r \le h$ is obtained for $r = 0$, i.e. it is equal to

$$\Delta^* F_0(0) = 8d\sqrt{2}.$$

Due to the symmetry of the function E_2, it follows that this is also the absolute maximum for every $r \ge 0$, i.e.

$$\left\|\Delta^* F_0\right\|_\infty = 8d\sqrt{2}.$$

Taking into account that for $r \le h$ the function $F_0(x)$ is a polynomial of degree 4 in r, we obtain

$$\begin{aligned}
\Delta^2 F_0(x) &= -4 \cdot (\Delta^*)^2\left(r^4\right)\\
&= -32d(d+2) \quad \text{for} \quad r = |x| \le h.
\end{aligned}$$

and due to the definition of the function f_0 we have

$$\left|\Delta^2 F_0(x)\right| = 32d(d+2) \quad \text{for every} \quad x \in \mathbb{R}^d,$$

i.e.

$$\left\|\Delta^2 F_0(x)\right\|_\infty = 32d(d+2).$$

In formula (4.3) we put $f = F_0$. Using the above formulas and the fact that the spherical mean of a radially symmetric function is the value of the function at that radius, we obtain

$$\mu(F_0, 0; h) = F_0(h).$$

In a similar way, since $\Delta^2 F_0(x)$ is a constant for $r = |x| \le h$, we obtain

$$\mu(\Delta^2 F_0, 0; h) = \Delta^2 F_0 = -32d(d+2)$$

and finally

$$F_0(h) = F_0(0) + a_{d1} h^2 \Delta F_0(0) + \Delta^2 F_0(0)\, J^2(1; h) \cdot c_{2,d}. \tag{4.4}$$

It is important to note that we have alternatingly in sign

$$\begin{aligned}
F_0(0) &= -1, \\
F_0(h) &= 1, \\
\Delta^2 F_0(0) &= -32d(d+2).
\end{aligned}$$

Putting these values into (4.4) we obtain the following key identity:

$$\begin{aligned}
a_{d1} h^2 \Delta F_0(0) &= -F_0(0) + F_0(h) - \Delta^2 F_0(0)\, J^2(1; h) \cdot c_{2,d} \\
&= 1 + 1 + \left\|\Delta^2 F_0\right\|_\infty J^2(1; h) \cdot c_{2,d}. \tag{4.5}
\end{aligned}$$

Let now f be an arbitrary function for which $\|f\|_\infty \le 1$, $\left\|\Delta^2 f\right\|_\infty \le \left\|\Delta^2 F_0\right\|_\infty = 32d(d+2)$. From formula (4.3) we express $\Delta f(0)$:

$$a_{d1} h^2 \Delta f(0) = -f(0) + \mu(f, 0; h) - J^2(\mu(\Delta^2 f, 0; \cdot); h) \cdot c_{2,d}.$$

Since $J^2(1; h) > 0$, and the kernel of the operator $J^2(\,\cdot\,, r)$ is evidently positive, we obtain the following inequality:

$$\begin{aligned}
a_{d1} h^2 |\Delta f(0)| &\le \left\|-f(0) + \mu(f, 0; h) - J^2(\mu(\Delta^2 f, 0; \cdot); h) \cdot c_{2,d}\right\|_\infty \\
&\le \|f\|_\infty + \|\mu(f, 0; h)\|_\infty + \left\|J^2(\mu(\Delta^2 f, 0; \cdot); h) \cdot c_{2,d}\right\|_\infty \\
&\le \|f\|_\infty + \|f\|_\infty + \left\|\mu(\Delta^2 f, 0; \cdot)\right\|_\infty J^2(1; h) \cdot c_{2,d} \\
&\le \|f\|_\infty + \|f\|_\infty + \left\|\Delta^2 f\right\|_\infty J^2(1; h) \cdot c_{2,d} \\
&\le \|f\|_\infty + \|f\|_\infty + \left\|\Delta^2 F_0\right\|_\infty J^2(1; h) \cdot c_{2,d} \\
&\le 1 + 1 + \left\|\Delta^2 F_0\right\|_\infty J^2(1; h) \cdot c_{2,d}.
\end{aligned}$$

By (4.5) we obtain

$$a_{d1} h^2 \left| \Delta f(0) \right| \leq a_{d1} h^2 \Delta F_0(0)$$
$$= a_{d1} h^2 \left\| \Delta F_0 \right\|_\infty .$$

It follows that

$$\left| \Delta f(0) \right| \leq \left\| \Delta F_0 \right\|_\infty .$$

We may apply the above arguments to the function $f(x + x_0)$, and prove the above inequality for every point $x_0 \in \mathbb{R}^d$, i.e.

$$\left| \Delta f(x_0) \right| \leq \left\| \Delta F_0 \right\|_\infty .$$

Hence it follows that

$$\left\| \Delta f \right\|_\infty \leq \left\| \Delta F_0 \right\|_\infty ,$$

i.e. under the hypotheses $\| f \|_\infty \leq 1$ and $\| \Delta^2 f \|_\infty \leq \| \Delta^2 F_0 \|_\infty$ we have proved that

$$\| \Delta f \|_\infty \leq \| \Delta F_0 \|_\infty ,$$

which is the statement of Theorem 4.1 in the formulation of Landau (see Remark 1.2).

If we put

$$K_d = \frac{\left\| \Delta F_0 \right\|_\infty}{\sqrt{\left\| \Delta^2 F_0 \right\|_\infty}} = \frac{8 d \sqrt{2}}{\sqrt{32 d (d + 2)}} = 2 \sqrt{\frac{d}{d + 2}}$$

then we obtain equality in (4.1) for the function $f = F_0$. That finishes the proof of the Theorem also in the formulation of Hadamard. $\qquad \square$

Acknowledgment. The present research was supported by the Volkswagen Foundation.

References

[1]   C. de Boor, I. J. Schoenberg: *Cardinal interpolation and spline functions, VIII. The Budan–Fourier theorem for splines and applications,* in: *Spline Functions,* K. Böhmer, G. Meinardus, W. Schempp (eds.)., Lect. Notes Math. **501**, 1–79, Springer, Berlin 1976.

[2]   K. N. Boyadzhiev: *A many variable Landau–Kolmogorov inequality,* Math. Proc. Camb. Philos. Soc. **101** (1987), 123–129.

[3]   Z. Ditzian: *Multivariate Landau–Kolmogorov–type inequality,* Math. Proc. Camb. Philos. Soc. **105** (1989), 335–350.

[4]   J. Hadamard: *Sur le module maximum d'une fonction et de ses dérivées,* C. R. Acad. Sci. Paris **41** (1914), 68–72.

[5]   W. Haussmann, O. Kounchev: *Peano theorem for linear functionals vanishing on polyharmonic functions,* in: *Approximation Theory, VIII, Vol. I,* C. K. Chui, L. L. Schumaker (eds.), 233–240, World Scientific, Singapore 1995.

[6]   L. Hörmander: *The Analysis of Linear Partial Differential Operators, Vol. II,* Springer, Berlin 1983.

[7]   A. N. Kolmogorov: *On inequalities for suprema of consecutive derivatives of an arbitrary function on an infinite interval,* in: *Selected Works of A.N. Kolmogorov, Vol. I,* V. Tikhomirov (ed.), 277–290, Kluwer, Dordrecht–Boston–London 1991.

[8]   E. Landau: *Einige Ungleichungen für zweimal differentiierbare Funktionen,* Proc. London Math. Soc. (2) **13** (1913), 43–49.

[9]   M. Nicolescu: *Sur les fonctions de n variables harmoniques d'ordre p,* Bull. Soc. Math. France **60** (1932), 129–151.

[10]   I. J. Schoenberg: *Cardinal Spline Interpolation,* SIAM, Philadelphia 1973.

[11]   L. L. Schumaker: *Spline Functions: Basic Theory,* Wiley, New York 1981.

[12]   V. G. Timofeev: *Landau inequality for function of several variables,* Mat. Zametki **37** (1985), 676–689.

Address:

OGNYAN KOUNCHEV
Department of Mathematics
University of Duisburg
D–47048 Duisburg
Germany

Nested Sequences of Triangular Finite Element Spaces

A. Le Méhauté

Abstract

The present paper is a continuation of the discussion of hierarchical bases of triangular finite elements. It is an attempt to determine in what sense it is possible to define nested sequences of finite element spaces and, more precisely, what are the triangular finite elements which are relevant to use in such nested sequences.

1   Introduction

In a recent paper, Dahmen, Oswald and Shi [9] investigate the construction of sequences of C^1 finite elements, following some ideas from Yserentant [21], where C^0 nested sequences of (piecewise linear) finite element approximations play a key role and from Oswald [18] where nested sequences of C^1–piecewise quadratic and cubic finite element spaces are considered which involve relatively complex macro–elements.

It is the aim of this note to complement and continue the discussion in [9] and to try to determine in what sense it is possible to define nested sequences of finite element spaces and more precisely what are the triangular finite elements which are relevant to use in such nested sequences.

Let us first explain what we understand by this idea of nested sequences of finite element spaces. Suppose $\Omega \subset \mathbb{R}^2$ is a bounded polygonal domain, *i.e.*, the boundary $\partial\Omega$ of Ω consists of straight line segments. Let Δ be a triangulation of Ω built upon a set $\mathcal{A}$ of given (initial) points, $\mathcal{A} = \{a_1, \ldots, a_N\}$. Let $\mathcal{S}_d^k(\Delta)$ be the set of piecewise polynomial spline functions with continuity k and degree d built upon Δ, *i.e.*, :

$$\mathcal{S}_d^k(\Delta) = \{f : f \in C^k(\overline{\Omega}), \; f|_T \in \mathbb{P}_d[\mathbb{R}^2], \; \forall\, T \in \Delta\}$$

where $\mathbb{P}_d[\mathbb{R}^2]$ denotes the set of bivariate polynomials of degree d.

Multivariate Approximation: Recent Trends and Results; W. Haußmann, K. Jetter and M. Reimer (eds.)
Mathematical Research, Vol. 101, pp. 133–145, ISBN 3-05-501770-6
© Akademie–Verlag, Berlin 1997

The initial set of approximating functions will be $V_1 = \mathcal{S}_d^k(\Delta)$ for some k and some d, and we want to describe a sequence of spaces $\{V_i\}_{i \geq 1}$ such that

$$V_1 \subset V_2 \subset \ldots \subset V_i \subset \ldots$$

As we will show, there exist different possibilities for the construction of such nested families and at least one could think of the two different approaches that are related to each of the two aspects of finite elements: the usual h–version and the p–version (see [3]), without mentioning the use of macro–elements generalizing Hsieh–Clough–Tocher or Fraeijs de Veubeke–Sanders elements ([7, 5, 8]).

2 Basic Aspect of Finite Element Approximation

Let us recall some (well known) basic aspects of the finite element method in $\mathbb{R}^2$, or equivalently of the construction of piecewise polynomial spline approximation of functions of two real variables.

Let Δ be a (proper) triangulation of a polygonal domain Ω, with no splitting of the triangles (which means that we are dealing with a unique polynomial piece defined on each triangle).

The approximation by simplicial polynomial finite elements is based on the following hypotheses:

(h1) Interpolation data are linearly independent;

(h2) Interpolation is local (*i.e.*, any alteration of a single datum should affect only the triangles containing the point supporting this datum);

(h3) Those degrees of freedom that are necessary to ensure k–continuity between adjacent triangles (called *first type data*) are given on the boundary of each triangle. Other data are needed to ensure unisolvency on each triangle T , and are given either on the boundary or in the interior of T (they are called *second type data*);

(h4) The first type data are "symmetric" relatively to the structure of each triangle (*i.e.*, the 3 vertices play the same role, just like the 3 edges);

(h5) First type data are linear functionals of the form $\delta_{a_i}^\alpha : f \mapsto \partial^\alpha f(a_i)$, where a_i are points on the boundary of T , $\alpha \in \mathbb{N}^n$, or belong to $\mathrm{span}\{\delta_{a_i}^\alpha\}$ (in order to use derivatives normal to the edges, for instance).

From Morgan–Scott [17] (in the specific case of $k = 1$), and Ženíšek [22] in $\mathbb{R}^2$, (more generally for any k in $\mathbb{R}^n$, see [12]), we obtain:

Theorem 2.1. *Assume that $(h1)$–$(h5)$ are fulfilled, then the space $\mathcal{S}_d^k(\Delta)$ is non–empty whenever $d \geq 4k+1$ (and more generally, $d \geq 2^n k+1$ for simplicial polynomial finite elements in $\mathbb{R}^n$).*

As a corollary, in the proof of [12], using results from $1D$ approximation for the restriction of the interpolant on each edge of the triangle (such as Pólya's and Ferguson's theorems for well poised approximation), one proves that the simplest triangular element that could be used in the plane for C^1 approximation is the Argyris triangle [2], or its reduced version, the Bell's interpolant [4].

Let us recall here that the data involved in the Argyris interpolant for a triangle $T = [a_1, a_2, a_3]$ are the following:

$$\partial^\alpha f(a_s), \ |\alpha| \leq 2, s = 1,2,3; \ \text{and the normal derivatives} \ \frac{\partial}{\partial \nu_s} f(Q^1_{1,s}),$$

where $Q^1_{1,s}$ is any point different from a vertex on the edge $\sigma_s = [a_{s-1}, a_{s+1}]$ opposite to a_s, (when dealing with finite elements approximation for solving PDE's, it is usual, but not mandatory, for $Q^1_{1,s}$ to be the midpoint of σ_s). Bell's interpolant is nothing but Argyris' one where the normal derivative at $Q^1_{1,s}$ is obtained from interpolation of the data given at the two vertices a_{s-1} and a_{s+1} [12].

Obviously, the easiest way to build a nested family of finite elements is to start with a given triangulation, decide what generic element will be used, then split each triangle in 4 isometric parts dividing each edge by the midpoint. Unfortunately, as already mentioned in [9], this subdivision is not directly permissible when dealing with Argyris' element and its generalization of higher order. This is due to the C^2 continuity at the vertices which could spead out to a dense set of points.

Due to lack of space, we cannot develop here the following results which will be published somewhere else and we apologize just to sketch them. One other corollary from the Morgan–Scott and Ženíšek approach is usually stated as [22, 12]:

Corollary 2.2. *The space $\mathcal{S}_d^k(\Delta)$ is non–empty whenever the data set contains all the derivatives up to order $2k$ at any of the vertices of the triangulation (and up to order $2^{n-1}k$ in $\mathbb{R}^n$).*

It was already noticed in [12] that those data are convenient to be used but are not mandatory; for instance in the plane and for $k \geq 3$, it is possible, at any vertex a of the triangulation, to replace the data $\partial^\alpha f(a)$, for all $\alpha \leq 2k$ by $\partial^\alpha f(a)$ for all $\alpha \leq 2k - 2$ plus $\partial^\alpha f(a)$, for $\alpha = p, p + 1$, where $2k + 1 \leq p \leq Int \left[\frac{1}{12}(k^2 + 25k + 12)\right]$, and other possibilities depending on that $k \geq 6, k \geq 9, \ldots$ (*Int* means "the largest integer strictly less than"). For example, when $k = 4$, one can replace the derivatives up to order 8 by the derivatives up to order 6 plus derivatives of order 8 and 9. The interest is that the dimension of the interpolating space is smaller.

A more careful look at the Morgan–Scott proof shows that the corollary is in fact a fast interpretation of the fundamentals. Their paper is essentially based on the fact that a piecewise polynomial function f which is C^1 at the vertices of the triangulation has to fulfill the Schwarz equality for cross derivatives of order 2 (and as a matter of fact, of order $2k$ for C^k continuity). Practically, let e_1, e_2, and e_3 be three successive edges (ordered clockwise for instance) meeting at the same vertex a and let $\frac{\partial}{\partial\sigma_i}$ and $\frac{\partial}{\partial\tau_i}$ be the directional derivatives along and accross the edge e_i. The core of the Morgan–Scott proof is to write that the equality $\frac{\partial^2 f}{\partial\sigma_2\partial\tau_2}(a) = \frac{\partial^2 f}{\partial\tau_2\partial\sigma_2}(a)$ is valid whatever the way τ_2 is written as a linear combination of $\{e_1, e_2\}$ or $\{e_2, e_3\}$. But another admissible conclusion is that those higher order derivatives (higher than the order k) should be available in such a way that even crossing an edge of the triangulation, the final interpolant would fulfill the Schwarz' equality, which leads us to a new formulation:

Proposition 2.3. *Given any integer $k \geq 0$, and in order to obtain C^k continuity between simplices, it is sufficient that, for any of the vertices a in the triangulation, the linear forms $v \mapsto \partial^\alpha v(a)$, for all α such that $|\alpha| \leq 2k$, are available from the data set.*

Of course, this does not mean that they are necessarily given in the data set, and in the latter case, we will have to pay a price for, not only in the evaluation of those degrees of freedom, but also from the fact that the hypotheses $(h2, h3)$ have to be enlarged: the evaluation of the interpolant on one triangle will involve data from the adjacent ones.

3 On the Polynomial Spline Spaces $\mathcal{S}_d^k(\Delta)$ and $\mathcal{Z}_d^k(\Delta)$

As seen in the previous section, the usual space of finite elements involves separately on each triangle an interpolant such as Argyris' or any of its generalization [12, 14] and is a strict subspace of $\mathcal{S}_d^k(\Delta)$. Let us denote by $\mathcal{Z}_d^k(\Delta)$ this space also called super splines space [6]. The data set includes the derivatives up to order $2k$ at any vertex, and we can write

$$\mathcal{Z}_d^k(\Delta) = \mathcal{S}_d^k(\Delta) \cap C^{2k}(a), \text{ for any vertex } a \text{ in } \Delta.$$

Given a triangulation Δ , let V_I (resp. E_I) be the number of interior vertices (resp. edges), V_B (resp. E_B) the number of boundary vertices (resp. edges), N_T the number of triangles, and $V = V_I + V_B$, $E = E_I + E_B$. These numbers are related via the classical Euler's formulas:

$$E_B = V_B, \quad E_I = 3V_I + V_B - 3, \quad \text{and} \quad N = 2V_I + V_B - 2.$$

For each vertex $a_i \in \Delta$, let E_i be the number of edges emanating from a_i and s_i be the number of different slopes of these edges. From [1], we know that

$$\dim \mathcal{S}_d^k(\Delta) = \binom{k+2}{2} + \binom{d-k+1}{2} E_I - \frac{1}{2}\Big(d(d+3) - k(k+3)\Big) V_I + \sigma \quad (3.1)$$

where $\sigma = \sum_{j=1}^{d-k}(k+j+1-js_i)_+$.

It is obvious that

$$\dim \mathcal{Z}_d^k(\Delta) = \binom{2k+2}{2} V + \binom{k+2}{2} E + \binom{k+2}{2} N \leq \dim \mathcal{S}_d^k(\Delta).$$

Before considering nested sequences of finite element spaces, let us spend a short time on the effective construction of the polynomial pieces of the interpolants.

Let T be one of the triangles in Δ, with the vertices a_1, a_2, a_3, with edges $\sigma_s = [a_{s-1}, a_{s+1}]$ opposite to a_s and ν_s the direction of the normal to σ_s, for $s = 1, 2, 3$. Let us assume first that we are given or we have already evaluated the following first type degrees of freedom:

$$\partial^\alpha f(a_s), |\alpha| \leq 2k, \quad s = 1, 2, 3,$$

$$\frac{\partial^i}{\partial \nu_s^i} f(Q_{r,s}^i) \quad \begin{cases} s = 1, 2, 3, \\ i = 1, \ldots, k \\ r = 1, 2, \ldots, i \end{cases} \quad (3.2)$$

where Q^i_{rs} are points on σ_s such that $Q^i_{rs} \neq a_j$, $j{=}1,\ 2,\ 3$, and $Q^i_{rs} \neq Q^i_{ts}$ whenever $t \neq r$.

The second type degrees of freedom, needed to ensure the polynomial unisolvency of degree $d = 4k + 1$ could be $\partial^\alpha f(a_0), |\alpha| \leq k - 2$, where a_0 is a strictly interior point, or any equivalent set (see [12]).

It was shown in [13, 15] that it is beneficial to use instead of (3.2) the equivalent set of degrees of freedom (which reduces drastically the number of elementary operations needed for the pointwise evaluation of the polynomial):

$$f(a_s),\ T^i_{a_s} f(a_{s-1}),\ T^i_{a_s} f(a_{s+1});\ i = 1,\dots,2k;\ \ s = 1,2,3,$$
$$D^{i+j} f(a_s) \cdot (a_s a^i_{s-1}, a_s a^j_{s+1}),\ \ i = 1,\dots,2k,\ j = 1,\dots,2k;\ i + j \leq 2k \tag{3.3}$$

for the data at each vertex, and keep the same for the others. The notation $T^i_{a_s} f(a_{s-1})$ is for the polynomial Taylor expansion of order i about the point a_s evaluated at a_{s-1}, and $D^{i+j} f(a_s) \cdot (a_s a^i_{s-1}, a_s a^j_{s+1})$ are directional derivatives at a_s evaluated along $a_s a_{s-1}$ and $a_s a_{s+1}$. More precisely, following [10], let us write

$$D^2 f(x) \cdot (\mu, \nu) = \partial^{2,0} f(x) \mu_x\, \nu_x + \partial^{1,1} f(x)[\mu_x\, \nu_y + \mu_y\, \nu_x] + \partial^{0,2} f(x) \mu_y\, \nu_y;$$

more generally, for any $\mathbf{T} = (t_1, t_2, \dots, t_p)$ with $t_i = (\xi_{i1}, \xi_{i2}, \dots, \xi_{in}) \in \mathbb{R}^n$,

$$\begin{aligned}
D^p f(x) \cdot (\mathbf{T}) &= D^p f(x) \cdot (t_1, t_2, \dots, t_p) = \\
&\quad \textstyle\sum_{\mathbf{J}}\ D_{j_1} D_{j_2} \dots D_{j_p} f(x)\, \xi_{1,j_1} \xi_{2,j_2} \dots \xi_{p,j_p}
\end{aligned} \tag{3.4}$$

where the sum $\sum_{\mathbf{J}}$ is to be understood for all the n^p different sequences $\mathbf{J} = (j_1, j_2, \dots, j_p)$ of integers in $\{1, 2, 3, \dots, n\}$, and $D^p f(x) \cdot (\mu^q, \nu^{p-q})$ denotes the particular case where q of the t_i are equal to the same vector μ, and $p - q$ of them are equal to ν.

Now we are in situation to consider a first family of nested finite element spaces. As we have already noticed, it is not possible to use the Argyris elements and its generalizations which lead to the spaces $\mathcal{Z}^k_d(\Delta)$. Instead, we have to use the full space $\mathcal{S}^k_d(\Delta)$. In order to get a basis, we use the following data:

1. On each edge e of Δ, the normal derivatives

$$\frac{\partial^i}{\partial \nu^i_s} f(Q^i_{r,s}) \quad \left\{ \begin{array}{l} s = 1,2,3, \\ i = 1,\dots,k \\ r = 1,\dots,i \end{array} \right. \tag{3.5}$$

where Q^i_{rs} are points on $e = \sigma_s$ such that Q^i_{rs} is not a vertex and $Q^i_{rs} \neq Q^i_{ts}$ whenever $t \neq r$.

2. Inside each triangle T, the second type degrees of freedom, needed to ensure the $(d = 4k+1)$–unisolvency, which could be $\partial^\alpha f(a_0), |\alpha| \leq k-2$, where a_0 is a strictly interior point, or any equivalent set.

3. Around each vertex a, two possibilities according to the fact that a is an interior or a boundary vertex.

3.a) For an interior vertex, let $e_1, e_2, \ldots, e_n$ be the different edges emanating from a, ordered clockwise for instance. If a is not a defective vertex [1], all the slopes are different and we take the data set

$$\partial^\alpha f(a), |\alpha| \leq k,$$
$$D^{i+j} f(a) \cdot (e_s^i, e_{s+1}^j), \quad i = 1, \ldots, k, j = 1, \ldots, k; s = 1, \ldots, n. \tag{3.6}$$

From these data, it is possible to evaluate locally, and separately in each of the triangles who share the vertex a the Taylor expansions of (3.3) in such a way that the final result will be C^k. When a is a defective vertex, we have to add another set of data, corresponding to the directional derivatives $D^{i+j} f(a) \cdot (e_s^i, e_{s+1}^j)$, $i = 1, \ldots, 2k, j = 1, \ldots, 2k$ which cannot be obtained from (3.6). (This corresponds to the terms in the sum σ in (3.1)).

3.b) For a boundary vertex a, let $e_1, e_2, \ldots, e_n$ be the different edges emanating from a, ordered clockwise for instance in such a way that the boundary edges are e_1 and e_n. If a is not a defective vertex, all the slopes are different and we take the data set

$$\begin{aligned}
&\partial^\alpha f(a), |\alpha| \leq k, \\
&D^{i+j} f(a) \cdot (e_s^i, e_{s+1}^j), \quad i = 1, \ldots, k, \ j = 1, \ldots, k, \ s = 1, \ldots, n, \\
&D^{i+j} f(a) \cdot (e_1^i, e_2^j), \quad\ \ i = 1, \ldots, 2k, \ j = 1, \ldots, k, \ i+j \leq 2k, \\
&D^{i+j} f(a) \cdot (e_{n-1}^i, e_n^j), \quad i = 1, \ldots, k, \ j = 1, \ldots, 2k, 1+j \leq 2k,
\end{aligned} \tag{3.7}$$

(and the corresponding changes for a defective vertex).

The resulting space is now $\mathcal{S}_d^k(\Delta)$. At this point, any triangle can then be splitted in 4 isometric subtriangles, using as new vertices the midpoints of the edges and it is easy to see that we can iterate this process. If we denote by Δ_1 the original triangulation and by Δ_2 the new one, and so on, we have

$$\mathcal{S}_d^k(\Delta_1) \subset \mathcal{S}_d^k(\Delta_2) \subset \ldots.$$

For example, and for C^1 continuity, we will start with $\mathcal{S}_5^1(\Delta_1)$ as in [9], then build $\mathcal{S}_5^1(\Delta_2), \mathcal{S}_5^1(\Delta_3)$ and so forth.

4   Another Nested Family of Finite Elements Spaces: The p–Version.

Another but possibly not so familiar approach for the finite element approximation is to use what Babuska [3] called the *p–version* instead of the usual *h–version*, which roughly means that instead of having the size of the elements going to 0 when refining the triangulation (for the *h*–version), it is the degree of the local polynomials which goes to ∞, keeping the same triangulation.

Now let us go back to the construction of the local interpolant. It is shown in [12, 16] that another way of generalizing Argyris' triangle is to define triangular finite elements of type $(k, p, 4k+p+1)$, for C^k continuity, where p is the number of time the function itself is given on each edge of a triangle, and $4k + p + 1$ is the degree of the polynomial. The general result is proved in [12]:

Theorem 4.1. *Let p , k , κ be three integers such that $p \geq 0, \kappa \geq k \geq 0$, and f a function in $C^{k+\kappa}(\bar{\Omega})$. Let $N = 2(k + \kappa) + p + 1$ and T be a triangle of the triangulation Δ of Ω with a_1, a_2, a_3 its vertices, σ_1, σ_2, σ_3 the edges of T, and ν_s the normal to σ_s, for $s = 1, 2, 3$.*

Let Π_Σ^N be a polynomial of degree N such that

$$\partial^\alpha \Pi_\Sigma^N(a_s) = \partial^\alpha f(a_s), \qquad |\alpha| \leq k + \kappa, \quad s = 1, 2, 3,$$

$$\frac{\partial^i}{\partial \nu_s^i} \Pi_\Sigma^N(Q_{rs}^i) = \frac{\partial^i}{\partial \nu_s^i} f(Q_{rs}^i), \qquad \begin{cases} s = 1, 2, 3, \\ i = 0, 1, \ldots, k, \\ r = 1, 2, \ldots, p + i, \end{cases} \qquad (4.1)$$

where Q_{rs}^i are points on σ_s such that $Q_{rs}^i \neq a_j$, $j = 1, 2, 3$, and $Q_{rs}^i \neq Q_{ts}^i$ whenever $t \neq r$.

Let g be the function defined on $\overline{\Omega}$ by $g_{|\Sigma} = \Pi_\Sigma^N$, for each $\Sigma \in \Delta$. Then $g \in C^k(\overline{\Omega})$.

Remark : This theorem defines only the first type data , which are necessary to obtain C^k triangular elements. One has to add the data which are necessary to ensure the N–unisolvency of the interpolation problem. They can be defined as derivatives in one interior point or from any equivalent set of linear forms. It is possible to use this result for the construction of a nested family of triangular elements, keeping the same vertices, the same order of continuity k, and $\kappa = k$ and going from a step to another step by changing p into $p+1$ and the relevant data at the appropriate points Q_{rs}^i. Of course, we start with a convenient data set concerning the derivatives at the vertices of Δ (as defined in the previous section).

The nested family will be (with $N = 4k + p + 1$, p initially given)

$$\mathcal{S}_N^k(\Delta) \subset \mathcal{S}_{N+1}^k(\Delta) \subset \mathcal{S}_{N+2}^k(\Delta) \subset \cdots \subset \mathcal{S}_{N+p}^k(\Delta) \subset \cdots$$

For example, and for $\mathcal{C}^1$ continuity, we will start with $\mathcal{S}_5^1(\Delta)$ as in [9], then build $\mathcal{S}_6^1(\Delta), \mathcal{S}_7^1(\Delta)$ and so forth.

5 Nested Families of Composite Finite Elements

In order to avoid the "super–spline" effect, *i.e.*, the high order of derivatives at the vertices of the triangulation, one should think of using composite finite elements. Those interpolants are built by splitting each triangle in Δ, usually called a macro–element, in smaller sub triangles in order to have some pieces of low degree instead of one piece of higher degree. These pieces realize the data linked to the macro–element which contribute to the dimension of the spline space, and also some other degree of freedom which are needed to ensure continuity of some order inside the macro–element (at least k, the global order of smoothness).

The most convenient macro–elements for $\mathcal{C}^1$ interpolation are the Hsieh–Clough–Tocher for triangles and the Fraeijs de Veubeke–Sanders for the quadrilaterals [7, 8]. Both use polynomial pieces of degree 3. Hierarchical $\mathcal{C}^1$ bases are available from both elements [9], obtained as in Section 3 by subdividing each element in isometric ones when introducing as new vertices the midpoints of the edges. Our work is still in progress but it is possible to generalize partially the results in [9] in both directions: from $k = 1$ to higher order of smoothness for the *h–version*, and also with the idea of *p–version* of the composite elements.

Let us introduce first an efficient way for the construction and the evaluation of H–C–T $\mathcal{C}^1$ elements (easily transposable to FdV–S). The degrees of freedom for an Hsieh–Clough–Tocher triangle $T = [a_1, a_2, a_3]$ are

$$\partial^\alpha f(a_s), \ |\alpha| \le 1, s = 1, 2, 3, \ \text{and the normal derivatives} \ \frac{\partial}{\partial \nu_s} f(Q_{1,s}^1),$$

where $Q_{1,s}^1$ is any point different from a vertex on the edge $\sigma_s = [a_{s-1}, a_{s+1}]$ opposite to a_s. Let a_0 be any strictly interior point. As a_0 is not a given data point, it can be chosen by the user; it is usually taken as the iso–barycenter of the triangle T. Instead of the classical approach involving a list of basis functions on a reference element, we recommend the following one, which allows to reduce the number of elementary operations in the pointwise evaluation of the polynomial, as was done in Section 2.

Let us denote by $\Sigma 1$ the subtriangle $[a_0, a_2, a_3]$, $\lambda_{i,\Sigma 1}$, $i = 1, 2, 3$, the corresponding barycentric coordinates (with $i = 1$ standing for a_0). Then the polynomial piece of degree 3 on $\Sigma 1$ is given by

$$\Pi^3_{\Sigma 1}[f](M) =$$
$$= \sum_{i=1}^3 \lambda_i^2 \Big\{ (2 - \lambda_{i,\Sigma 1}) f(A_i) + \lambda_{i+1,\Sigma 1} T^1_{A_i} f(A_{i+1}) + \lambda_{i-1,\Sigma 1} T^1_{A_i} f(A_{i-1}) \Big\}$$
$$+ a_{1,\Sigma 1} \lambda_{1,\Sigma 1} \lambda_{2,\Sigma 1} \lambda_{3,\Sigma 1}$$

where $a_{1,\Sigma 1}$ is obtained from the initial data writing that $\frac{\partial}{\partial \tau_s} \Pi^3_{\Sigma 1}[f](Q^1_{1,s}) = \frac{\partial}{\partial \tau_s} f(Q^1_{1,s})$, τ_s denoting the normal to the edge $[a_2, a_3]$ (a_0 does not interfere here).

The Taylor expansions $T^1_{a_0} f(a_i)$, $i = 1, 2, 3$, provide $f(a_0) = T^1_{a_0} f(a_0) = \frac{1}{3} \sum_{i=1}^3 T^1_{a_0} f(a_i)$ but are unknown. They are solutions of a 3×3 linear system obtained in writing that two of the three pieces match together with C^1 continuity. This gives us 3 conditions, such as $\frac{\partial}{\partial \nu_3} \Pi^3_{\Sigma 1}[f](B_3) = \frac{\partial}{\partial \nu_3} \Pi^3_{\Sigma 2}[f](B_3)$ where B_3 is any point on $[a_0, a_3]$ and ν_3 the normal derivative along $[a_0, a_3]$, and the two other conditions along $[a_0, a_i]$, $i = 1, 2$, respectively.

The $(C^1, \mathbb{P}_3)$ Fraeijs de Veubeke–Sanders quadrilateral element with vertices a_1, a_2, a_3, a_4 is essentially the same. Here a_0 is the intersection of the two diagonals. Thus there are 4 edges with 2 different slopes at a_0 which is a defective vertex. The set of C^1 continuity conditions between the 4 polynomial pieces is now a system of 4 linear equations, but its rank is 3, and therefore provides the values of $T^1_{a_0} f(a_i)$, $i = 1, 2, 3, 4$.

The h–version of the nested family of H–C–T triangles can be directly obtained by splitting each macro–triangle by the midpoints of the edges, as in [9]. A higher order of continuity can be obtained using the generalization of this element due to Sablonnière [19], and we do not emphasize here on this aspect.

For the p–version, which works also for the Fraeijs de Veubeke–Sanders quadrilateral element, let us consider elements which are $(C^1, \mathbb{P}_4)$. The triangle uses the following data:

$$\partial^\alpha f(a_s) \quad , \qquad |\alpha| \leq 1 \quad , \qquad s = 1, 2, 3,$$

$$\frac{\partial^i}{\partial \nu_s^i} f(Q_{rs}^i), \quad \begin{cases} s = 1, 2, 3, \\ i = 0, 1; r = 1, \dots, i+1, \end{cases} \tag{4.2}$$

where Q_{rs}^i are points on the edges of the macro–triangle.

The construction is essentially the same. Writing that the interpolant is $\mathcal{C}^2$ around a_0, we get 6 unknowns $\partial^\alpha f(a_0), |\alpha| \leq 2$, which are obtained from the continuity conditions

$$\frac{\partial^2}{\partial^2 \nu_3} \Pi_{\Sigma 1}^3 [f](a_0) = \frac{\partial^2}{\partial^2 \nu_3} \Pi_{\Sigma 2}^3 [f](a_0),$$

$$\frac{\partial}{\partial \nu_3} \Pi_{\Sigma 1}^3 [f](B_3) = \frac{\partial}{\partial \nu_3} \Pi_{\Sigma 2}^3 [f](B_3),$$

where B_3 is any point on $[a_0, a_3]$ and ν_3 the normal derivative along $[a_0, a_3]$, and the other conditions along $[a_0, a_i]$, $i = 1, \dots, 4$, respectively (the rank of the linear system is 3).

References

[1] P. Alfeld, L. L. Schumaker, *The dimension of bivariate splines spaces of smoothness r for degree $d \geq 4r + 1$*, Constr. Approx. **3** (1987), 189–197.

[2] J. H. Argyris, I. Fried, D. W. Scharpf, *The TUBA family of plate elements for the matrix displacement method*, Aero. J. Royal Aeronautic Society **72** (1968), 701–709.

[3] I. Babuska, B. A. Szabo, I. N. Katz, *The p version of the finite element method*, SIAM J. Numer. Anal. **18** (1981), 516–545.

[4] K. Bell, *A refined triangular plate bending element*, Internat. J. Numer. Methods Engrg. **1** (1969), 101–122.

[5] M. Bernadou, J. M. Boisserie, *The Finite Element Method in Thin Shell Theory*, Birkhäuser, Boston–Basel 1982.

[6] C. K. Chui, M. J. Lai, *On bivariate super vertex splines*, Constr. Approx. **6** (1990), 399–419.

[7]   P. G. Ciarlet, *The Finite Element Method for Elliptic Problems*, North–Holland, Amsterdam 1978.

[8]   J. F. Ciavaldini, J. C. Nedellec, *Sur lélément fini de Fraeijs de Veubeke & Sanders*, RAIRO Analyse Numérique R2 , August 1974.

[9]   W. Dahmen, P. Oswald, X. Q. Shi, C^1*-hierarchical bases*, J. Comp. Appl. Math. **51** (1994), 37–56.

[10]   J. Dieudonné, *Eléments d'Analyse, Tome 1, Fondements de l'Analyse Moderne*, Gauthier–Villars, Paris 1968; (English translation: Foundations of Modern Analysis, Academic Press, New York 1960).

[11]   A. Le Méhauté, *Taylor interpolation of order n at the vertices of a triangle*, in: *Approximation Theory and Applications*, Z. Ziegler (ed.), 171–185, Academic Press, New York 1981.

[12]   A. Le Méhauté, *Interpolation et Approximation par des Fonctions Polynomiales par Morceaux dans* $\mathbb{R}^n$, Thèse d'Etat, Université de Rennes, 1984.

[13]   A. Le Méhauté, *An efficient algorithm for C^k-simplicial finite element interpolation in* $\mathbb{R}^n$, Annex, CAT Report #111, Texas A&M University, March 1986.

[14]   A. Le Méhauté, *Unisolvent interpolation in* $\mathbb{R}^n$ *and the simplicial polynomial finite element method*, in: *Topics in Multivariate Approximation*, C. K. Chui, L. L. Schumaker and F. I. Utreras (eds.), 141–152, Academic Press, New York–London 1987.

[15]   A. Le Méhauté, *An efficient algorithm for C^k-simplicial finite element interpolation in* $\mathbb{R}^d$, in: *Multivariate Approximation and Interpolation*, W. Haussmann and K. Jetter (eds.), Internat. Ser. Numer. Math. **94**, 179–192, Birkhäuser, Basel–Stuttgart–Boston 1990.

[16]   A. Le Méhauté, *A finite element approach to surfaces reconstruction*, in: *Computation of Curves and Surfaces*, W. Dahmen, M. Gasca, C. A. Micchelli (eds.), 237–274, Kluwer, Dordrecht 1990.

[17]   J. Morgan, R. Scott, *A nodal basis for C^1 piecewise polynomials of degree* ≥ 5, Math. Comp. **29** (1975), 736–740.

[18]   P. Oswald, *Hierarchical conforming finite element methods for the biharmonic equation,* SIAM J. Numer. Anal. **29** (1992), 1610–1625.

[19]   P. Sablonnière, *Composite finite elements of class C^2,* in: *Topics in Multivariate Approximation,* C. K. Chui, L. L. Schumaker, F. I. Utreras (eds.), 207–217, Academic Press, New York–London 1987.

[20]   L. L. Schumaker, *Numerical aspects of spaces of piecewise polynomials on triangulation,* CAT Report #91, Texas A&M University, Oct. 1985.

[21]   H. Yserentant, *On the multilevel splitting of finite element spaces,* Numer. Math. **49** (1986), 379–412.

[22]   A. Ženíšek, *A general theorem on triangular C^n elements,* RAIRO Anal. Numer. R2, **4** (1974), 119–127.

Address:

ALAIN LE MÉHAUTÉ
Département de Mathématiques
Faculté des Sciences et Techniques
Université de Nantes
2 rue de la Houssinière, BP 92208
44322 Nantes Cedex 3
France

Charge Distribution
of Points on the Sphere and
Corresponding Cubature Formulae

U. Maier and J. Fliege

Abstract

In the univariate case most satisfactory formulae for numerical integration concerning degree of exactness and convergence are obtained if the nodes are defined by the roots of orthogonal polynomials. We interprete these nodes as charged particles and generalize this idea to the multivariate case to get appropriate points for cubature formulae on the unit sphere. With the aid of a two–stage optimization algorithm we calculate (for $r = 3$) nodal systems on the sphere which are nearly equally spaced. Some of these nodal systems turn out to be spherical designs.
In case the number of nodes equals the dimension of the space of polynomials of certain degree, we calculate weights for corresponding quadrature formulae. Numerical examples show that the constructed formulae yield impressively small absolute integration errors of approximately 10^{-12} in certain test calculations.

1 Introduction

In applications (e.g. simulation of interstellar dust clouds [12]) multivariate quadrature formulae for the approximation of integrals over the unit sphere are needed. Formulae commonly used are *product trapezoidal* or *product Gaussian formulae* where the nodes lie on a grid such that they concentrate near the poles. This particular property is often not wanted because the poles are emphasized too much. Thus, a more regular distribution of points on the sphere is advisable.

On the other hand quadrature formulae with highest possible degree of exactness and equal weights to minimize the numerical effort of evaluation are desired. This leads to the concepts of multivariate Gaussian quadratures and

Multivariate Approximation: Recent Trends and Results; W. Haußmann, K. Jetter and M. Reimer (eds.)
Mathematical Research, Vol. 101, pp. 147–159, ISBN 3-05-501770-6
© Akademie-Verlag, Berlin 1997

spherical designs, respectively. Besides, multivariate Gaussian quadrature formulae are related to formulae with equal weights (Reimer [10]). Unfortunately, multivariate Gaussian formulae or tight spherical designs, i.e. spherical designs with the minimal number of nodes, rarely exist (Bannai, Damerell [1, 2], Bos [3], Reimer [10, 11]).

To overcome the difficulties mentioned above, we approach the multivariate quadrature problem by first computing nearly equally distributed points on the sphere and subsequently calculating corresponding weights which only differ slightly from equal weights.

The paper is organized as follows. In Section 2 we describe the first stage of our approach: the idea for finding the points, the computation of them and numerical results. In Section 2.4 we point out a connection between some of our node systems and spherical designs. Section 3 treats the second stage of our approach, namely the calculation of the weights and in Section 4 we report on tests of our quadrature formulae applied to different functions. A more detailed description of our algorithms and their implementation can be found in [6].

The following notation is used. For $r \in \mathbb{N}$ the unit sphere in $\mathbb{R}^r$ is denoted by S^{r-1}. For $m \in \mathbb{N}_0$ let $I\!\!P_m^r(S^{r-1})$ denote the linear space of restrictions of polynomials onto S^{r-1} in r variables of degree $\leq m$ and let $\overset{*}{I\!\!P}{}_m^r$ denote the corresponding space of homogeneous polynomials.

Further, let the inner product $\langle .,. \rangle$ on $I\!\!P_m^r(S^{r-1})$ be defined by

$$\langle f, g \rangle := \int_{S^{r-1}} f(x)g(x)dx.$$

The reproducing kernel of $I\!\!P_m^r(S^{r-1})$ is denoted by G_m and has the representation

$$G_m(x,y) = \frac{1}{\mu(S^{r-1})}[C_m^{r/2}(x^\top y) + C_{m-1}^{r/2}(x^\top y)], \tag{1.1}$$

where $\mu(S^{r-1}) = \langle 1, 1 \rangle$ is the surface area of S^{r-1} and $C_m^{r/2}$ denotes the *Gegenbauer polynomial* of degree m with index $r/2$.

The reproducing kernel of $\overset{*}{I\!\!P}{}_m^r$ is denoted by $\overset{*}{G}_m$ and has the representation

$$\overset{*}{G}_m(x,y) = \frac{1}{\mu(S^{r-1})} \cdot C_m^{r/2}(x^\top y). \tag{1.2}$$

2 The First Stage: The Points

2.1 Finding the Points

In the univariate case, the zeros of Jacobi–polynomials define the well–known Gauss–Jacobi quadratures which are distinguished by highest possible degree of exactness. Moreover, these quadratures have positive weights which guarantees convergence.
For arbitrary indices, the zeros of these orthogonal polynomials can be interpreted as the loci of movable charges on a metal rod which is represented by the interval $[-1, 1]$. In the state of equilibrium the potential energy of the charge distribution is minimized and their locations coincide with the zeros of the Jacobi polynomials. For details see Stroud and Secrest [13, p. 17].

Thus, minimization of the potential energy of a point distribution on the sphere is a good criterion for an appropriate node distribution.

With $\|.\|_2$ the Euclidian norm in $\mathbb{R}^r$ we get for electrically charged point particles at locations $x_1, \ldots, x_N \in S^{r-1}$ the potential energy

$$E(x_1, \ldots, x_N) = \sum_{i=1}^{N} \sum_{j=i+1}^{N} \frac{1}{\|x_i - x_j\|_2} \tag{2.3}$$

and thus the nonlinear optimization problem

$$\text{minimize} \qquad E(x_1, \ldots, x_N) \tag{2.4}$$
$$\text{subject to} \qquad x_i \in S^{r-1} \qquad (1 \leq i \leq N). \tag{2.5}$$

In optimization theory this is called a *facility dispersion problem* on the sphere.

The nonlinear optimization problem (2.4)–(2.5) has as constraints the relations $x_i \in S^{r-1}$ $(1 \leq i \leq N)$, which can be written as $\|x_i\|_2 = 1$. However, these nonlinear equality constraints are computationally not so easy to handle. A reformulation of the objective function involving spherical coordinates leads to an optimization problem with far less unknowns, e.g. $2N$ instead of $3N$ unknowns for $r = 3$. Fixing the north pole and the longitude of a further point, the number of unknowns can be dropped to $2N - 3$ for $r = 3$.

2.2 Specific Algorithms

We have to solve a difficult nonlinear nonconvex constrained global optimization problem, which has a high number of local minima. In order to solve this problem we propose a two–phase approach. In Phase I, only a rough approximation of a global optimum is computed. This is done with a stochastic global optimization method. In Phase II, the previously computed approximation is refined with a highly accurate nonlinear local optimization method. Our computational study suggests that a two–phase approach results in a better performance than using only one of the two phases. This is confirmed by results of Steinacker, Thamm and Maier [12] and Zhou [15].

For Phase I we chose the code of Goffe, Ferrier and Rogers [8], a standard simulated annealing approach for continuous global optimization problems. There, some parameters had to be adapted to our special situation. For Phase II we used a limited memory BFGS–method (L–BFGS) implemented by Zhu [16]. A detailed discussion of our implementation can be found in [6].

2.3 Computational Results

The optimization codes were compiled with the `f77` compiler using the `-O3` (aggresive code optimization) compilation flag on a Silicon Graphics Power Indigo^2 with Mips R8000 Chip, 192 MB main memory and IRIX 6.0.1 operating system. The machine precision on this machine is approximately 2.22×10^{-16}. All point distributions were computed on this machine. The computation times were obtained by the IRIX `time` command.

We computed point systems for problems on S^2 with $N = (m + 1)^2$ points, $m = 1, \ldots, 39$, and, additionally, $N = 1, \ldots, 100$. The last set of distributions was computed for benchmarking reasons. The biggest problem we solved had therefore $N = 1600$ nodes, which resulted in a nonlinear nonconvex global optimization problem with 3197 unknowns and 6394 constraints. Our technique can easily be extended to higher dimensional problems.

Figure 1 shows the computation time for the first (simulated annealing, SA) and the second (limited-memory BFGS, L-BFGS) phase in Stage 1. A detailed table of all computation times can be found in [6]. Note that only one run of the simulated annealing algorithm of Phase I was needed to produce a good starting distribution for Phase II. For the L-BFGS algorithm only one run was necessary for good results, as well.

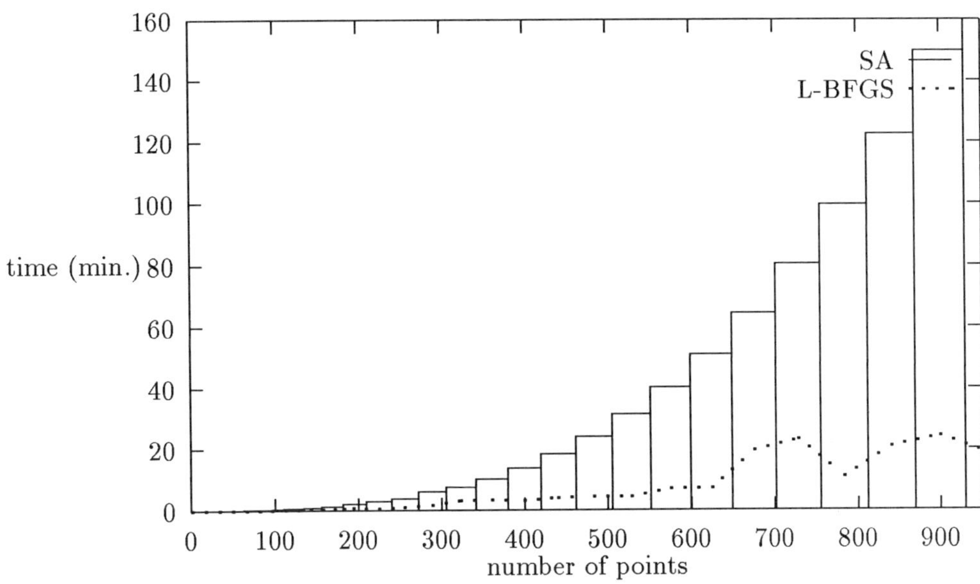

Figure 1: Time needed in Stage 1

Figure 2 shows, as an example, one of our node distributions on the sphere. All other cases considered give similar configurations.

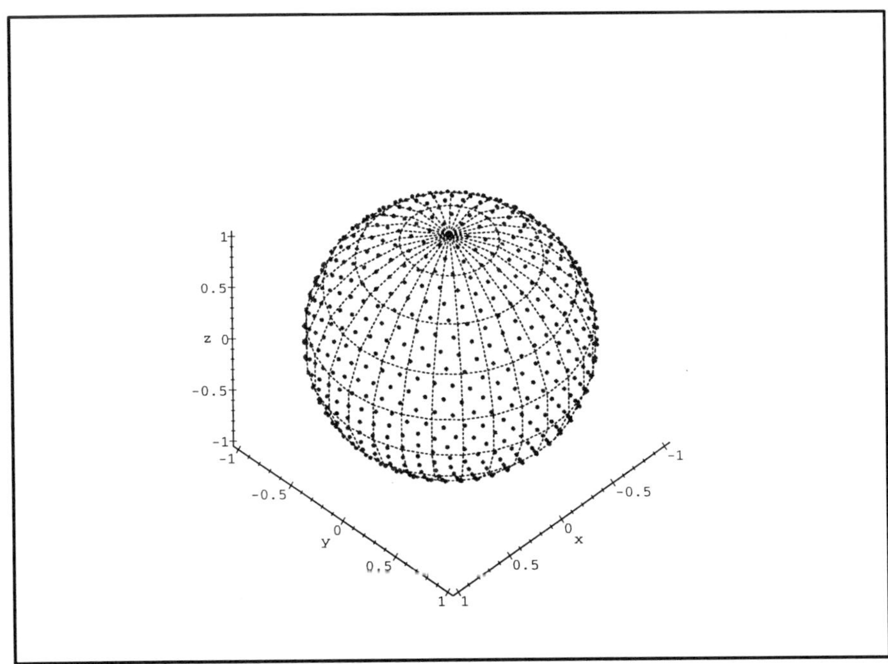

Figure 2: 900 nodes on a sphere

2.4   A Connection with Spherical Designs

One possibility to test the quality of the nodes is to compare them with special point systems on S^{r-1}, socalled spherical designs.

Definition 2.1. *Let* $X \subset S^{r-1}$ *and* $A(X) := \{\langle \xi, \eta \rangle, \xi, \eta \in X, \xi \neq \eta \} \subset [-1, 1]$. *Then* X *is called a* spherical A-code. *A finite subset* $X \subset S^{r-1}$ *is a* spherical design *of strength* m *(for short* m-design*) if*

$$\frac{1}{|X|} \sum_{x \in X} f(x) = \frac{1}{\mu(S^{r-1})} \int_{S^{r-1}} f(\xi) d\mu(\xi) \quad \text{for every} \quad f \in I\!\!P_m^r(S^{r-1}).$$

Here $|X|$ *denotes the number of elements of* X *and* μ *the surface measure of the sphere.*

Quadrature formulae resulting from spherical m-designs have the advantage of minimal numerical effort of evaluation and maximal degree of exactness, namely $m = 2n$ or $m = 2n + 1$, $n \in I\!\!N$. In contrast to this, interpolatory quadrature formulae in general only have a degree of exactness equal to n.

Let $m = 2n$ or $m = 2n + 1$, $n \in I\!\!N$. In [5] Delsarte, Goethals and Seidel proved the following theorems.

Theorem 2.2. *Let* X *be a* $(2n)$-*design. Then* $|X| \geq \dim I\!\!P_m^r(S^{r-1})$. *Equality holds if and only if* $A(X)$ *consists of the zeros of* G_m.

Theorem 2.3. *Let* X *be a* $(2n + 1)$-*design. Then* $|X| \geq 2 \dim \overset{*}{I\!\!P}_m^r (S^{r-1})$. *E-quality holds if and only if* $A(X)$ *consists of* -1 *and the zeros of* $\overset{*}{G}_m$. *Moreover, in the case of equality the design* X *is antipodal (i.e.* $X = -X$).

A spherical design is called *tight* if any of the bounds is attained.

Unfortunately, tight spherical m-designs scarcely exist for $r \geq 3$ (Bannai and Damerell [1, 2]). A general principle of construction for spherical designs is not known either.

We now compare our node systems with spherical m-designs. To give a uniform representation we consider the connection between the Jacobi–polynomials $P_m^{(\alpha,\beta)}$ and the Gegenbauer–polynomials C_m^{λ} (Tricomi [14]).

$$G_m = \text{const} \cdot P_m^{(1,0)} = \text{const} \cdot (C_m^{3/2} + C_{m-1}^{3/2}),$$

$$\overset{*}{G}_m = \text{const} \cdot P_m^{(1,1)} = \text{const} \cdot C_m^{3/2}$$

and

$$\text{const} \cdot P_m^{(1/2,-1/2)} = C_m^1 + C_{m-1}^1,$$
$$\text{const} \cdot x \cdot P_m^{(1/2,-1/2)} = C_{m+1}^1 + C_m^1 + C_{m-1}^1 + C_{m-2}^1.$$

A comparison of the scalar products of the calculated nodes for $1 \leq N \leq 12$ with the zeros of G_m and $\overset{*}{G}_m$ gives the following table.

N	tight design	-1 is zero	zeros of
2	1-design	yes	$P_0^{(1,1)}$
3	–	no	$P_1^{(1/2,-1/2)}$
4	2-design	no	$P_1^{(1,0)}$
5	–	yes	$x \cdot P_1^{(1/2,-1/2)}$
6	3-design	yes	$P_1^{(1,1)}$
7	–	no	$x \cdot P_2^{(1/2,-1/2)}$
8	–	no	?
9	–	no	?
10	–	yes	?
11	–	no	?
12	5-design	yes	$P_2^{(1,1)}$

Table 1: Comparison of nodes with Theorems 2.1–2.2

In the cases $N = 2, 4, 6, 12$ the nodes are tight spherical designs. Thus the weights of such a quadrature formula can be chosen to be $4\pi/N = \mu(S^2)/N$. The cases $N = 3, 5, 7$ also show a regular behaviour, but up to now it could not be decided whether these nodes were spherical designs.

3 The Second Stage: The Weights

3.1 Computing the Weights

In the cases where the calculated nodes can be shown to be spherical designs ($N = 2, 4, 6, 12$), equal weights are optimal. In all other cases appropriate weights have to be calculated. With the aid of multivariate interpolation theory this is at least possible if the number N of nodes equals the dimension of the linear space of polynomials $I\!\!P_m^r(S^{r-1})$.

Let $G := (G_m(x_j, x_k))_{j,k=1}^N$ be the symmetric matrix being composed by the aid of the kernel G_m and the nodes $x_i \in S^{r-1}$, $i = 1, \ldots, N$ (see (1.1)). If $\det G \neq 0$, the associated *Lagrange fundamental polynomials* l_j, $j = 1 \ldots, N$, are well defined by $l_j(x_k) = \delta_{jk}$ (δ_{jk} the Kronecker symbol). Following Reimer [9], as a consequence of the reproducing property of G_m,

$$l_j(x) = \int_{S^{r-1}} l_j(y) \cdot \left(\sum_{k=1}^N G_m(x_k, x) \cdot l_k(y) \right) dy, \tag{3.6}$$

which leads to

$$\sum_{k=1}^N \left(\int_{S^{r-1}} l_j(y) \cdot l_k(y) dy \right) \cdot G_m(x_k, x_j) = \delta_{jk}. \tag{3.7}$$

This means that the matrices $L := (< l_j, l_k >)_{j,k=1}^N$ and $G := (G_m(x_j, x_k))_{j,k=1}^N$ are inverse to each other.

Consider now the interpolatory cubature formula on $I\!\!P_m^r(S^{r-1})$

$$\int_{S^{r-1}} f(x) dx = \sum_{j=1}^N \omega_j f(x_j), \quad f \in C(S^{r-1}). \tag{3.8}$$

Then, we get $\omega_j = \int_{S^{r-1}} l_j(x) dx$. Because $1 \in I\!\!P_m^r(S^{r-1})$ and $\sum_{k=1}^N l_k(x) = 1$ this gives

$$\omega_j = \sum_{k=1}^N \left(\int_{S^{r-1}} l_j(x) l_k(x) dx \right), \tag{3.9}$$

that is each weight ω_j is a row–sum of the matrix L (Reimer [9]).

Furthermore, the inversion of the matrix $G := (G_m(x_j, x_k))_{j,k=1}^N$ can be avoided. With $e := (1, \ldots, 1)^\top \in I\!\!R^N$ we obtain the vector $\omega := (\omega_1, \ldots, \omega_N)^\top$ of weights as the solution of the linear system of equations

$$G \cdot \omega = e. \tag{3.10}$$

Remark. To ensure a numerically stable computation of the matrix G the Gegenbauer polynomials are evaluated with the aid of their recurrence relation. Since G is symmetric and positive definite, we have used Cholesky's method to solve the system of equations (3.10).

3.2 Computational Results

In this section we restrict our attention to the case $r = 3$. An extension of our considerations to higher dimensions causes no problems.

The code for computing the weights was developed with SPARCompiler Pascal 3.0 and cross–compiled to C with `p2c`. The actual compilation took place with the `cc` compiler (again using the `-03` flag) on a SPARCstation 20 with 64 MB main memory and SunOS 5.3 operating system.

On this machine we computed the optimal weights for the node distributions from the first stage with $N = (m + 1)^2$ points ($m = 1, \ldots, 29$).

The following Figures 3 and 4 show the minimal and maximal weights compared with the equal weights $4\pi/N = \mu(S^{r-1})/N$. The time needed to compute the weights was in all cases considered less than 6 minutes even for the problem of 900 nodes.

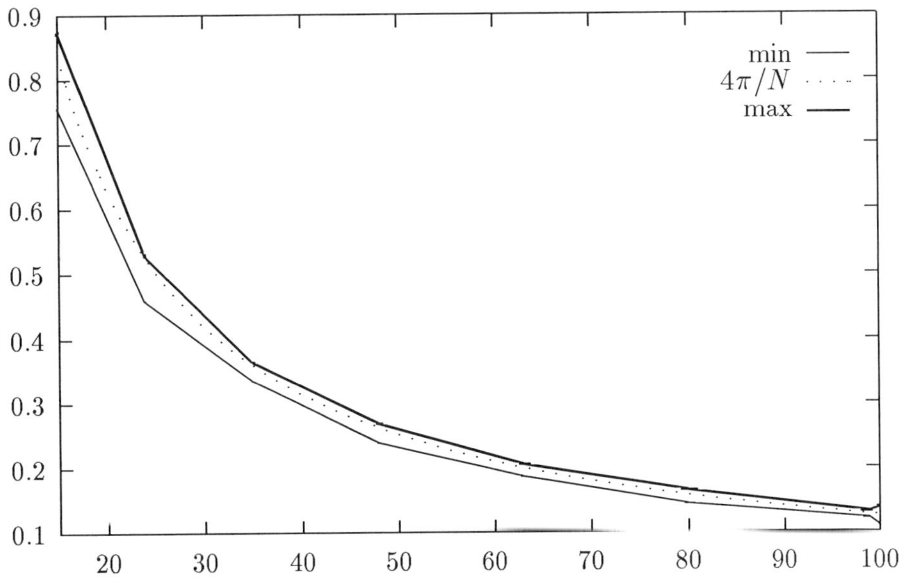

Figure 3: Weights compared with equal weights, $N \leq 100$

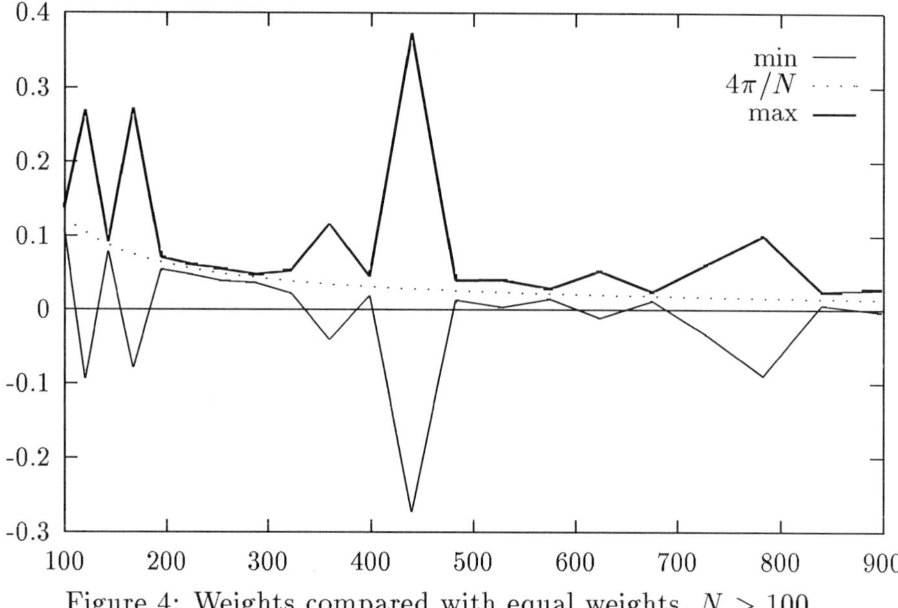

Figure 4: Weights compared with equal weights, $N > 100$

4   Testing the Formulae

We tested our integration formulae on six different functions $f_j : \mathbb{R}^3 \to \mathbb{R}$ described in Table 2.

j	$f_j(x)$	$\int_{x \in S^2} f_j(x)dx$
1	$\exp(x_1 + x_2 + x_3)/10$	$1.98622365458551\ldots$
2	$\|x\|_1 /10$	$1.88495559215388\ldots$
3	$-5\sin(1 + 10x_3)$	$2.87630387748651\ldots$
4	$1/(10.1 - 10x_3)$	$3.33216474778102\ldots$
5	$\exp(x_1)$	$14.7680137457653\ldots$
6	$x_1 x_2 x_3$	0

Table 2: Test functions and exact integration values

The absolute integration error for an N–node formula is now defined as

$$\left| \int_{x \in S^2} f_j(x)dx - \sum_{i=1}^{N} \omega_i f_j(x_i) \right|, \quad j = 1,\ldots,6.$$

Figures 5 and 6 show these integration errors for all computed formulae and all test functions.

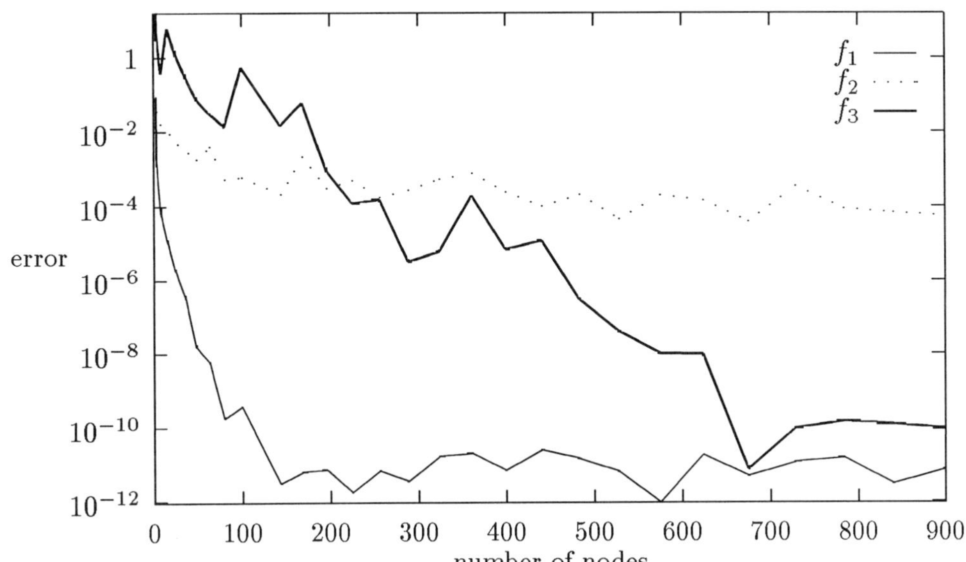

Figure 5: Absolute error for test functions f_1, f_2 and f_3

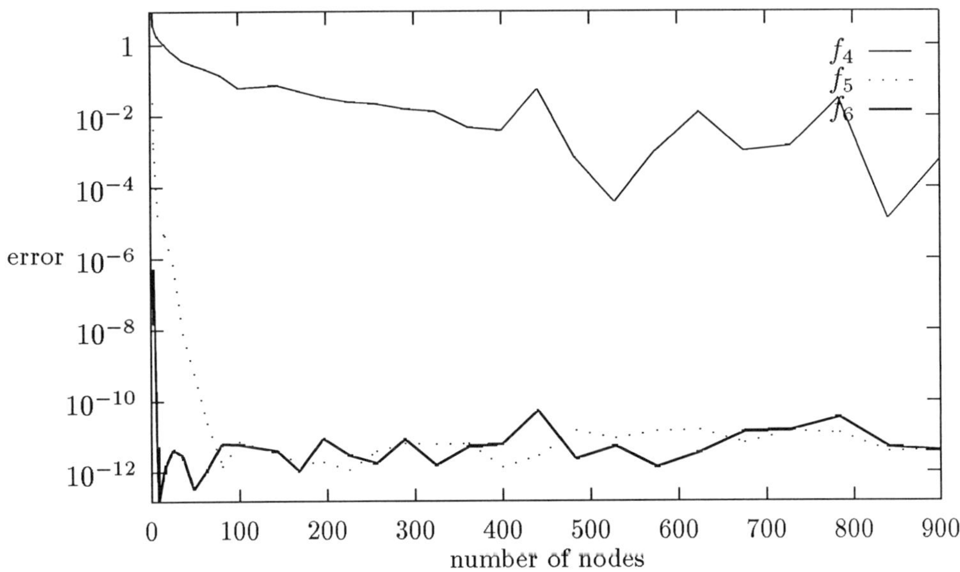

Figure 6: Absolute error for test functions f_4, f_5 and f_6

Since the machine precision is about 10^{-16}, the computed formulae have a "threshold" at about 10^{-12}. Without a doubt, the errors will get smaller when the numerical integration is done on a machine with higher machine precision (or which simulates more accurate floating point numbers by software).

Our formulae are exact for polynomials of a certain degree. Thus, the absolute integration error will be the smaller the better a function can be approximated by multivariate polynomials. In other words: functions with rapidly converging Taylor series will give extremely small absolute errors even for small numbers of nodes. Since the function f_2 is not differentiable, it cannot be expected to have a reasonable absolute integration error. The same is true for the function f_4 which has a pole near $(0,0,1)^\top$. Note that we have fixed the north pole to reduce the numerical effort for the computation of our point systems.
Unfortunately, to the best of our knowledge, comparable numerical results for multivariate quadrature formulae on the sphere do not seem available in the literature.

Our computed nodal systems and weights are available at our WWW–server with the
URL http://www.mathematik.uni-dortmund.de/lsx/fliege/nodes.html.

References

[1]   E. Bannai, R. M. Damerell: *Tight spherical designs I*, J. Math. Soc. Japan **31** (1979), 199–207.

[2]   E. Bannai, R. M. Damerell: *Tight spherical designs II*, J. London Math. Soc. (2) **21** (1980), 13–30.

[3]   L. Bos: *Some remarks on the Fejér–problem for Lagrange interpolation in several variables*, J. Approx. Theory **60** (1990), 133–140.

[4]   R. H. Byrd, P. Lu, J. Nocedal, C. Zhu: *A limited memory algorithm for bound constrained optimization*, SIAM J. Sci. Comput. **16** (5), 1995. Also published as Technical Report NAM–08, Department of Electrical Engineering and Computer Science, Northwestern University, Evanston, Illinois, USA. May 1994. Electronically distributed as ftp://eecs.nwu.edu/pub/lbfgs/lbfgs_bcm/byrd-lu-nocedal-zhu.ps

[5]   P. Delsarte, J. M. Goethals, J. J. Seidel: *Spherical codes and designs*, Geom. Dedicata **6** (1977), 363–388.

[6] J. Fliege, U. Maier: *A Two-Stage Approach for Computing Cubature Formulae for the Sphere*, Ergebnisberichte Angewandte Mathematik 139–T, Universität Dortmund, Fachbereich Mathematik, Universität Dortmund, D–44221 Dortmund, August 1996.

[7] W. L. Goffe, G. D. Ferrier, J. Rogers: *Global optimization of statistical functions with simulated annealing*, J. Econometrics **60** (1994), 65–99.

[8] W. L. Goffe, G. D. Ferrier, J. Rogers: SIMANN code, Electronically distributed as `ftp://netlib2.cs.utk.edu/opt/simann.f`

[9] M. Reimer: *Constructive theory of multivariate functions*, BI Wissenschaftsverlag, Mannheim–Wien–Zürich 1990.

[10] M. Reimer: *On the existence-problem of Gauss–quadrature on the sphere*, In: Approximation by Solutions of Partial Differential Equations, B. Fuglede et al. (eds.), 169–184, Kluwer, Dordrecht 1992.

[11] M. Reimer: *On the existence of Gauss–like node–distributions on high–dimensional spheres*, In: Multivariate Approximation, K. Jetter and F. Utreras (eds.), 281–291, World Scientific, Singapore 1993.

[12] J. Steinacker, E. Thamm, U. Maier: *Efficient integration of intensity functions on the sphere*, J. Quant. Spec. Rad. Transfer **56**, No. 1, (1996), 97–107.

[13] A. H. Stroud, D. Secrest: *Gaussian quadrature formulas*, Prentice Hall, Series in Automatic Computation, 1966.

[14] F. G. Tricomi: *Vorlesungen über Orthogonalreihen*, Springer, Berlin–Göttingen–Heidelberg, 1955.

[15] Yanmu Zhou: *Arrangements of points on the sphere*, Ph. D. Thesis, Tampa, Florida 1995.

[16] C. Zhu: *L–BFGS FORTRAN code*, November 1994. Electronically distributed via `ftp://eecs.nwu.edu/pub/lbfgs/lbfgs_bcm/` and `http://www.eecs.nwu.edu/ ciyou/code/lbcode.html`

Address:

JÖRG FLIEGE, ULRIKE MAIER
Department of Mathematics
University of Dortmund
D–44221 Dortmund
Germany

Solving of Algebraic Equations –
An Interplay of Symbolical and Numerical Methods

H. M. Möller

Abstract

For the numerical solving of systems of algebraic equations, tools from computer algebra can be useful. We give a survey on available symbolical methods for a preprocessing of such systems, for obtaining informations on the solution set, and for transforming them into simpler, numerically solvable systems of equations or into eigenproblems. Using invariant theory, we show by two examples how to reformulate and solve otherwise unsolvable problems with symmetries.

1   Introduction

Usually, when authors describe how to solve systems of algebraic equations, they concentrate either on numerical algorithms for finding approximate solutions, see for instance [4], or they exhibit symbolic methods for solving algebraically the equations, for instance [3]. Depending on the authors' point of view, the symbolical manipulations for simplifying the equations are considered at most as a preprocessing for the hard work of finding good approximations or the numerical computation is considered as just a postprocessing after having found the solutions in an algebraic sense.

The intention of the present paper is to encourage users of numerical software to employ algebraic methods and to use symbolic software as integrated in computer algebra systems like AXIOM, MAPLE, MATHEMATICA, and RE-DUCE. For a better understanding of the algebraic background, we try to give a low level access to the needed algebraic facts. We will show the limitations of the pure access (either numerical or symbolical) and present actually available tools from computer algebra which help to find a better understanding of the solution set and possible alternative formulations of the given system which admit hopefully an easier numerical solution.

Multivariate Approximation: Recent Trends and Results; W. Haußmann, K. Jetter and M. Reimer (eds.)
Mathematical Research, Vol. 101, pp. 161–176, ISBN 3–05–501770–6
© Akademie–Verlag, Berlin 1997

Let us illustrate these limitations by two typical problems arosen from questions in multivariate approximation and interpolation.

Example 1. The first set of equations was given by [9]. The minimization of zonal spherical polynomials in L_1–norm on the unit sphere S^2 amounts to the well known (univariate) minimizing problem

$$\gamma_n := \min\{\int_0^1 |f(x)|dx \ : \ f \in \mathcal{P}_n, \ f(0) = 1 \}.$$

The zeros $x_1, \ldots, x_n$ of the optimal f satisfy

$$x_1^k - x_2^k \pm \ldots - (-1)^n x_n^k = \frac{(-1)^{n+1}}{2}, \quad k = 2, \ldots, n+1. \tag{1.1}$$

The solution with $0 < x_1 < \ldots < x_n < 1$ is unique. The observation, that the components of this unique solution for the n–th problem interlace with those for the $(n+1)$–st one gives good starting points for the computation of the unique $(n+1)$–st solution by Newton's method. This allows an easy numerical calculation of the floating point solutions up to very high degrees [10]. However, if a quantity depends on the unique solution $x_1, \ldots, x_n$, for instance their alternating sum, [9], then it is impossible to find an algebraic dependence of these quantities for some consecutive n by numerical methods.

These systems were imported into the PoSSo benchmark suite and tested on different platforms with different symbolic software. Usually the solution set was only found for $n \leq 5$. The optimum was the solution for $n = 6$, see the PoSSo benchmark suite.[1]

Example 2. In her thesis, U. Maier [5] investigated sets of points on S^2 in which the Kergin interpolation polynomial has a certain best L_2–approximation property. In the case $m = (3, 1, 0)$, four points (x_i, y_i, z_i), $i = 1, 2, 3, 4$, on the sphere are involved. The geometrical configuration of the $x-$ and $y-$ coordinates is described by the system

---

[1] available by ftp from barsotti.dm.unipi.it

$$x_1 + x_2 + x_3 + x_4 = 0,$$
$$y_1 + y_2 + y_3 + y_4 = 0,$$
$$3x_1x_2 + 3x_1x_3 + 3x_1x_4 + 3x_2x_3 + 3x_2x_4 + 3x_3x_4 + 2 = 0,$$
$$x_1(y_2 + y_3 + y_4) + x_2(y_1 + y_3 + y_4) + x_3(y_1 + y_2 + y_4) + x_4(y_1 + y_2 + y_3) = 0,$$
$$y_1(x_2x_3 + x_2x_4 + x_3x_4) + y_2(x_1x_3 + x_1x_4 + x_3x_4) +$$
$$y_3(x_1x_2 + x_1x_4 + x_2x_4) + y_4(x_1x_2 + x_1x_3 + x_2x_3) = 0,$$
$$x_1x_2x_3 + x_1x_2x_4 + x_1x_3x_4 + x_2x_3x_4 = 0,$$
$$x_1x_2x_3y_4 + x_1x_2y_3x_4 + x_1y_2x_3x_4 + y_1x_2x_3x_4 = 0.$$

$$(1.2)$$

These are seven equation for eight unknowns. Hence the system is numerical unsolvable. But, as we will see later, it is easily solvable by standard computer algebra systems.

2 Algebraical and Numerical Solving

We consider $\mathcal{P} := \mathbb{K}[x_1, \ldots, x_n]$, the ring of polynomials

$$f(x_1, \ldots, x_n) := \sum_{i_1,\ldots,i_n=0}^{\infty} a_{i_1\ldots i_n} x_1^{i_1} \cdots x_n^{i_n} \tag{2.1}$$

with at most finitely many nonzero coefficients $a_{i_1\ldots i_n} \in \mathbb{K}$. Here $\mathbb{K}$ denotes a field between the rationals and the complex numbers, $\mathbb{Q} \subseteq \mathbb{K} \subseteq \mathbb{C}$, which can be approximated by floating point numbers. By the leading form $l(f)$ of the polynomial f in (2.1) of degree d we mean

$$l(f) := \sum_{i_1+\ldots+i_n=d} a_{i_1\ldots i_n} x_1^{i_1} \cdots x_n^{i_n} \ .$$

We fix an algebraic system

$$f_1(x_1, \ldots, x_n) \;=\; 0 \ ,$$
$$\vdots \tag{2.2}$$
$$f_r(x_1, \ldots, x_n) \;=\; 0 \ ,$$

with polynomials $f_1, \ldots, f_r$ and not necessarily $n = r$.

Solving this system means finding a description (in a sense to be specified later) of the variety, i.e. of the solution set

$$V := \{y \in \mathbb{C}^n \mid f_k(y) = 0, \ k = 1, \ldots, r\} \ . \tag{2.3}$$

This variety may be either a finite or an infinite set of points.

If we admit also common zeros 'at infinity', which means the common projective zeros of the leading forms $l(f_1), \ldots, l(f_r)$, then

- V empty only for $r > n$,
- V finite only for $r \geq n$.

If $r \leq n - k$, then V is a variety of dimension $\geq k$, which roughly spoken means, that V contains points, for which a parametric description needs at least k free parameters.

In the following, we restrict our considerations to the case that V is finite, if not explicitly stated otherwise.

In order to describe the notion *multiplicity* of a point y of the variety V, let us assume here, for simplicity, $V \subset \mathbb{K}$. We introduce the space of differentials at y by

$$\mathbb{D}_y := \{ L \in \mathcal{P}' \mid \exists c_\alpha \in \mathbb{K} : \ L : f \mapsto \sum_\alpha c_\alpha \frac{\partial^{|\alpha|} f}{\partial x^\alpha}(y) \} \ . \tag{2.4}$$

If we denote by I the ideal generated by $f_1, \ldots, f_r$, i.e.

$$I := \{ f \in \mathcal{P} \mid \exists h_1, \ldots, h_r \in \mathcal{P} : \ f = \sum_{i=1}^r h_i f_i \} \ , \tag{2.5}$$

then the linear space

$$V_y := \{ L \in \mathbb{D}_y \mid L(f) = 0 \ \forall f \in I \} \tag{2.6}$$

has a dimension $\mu \geq 1$ for $y \in V$. This μ is the multiplicity of $y \in V$. $\mu \geq 1$ holds because the point evaluation functional at y belongs to V_y. For a more detailed discussion of V_y see [6].

Now we can explain the meaning of solving the equations (2.2).

In the *numerical* sense it means to find for every $y \in V$ and every given precision an approximation to y within this precision. In the *algebraic* sense it means to find the right finite algebraic field extension $\mathbb{K}_1$ of $\mathbb{K}$ such that each component y_i of $y \in V$ is given in terms of elements of $\mathbb{K}_1$. If K_1 is considered as a simple extension, then this amounts to finding univariate polynomials $p, p_1, \ldots, p_n \in \mathbb{K}[t]$ such that for every component y_i of y the equation $y_i = p_i(a)$ holds and a is an arbitrary zero of the irreducible polynomial $p \in \mathbb{K}[t]$.

The task of finding all solutions to (2.2) can be restricted to find all real solutions or more generally to all solutions in a set $D \subset \mathbb{C}^n$ or $\subseteq \mathbb{R}^n$. It may turn out that points y in V can not be distinguished numerically (cluster) or algebraically (multiple points). Then it may be sufficient to estimate the size of the cluster or to find the multiplicity instead of a basis of V_y.

3 Gröbner Bases

A powerful tool for dealing with polynomials symbolically is the Gröbner basis concept. It is integrated in most computer algebra systems (CAS). Nearly all textbooks on computer algebra contain this concept. We collect some properties of Gröbner bases useful for solving systems and refer for details to [1].

In the Gröbner concept, the ordering of power products (terms) $x_1^{i_1} \cdots x_n^{i_n}$ is essential. Let T be the set of power products and $<_T$ a linear ordering of T. $<_T$ is called admissible, if

$$1 \leq_T t \quad \forall t \in T \quad \text{and} \quad t_1 <_T t_2 \implies t \cdot t_1 <_T t \cdot t_2 \quad \forall t, t_1, t_2 \in T. \qquad (3.1)$$

Then each nonzero polynomial f being a linear combination of power products contains a greatest one with respect to $<_T$, the leading term, $lt(f)$. Its coefficient is the leading coefficient of f. We say a polynomial f is reduced modulo a finite set $G \subset \mathcal{P}$ if no power product of f is a multiple of an $lt(g)$, $g \in G$.

Definition 3.1. *A finite set* $G \subset \mathcal{P} \setminus \{0\}$ *is called a* Gröbner basis *for an ideal* I *if* $G \subset I$ *and if*

$$\forall \, 0 \neq f \in I \; \exists g \in G \; : \; lt(g)/lt(f). \qquad (3.2)$$

A Gröbner basis is called minimal, *if the leading coefficients are normalized to* 1 *and each* $g \in G$ *is reduced modulo* $G \setminus \{g\}$.

For every admissible term ordering, every ideal $I \neq (0)$ has a minimal Gröbner basis. It is uniquely determined (by the ordering and I). The Gröbner basis is an ideal basis, i.e. if $G = \{g_1, \ldots, g_m\}$, then

$$I = \{f \in \mathcal{P} \mid \exists h_1, \ldots, h_m \in \mathcal{P} \; : \; f = \sum_{i=1}^{m} h_i g_i\} \; .$$

Given an arbitrary (ideal) basis of an ideal I, there is an algorithm which computes the minimal Gröbner basis with respect to a given admissible term ordering, the so called Buchberger algorithm. It is integrated in most CAS. Unfortunately, its complexity is bad, such that in many cases the computation fails because of storage or time problems. Much effort is made to improve its performance. One of the last efforts is made within the PoSSo project, an EU project (ESPRIT Basic Research Action), see the documentation on the server *barsotti.dm.unipi.it* .

Crucial for the determination of the variety V is the fact, that it does not depend on the specific basis $\{f_1, \ldots, f_r\}$ of the ideal I given by (2.5), but merely on I itself, since

$$V = \{y \in \mathbb{C}^n \mid f(y) = 0 \ \forall f \in I\} \ .$$

Hence turning over from one ideal basis to the (minimal) Gröbner basis of the same ideal does not change V, but allows insight into the structure of I, for instance the computation of the Hilbert function.

Definition 3.2. *Let* $\mathcal{P}_k := \{f \in \mathcal{P} \mid deg(f) \leq k\}$ *. Then the mapping* $H(\,.\,, I):$ $k \mapsto dim(\mathcal{P}_k / I \cap \mathcal{P}_k)$ *is called the* Hilbert function *of* I *and*

$$\sum_{k=0}^{\infty} H(k, I) \cdot t^k$$

the Hilbert–Poincaré series.

Another feature of ideals I is $\mathcal{P}/I$ the vector space of equivalence classes modulo I. Its basis can be constructed by means of a Gröbner basis.

Theorem 3.3. *Let* V *be finite. Then* $\mathcal{P}/I$ *is a finite vector space with basis*

$$B := \{x_1^{i_1} \cdots x_n^{i_n} + I \mid \ \not\exists g \in G \ : \ lt(g)/x_1^{i_1} \cdots x_n^{i_n}\}, \tag{3.3}$$

where G *is a Gröbner basis of* I.

4   The Number of Solutions

The best known bound for the number of solutions deals with the numerical solvable system (2.2), i.e. with the special case $r = n$.

Theorem of Bézout 4.1. *If* $f_1, \ldots, f_n$ *have only finitely many zeros in common, then their number is* $d_1 \cdots d_n$, *where* $d_i := deg(f_i)$. *Points at infinity are included and all points are counted corresponding to their multiplicity.*

If the multiplicity of a zero is not of interest, then the polynomial set of Theorem 4.1 can be modified to an other polynomial set with the same variety, but each zero has multiplicity one. This requires the knowledge of one Gröbner basis for the ideal I generated by $\{f_1,\ldots,f_n\}$, independent of the term ordering $<_T$. Then as implemented in most CAS, for each variable x_i, $i = 1,\ldots,n$, a polynomial $p_i \in I$, depending only on x_i can be constructed. If multiple factors are removed from p_i (divide p_i by $gcd(p_i,p_i')$) resulting in, say, q_i, then consider the system $\{f_1,\ldots,f_n,q_1,\ldots,q_n\}$. Now all common zeros are simple. However, the increased number of equations requires an other symbolic preprocessing.

The points at infinity cause an other difficulty, especially if only the affine zeros are needed. Here a better counting based on multigrading gives a better bound for the number of affine zeros, the so called BKK–bound. A practical procedure for calculating this bound is described in [12].

Example. Consider the Eigenproblem as a polynomial system

$$\sum_{j=1}^{n} a_{ij}x_j - \lambda \cdot x_i = 0, \quad i = 1,\ldots,n. \tag{4.1}$$

where without restriction $x_1 = 1$. Then it is a system of n equations for the n unknowns $x_2,\ldots,x_n,\lambda$. The Bézout bound is 2^{n-1}, the BKK–bound gives the sharp estimate n.

The exact information for the number of (affine) solutions of the system (2.2) is obtained by the Hilbert function for the ideal I in (2.5) (and V as in (2.3)).

Theorem 4.2. *The variety V is finite if and only if*

$$\exists k_0, M \in \mathbb{N}_0 \; \forall k \geq k_0 \; : \; H(k,I) = M. \tag{4.2}$$

Then $card(V) \leq M$ and, if $V = \{y_1,\ldots,y_s\}$ and $m_i := dim(V_{y_i})$, then $M = \sum_{i=1}^{s} m_i$, see [6].

The Hilbert function is also available in many CAS. The number M can be computed in a different way using Theorem 3.3 and $dim(\mathcal{P}/I) = M$, see [1].

There is also a method dating back to Ch. Hermite for counting the number of zeros in a semi–algebraic set D, i.e. in a set

$$D := \left\{ X \in \mathbb{R}^n \middle| \begin{array}{ll} \varphi_i(X) > 0, & i = 1,\ldots,s_1, \\ \varphi_i(X) \geq 0, & i = s_1+1,\ldots,s_2, \\ \varphi_i(X) = 0, & i = s_2+1,\ldots,s_3, \end{array} \right\} \tag{4.3}$$

with polynomials $\varphi_1, \ldots, \varphi_{s_3} \in \mathbb{R}[x_1, \ldots, x_n]$.

With $f_1, \ldots, f_r, I$ and V as above, let G denote a Gröbner basis of I and let V be a finite set. Take the basis $B =: \{b_1 + I, \ldots, b_s + I\}$ from thm. 3.3. For arbitrary $q \in \mathbb{R}[x_1, \ldots, x_n]$ let M_q denote the matrix for the endomorphism

$$\Phi_q : \mathcal{P}/I \longrightarrow \mathcal{P}/I, \quad h + I \mapsto q \cdot h + I. \tag{4.4}$$

Construct for a fixed $\varphi \in \mathbb{R}[x_1, \ldots, x_n]$ the matrix Q_φ with entries

$$q_{ij} := trace(M_{b_i \varphi b_j}), \quad i, j = 1, \ldots, s.$$

Then the signature of Q_φ equals

$$card(\{X \in V \cap \mathbb{R}^n \mid \varphi(X) > 0\}) - card(\{X \in V \cap \mathbb{R}^n \mid \varphi(X) < 0\}). \tag{4.5}$$

Calculating the signature for Q_1 we get the number of real zeros in V, and the signature for Q_{φ^2} gives $card(\{X \in V \cap \mathbb{R}^n \mid \varphi(X) \neq 0\})$. Together with (4.5), we get the number of real points in V, where φ is > 0, $= 0$ or < 0. By tensoring, this method can be extended to semi–algebraic sets which are given by more than one inequality. This method is described in [8], where further references can be found.

5   Symbolic Solving of Algebraic Systems

Several attempts have been made for solving symbolically algebraic systems. There are methods which are based on recursion on the number of variables, using that $\mathbb{K}[x_1, \ldots, x_n] = (\mathbb{K}[x_1, \ldots, x_{n-1}])[x_n]$ holds, like the *Wu–Ritt method*, see [13], or the *resultant method*, see [1]. In these methods, one tries to find at an early stage of computation univariate polynomials vanishing in the solution set. A factorisation of these univariate polynomials leads to a splitting of the solution set. The computation of these subsets is easier, but the procedure to get the univariate polynomials is involved. Tests with the Wu–Ritt method within the PoSSo project lead to the recommendation to apply it only for moderate degree polynomial input. The resultant method is recommended for small n, say $n \leq 3$.

Gröbner bases together with special term orderings allow also the solution of algebraic systems of equations.

Definition 5.1. *A term ordering is called an* elimination ordering *for* $x_1, \ldots, x_s$, $s < n$, *if*

$$x_{s+1}^{i_{s+1}} \cdots x_n^{i_n} <_T x_1^{j_1} \cdots x_n^{j_n} \quad if \; \exists k \le s : j_k \neq 0.$$

If G is a Gröbner basis with respect to such ordering for the ideal I, then $G \cap \mathbb{K}[x_{s+1}, \ldots, x_n]$ is a Gröbner basis of $I \cap \mathbb{K}[x_{s+1}, \ldots, x_n]$. Especially, the lexicographical term ordering is an elimination ordering for $x_1, \ldots, x_s$ for all $s < n$. Therefore, a Gröbner basis with respect to the lexicographical ordering contains one polynomial φ in x_n representing the basis of the ideal $I \cap \mathbb{K}[x_n]$, some polynomials in x_{n-1}, x_n representing a basis of the ideal $I \cap \mathbb{K}[x_{n-1}, x_n]$ etc. Hence the computation of the variety (2.3) amounts to computing a lexicographical Gröbner basis for the ideal (2.5) and the numerical computation of the possible n–th components by computing the zeros of φ substituting them into the other Gröbner basis polynomials, computing then the common zeros of the polynomials in x_{n-1} etc.

This procedure has two major drawbacks. At first, the numerical computation produces floating point approximations. By the successive substitutions, one gets instead of exact polynomials only polynomials with disturbed floating point coefficients. The common zeros of several univariate (not exactly known) polynomials first in x_{n-1}, then in x_{n-2} etc., have to be computed. There is just one polynomial for each variable only if the Gröbner basis has triangular shape, i.e. if it consists in exactly n polynomials where, after an ordering, the k–th polynomial depends exactly on $x_k, \ldots, x_n$.

The second drawback is the bad theoretical and practical complexity of Buchberger's algorithm. Its termination can not be observed in several uncritical looking examples. This empirical fact holds especially true for the computation of lexicographical Gröbner bases.

One remedy is factorisation as it is used by many CAS. During the computation each new polynomial is tested for factorisation. If a polynomial f factors, then the computation is branched. Each branch continues the computation with a factor instead of f. Since the factors have lower degrees and vanish usually only in a proper subset of the variety, the algorithm terminates usually for each branch earlier than the unbranched algorithm. In this way, one obtains Gröbner bases for subvarieties and hence the solutions grouped by the branching. However, the solution subsets are not necessarily disjoint. Some CAS dont even check subsets for being identical !

A possible hint to get at least different subsets is the observation, that a variety V for the set $\{f_1, \ldots, f_r, f\}$, for short $V(f_1, \ldots, f_r, f)$, can be split in case

$f = h_1 \cdot h_2$ into two point sets. In obvious notation

$$V(f_1, \ldots, f_r, f) = V(f_1, \ldots, f_r, h_1) \cup V(f_1, \ldots, f_r, h_2; h_1 \neq 0) . \qquad (5.1)$$

The least variety containing a set $V(f_1, \ldots, f_r; h \neq 0)$ is the variety of the ideal

$$A := \{f \in \mathcal{P} \mid \exists k \in \mathbb{N} : f \cdot h^k \in I\}. \qquad (5.2)$$

Using an elimination order for x_0 w.r.t. $\{x_0, x_1, \ldots, x_n\}$, and the representation

$$A = Ideal(f_1, \ldots, f_r, x_0 \cdot h + 1) \cap \mathbb{K}[x_1, \ldots, x_n], \qquad (5.3)$$

see [2], a Gröbner basis for A and hence a decomposition for (5.1) can be found.

There is an other decomposition technique, which splits the variety of an ideal given by a lexicographical Gröbner basis into disjoint varieties of ideals A_k, each generated by a triangular set

$$\{g_{1k}(x_1, \ldots, x_n), g_{2k}(x_2, \ldots, x_n), g_{nk}(x_n)\}.$$

Hence if the last j components of the solutions are already known and substituted into the remaining g_{ik}, then there is exactly one basis polynomial for fixing the next coordinate. This decomposition technique is integrated into REDUCE since version 3.3 and is described in [7].

In many instances, it is impossible to calculate by Buchberger's algorithm a lexicographical Gröbner basis but one with respect to an other admissible term ordering. In this situation, the algorithm of Faugère–Gianni–Lazard–Mora (FGLM–algorithm), see [1], allows to reconstruct the lexicographical Gröbner basis. This useful tool is also installed in CAS like MAPLE or REDUCE.

In summary, the Gröbner basis techniques can be used for solving systems of algebraic equations (with finitely many solutions) in several ways.

Polynomial input	Computation	Result
arbitrary G. basis	Hilbert function bound	Number of solutions
finite pol. set	Buchb. alg. for lex. ordering	solution set
finite pol. set	fact. Buchb. alg. for lex. ord.	union of sol. sets
arbitrary G. basis	FGLM	lexicogr. Gröbner basis
lexicogr. G. basis	decomposition technique	union of triang. sol. sets

There is an additional method which transforms the problem of solving a system of equations into an Eigenproblem. This method needs the matrix M_q

corresponding to the endomorphism Φ_q of (4.4). Hence for establishing M_q, one Gröbner basis w.r.t. an arbitrary term ordering is required first. This matrix is in case $n = 1$ and $q = x_1$ the Frobenius companion matrix. Therefore, for arbitrary n it may be called a *companion matrix*. The Eigenvalues of M_q are the numbers $q(y)$, $y \in V$. To $q(y)$ belongs the Eigenvector $(b_1(y), \ldots, b_s(y))^T)$, where $b_1, \ldots, b_s$ are the polynomials from which the basis B of $\mathcal{P}/I$, see (3.3), is built. One properly chosen q (often sufficient: $q = x_1$) allows the computation of the specific Eigenvectors from which the components of $y \in V$ can be read off. For details see [8].

6 Simplification by Symmetries

It is a truism, that an algebraic system (2.2) should be transformed first into a system which is as simple as possible. This can be done by clever substitutions, by cancelling obvious redundant equations etc. Here the use of CAS is helpful. We will illustrate it by concentrating on the two examples of section 1 and on simplifications by symmetries, more precisely on using invariant theory. For the general technique using invariant theory, we refer to the exposition of [11].

Let us consider example 1. The system (1.1) is invariant under all permutations of the odd indexed variables and under all of the even indexed ones. For simplicity let first n be even, $n = 2m$. Hence consider the invariants

$$\tau_k(a_1, \ldots, a_m) := a_1^k + \ldots + a_m^k, \quad k \in \mathbb{N}. \tag{6.1}$$

$\tau_1, \ldots, \tau_m$ are called fundamental invariants. They are algebraically independent, i.e. there is no polynomial $0 \neq p \in \mathbb{Q}[x_1, \ldots, x_m]$ satisfying $p(\tau_1, \ldots, \tau_m) = 0$, and the further τ_k depend on them algebraically,

$$\tau_k = p_k(\tau_1, \ldots, \tau_m), \quad k > m, \tag{6.2}$$

with suitable $p_k \in \mathbb{Q}[x_1, \ldots, x_m]$. These p_k are known by recursions, which can be found for instance in textbooks on combinatorics. But also Gröbner techniques can be used.

Compute a Gröbner basis for the ideal in $\mathbb{Q}[a_1, \ldots, a_m, s_1, \ldots, s_m, s]$ generated by $\{\tau_1 - s_1, \ldots, \tau_m - s_m, \tau_k - s\}$ using an elimination order for $a_1, \ldots, a_m$. Then just one polynomial depending only on $s_1, \ldots, s_m, s$ belongs to the Gröbner basis. It is up to normalization $s - p_k(s_1, \ldots, s_m)$.

Using u_k for $\tau_k(x_1, x_3, \ldots, x_{n-1})$ and v_k for $\tau_k(x_2, x_4, \ldots, x_n)$, the system (1.1) reduces to

$$
\begin{aligned}
u_k - v_k &= -\tfrac{1}{2}, & 2 \leq k \leq m, \\
p_k(u_1, \ldots, u_m) - p_k(v_1, \ldots, v_m) &= -\tfrac{1}{2}, & m < k \leq 2m + 1.
\end{aligned}
\tag{6.3}
$$

The first $m - 1$ equations can be used to eliminate $u_2, \ldots, u_m$. Then (6.3) is just a system of $m + 1$ equations for the unknowns $u_1, v_1, \ldots, v_m$. Thus the number of equations and unknowns is roughly halved compared to (1.1). Similarly one gets for odd n a system of $m + 1$ equations for the unknowns $u_1, v_1 \ldots, v_m$ where $n = 2m + 1$.

An ad hoc test with MAPLE gives for the up to now unsolved system (1.1) with $n = 7$ after transformation into a system of type (6.3) a lexicographical Gröbner basis $\{v_1 - q_1(u_1), v_2 - q_2(u_1), v_3 - q_3(u_1), q(u_1)\}$ with

$$
\begin{aligned}
q(u_1) := \ & 68719476736 * u_1^{20} - 1030792151040 * u_1^{19} \\
& + 6453188362240 * u_1^{18} - 21198347960320 * u_1^{17} \\
& + 38611546275840 * u_1^{16} - 54626829205504 * u_1^{15} \\
& + 158060808110080 * u_1^{14} - 338599298826240 * u_1^{13} \\
& - 531338425690880 * u_1^{12} + 4110328170800640 * u_1^{11} \\
& - 3903780825454752 * u_1^{10} - 27564203704883520 * u_1^{9} \\
& + 126353769451007775 * u_1^{8} - 283876384863715200 * u_1^{7} \\
& + 415113006662428200 * u_1^{6} - 424853487688855680 * u_1^{5} \\
& + 310161815375466600 * u_1^{4} - 158997360654720000 * u_1^{3} \\
& + 53346186217447200 * u_1^{2} - 9475825603507200 * u_1 \\
& + 324602161979760;
\end{aligned}
$$

and polynomials q_1, q_2, q_3 of degree 19. q has only four real zeros, only one of them leads to positive $v_1, v_2, v_3, u_1, u_2, u_3, u_4$. Using the connection of the τ_k to the elementary symmetric functions σ_k, we find the numerical values for $\sigma_k(x_1, x_3, x_5, x_7)$, $k = 1, 2, 3, 4$, and $\sigma_k(x_2, x_4, x_6)$, $k = 1, 2, 3$. Using Viëta's formulas, we find the values of x_1, x_3, x_5, x_7 as zeros of a fourth degree polynomial and x_2, x_4, x_6 as zeros of a third degree one. Calculation to 40 decimal places gives the only ordered solution set $0 < x_1 < \ldots < x_7 < 1$

$$\begin{aligned}
x_1 &= &.10192148148806277266844134241941323774 81,\\
x_2 &= &.23753969064964722140382247517830647291 27,\\
x_3 &= &.40350652965931782719319372854012433513 49,\\
x_4 &= &.58004057967585837284550641158812099671 89,\\
x_5 &= &.74590307806139906028038228632880788864 57,\\
x_6 &= &.88110712816952858876560984857695762458 43,\\
x_7 &= &.96935289770438760724942554535575385299 94;
\end{aligned}$$

Using invariant theory and standard CAS, the systems (1.1) can be solved easily up to $n = 9$. The power of this access results in the subdivision of a big system into one for the $u'_i s$ and $v'_i s$ and one for the $x'_i s$, once the numerical values of the $u'_i s$ and $v'_i s$ are known. However, the solutions by Gröbner basis techniques describe the whole variety. Therefore, numerical methods are superior, if only one single solution or only a few are needed.

Example 2 can be solved by standard software. For instance MAPLE computes, using the polynomials of (1.2) as input, a lexicographical Gröbner basis $\{g_1, \ldots, g_{13}\}$:

$$\begin{aligned}
g_1 =\ & y_1 + y_2 + y_3 + y_4,\\
g_2 =\ & 8y_2y_3 + 8y_4y_2 + 189x_4^5x_2y_3^2 - 117x_4^3x_2y_3^2 + 6x_4x_2y_3^2 + 45x_4^4y_3^2\\
& -33x_4^2y_3^2 + 10y_3^2 + 8y_4y_3 + 9x_4y_4^2x_2x_3^2 - 36x_2x_3x_4^2y_4^2 + 12y_4^2x_2x_3\\
& +9x_4x_2y_4^2 - 54x_3^2x_4^2y_4^2 + 15x_3^2y_4^2 + 21y_4^2x_4^2 - 2y_4^2,\\
g_3 =\ & 6y_4y_2x_4^2 - 2y_4y_2 + 9x_2x_3x_4^2y_4^2 - 3y_4^2x_2x_3 + 9x_3^2x_4^2y_4^2 - 3x_3^2y_4^2,\\
g_4 =\ & 2y_2x_2 - 9x_4^5y_2 + 3y_2x_4^3 + 2x_4y_2 - 3x_3x_2y_4x_4 + 3x_2y_4x_4^2 - 3x_3^2x_4y_4\\
& -3x_3x_4^2y_4 + 2y_4x_4,\\
g_5 =\ & x_3y_2 + 9x_4^5y_2 - 3y_2x_4^3 - x_4y_2 + x_2y_3 + 9x_4^5y_3 - 3x_4^3y_3 - x_4y_3\\
& +3x_3x_2y_4x_4 - 3x_2y_4x_4^2 + x_2y_4 - 3x_3x_4^2y_4 + x_3y_4 - y_4x_4,\\
g_6 =\ & 27x_4^6y_2 - 36x_4^4y_2 + 15y_2x_4^2 - 2y_2 + 9x_2x_3x_4^2y_4 - 3y_4x_2x_3 + 9x_3^2x_4^2y_4 - 3x_3^2y_4,\\
g_7 =\ & 6y_4x_4^2y_3 - 2y_4y_3 - 9x_3^2x_4^2y_4^2 + 3x_3^2y_4^2 + 6y_4^2x_4^2 - 2y_4^2,\\
g_8 =\ & 2y_3x_3 - 9x_4^5y_3 + 3x_4^3y_3 + 2x_4y_3 + 3x_3^2x_4y_4 + 6x_3x_4^2y_4 + y_4x_4,\\
g_9 =\ & 27x_4^6y_3 - 36x_4^4y_3 + 15x_4^2y_3 - 2y_3 - 9x_3^2x_4^2y_4 + 3x_3^2y_4 + 6y_4x_4^2 - 2y_4,\\
g_{10} =\ & 3y_4x_4^3 - y_4x_4,\\
g_{11} =\ & x_1 + x_2 + x_3 + x_4,\\
g_{12} =\ & 3x_2x_3 + 3x_2x_4 + 3x_3x_4 - 2 + 3x_2^2 + 3x_3^2 + 3x_4^2,\\
g_{13} =\ & 3x_3^2x_4 + 3x_3x_4^2 - 2x_3 - 2x_4 + 3x_3^3 + 3x_4^3.
\end{aligned}$$

And as result of a factorising Buchberger algorithm, MAPLE produces 21 output varieties which does not really illuminate the configuration of four points on a sphere, which is described by the equations (1.2).

A more careful look at the equations (1.2) shows, that the equations containing only x_1, x_2, x_3, x_4,

$$
\begin{aligned}
x_1 + x_2 + x_3 + x_4 &= 0, \\
x_1x_2 + x_1x_3 + x_1x_4 + x_2x_3 + x_2x_4 + x_3x_4 &= -\tfrac{2}{3}, \\
x_1x_2x_3 + x_1x_2x_4 + x_1x_3x_4 + x_2x_3x_4 &= 0
\end{aligned}
$$

just mean $\sigma_1(x_1, x_2, x_3, x_4) = \sigma_3(x_1, x_2, x_3, x_4) = 0$, $\sigma_2(x_1, x_2, x_3, x_4) = -\tfrac{2}{3}$. Hence, by means of Vïeta's formulas, the x_i are the zeros of a polynomial

$$
x^4 - \frac{2}{3}x^2 + c, \quad c \quad \text{a free parameter.}
$$

The y_i satisfy the remaining 4 (out of 7) equations. Easy symbolic calculation gives the equivalent system

$$
\begin{aligned}
y_1 + y_2 + y_3 + y_4 &= 0 \\
x_1y_1 + x_2y_2 + x_3y_3 + x_4y_4 &= 0 \\
x_1^2y_1 + x_2^2y_2 + x_3^2y_3 + x_4^2y_4 &= 0 \\
x_1x_2x_3y_4 + x_1x_2y_3x_4 + x_1y_2x_3x_4 + y_1x_2x_3x_4 &= 0 \, .
\end{aligned}
$$

This is a linear system for the y_i. The determinant of the coefficient matrix can either be computed by reducing it to a Vandermonde matrix or by using symbolic software for computing determinants and factorisation. The resulting determinant is

$$
-(x_1 - x_2)(x_1 - x_3)(x_1 - x_4)(x_2 - x_3)(x_2 - x_4)(x_3 - x_4).
$$

Degeneration happens if and only if

$$
c = 0 \quad \text{or} \quad c = \frac{1}{9}.
$$

In the nondegenerate case, the $x_i's$ are real exactly for $0 < c < \tfrac{1}{9}$. In this case, the solution points $(x_i, y_i), i = 1, \ldots, 4$, are

$$
\left(-\sqrt{\tfrac{1}{3} + \sqrt{\tfrac{1}{9} - c}}, 0\right), \quad \left(-\sqrt{\tfrac{1}{3} - \sqrt{\tfrac{1}{9} - c}}, 0\right),
$$
$$
\left(\sqrt{\tfrac{1}{3} - \sqrt{\tfrac{1}{9} - c}}, 0\right), \quad \left(\sqrt{\tfrac{1}{3} + \sqrt{\tfrac{1}{9} - c}}, 0\right).
$$

The degenerate case $c = 0$ means

$$
\{x_1, x_2, x_3, x_4\} = \{\sqrt{\tfrac{2}{3}}, -\sqrt{\tfrac{2}{3}}, 0, 0\}.
$$

Let us assume $x_3 = x_4 = 0$. Then the system for the y_i's reduces to $y_1 = y_2 = 0$, $y_3 + y_4 = 0$. Hence the four points are $(\pm\sqrt{\frac{2}{3}}, 0)$, $(0, \pm y_3)$.

The last case $c = \frac{1}{9}$ means, that after a possible renumbering of the four points

$$x_1 = x_2 = \sqrt{\frac{1}{3}}, \quad x_3 = x_4 = -\sqrt{\frac{1}{3}}$$

holds. Then (6.5) is equivalent to $y_1 + y_2 = 0$, $y_3 + y_4 = 0$. Hence the four points are

$$\left(\sqrt{\frac{1}{3}}, \pm y_1\right), \quad \left(-\sqrt{\frac{1}{3}}, \pm y_3\right), \qquad 0 < y_1, y_3 \leq \sqrt{\frac{2}{3}}.$$

This example showed, that symbolic software gives without human interaction nearly useless results. Together with some mathematical knowledge however, it is a valuable tool.

References

[1] D. Cox, J. Little, D. O'Shea: *Ideals, Varieties, and Algorithms*, Undergraduate Texts in Mathematics, Springer, New York 1992.

[2] P. Gianni, B. M. Trager, G. Zacharias: *Gröbner bases and primary decomposition of polynomial ideals*, J. Symb. Comp. **6** (1988), 149–167.

[3] D. Lazard: *Solving zero–dimensional algebraic systems*, J. Symb. Comp. **13** (1992), 117–131.

[4] T. Y. Li: *Solving polynomial systems*, Math. Intelligencer **9** (1987), 33–39.

[5] U. Maier: *Approximation durch Kergin–Interpolation*, Dissertation, Dortmund 1994.

[6] M. G. Marinari, H. M. Möller, T. Mora: *On multiplicities in polynomial system solving*, Trans. Amer. Math. Soc. **348** (1996), 3283–3321.

[7] H. M. Möller: *On decomposing systems of polynomial equations with finitely many solutions*, J. AAECC 4 (1993), 217–230.

[8]   H. M. Möller: *Systems of algebraic equations solved by means of endomorphisms*, in : *AAECC-10*, G. Cohen, T. Mora, O. Moreno (eds.), 43–56, Springer LNCS **673**, Berlin–Heidelberg 1993.

[9]   M. Reimer: *Zonal spherical polynomials with minimal L_1-norm*, Approx. Theory Appl. **11** (1995), 25–35.

[10]   M. Reimer: Private communication, 1996.

[11]   B. Sturmfels: *Algorithms in Invariant Theory*, Texts and Monographs in Symbolic Computation, Springer, Wien–New York 1993.

[12]   C. W. Wampler, A. P. Morgan, and A. J. Sommese: *Numerical continuation methods for solving polynomial systems arising in kinematics*, ASME Journal of Mechanical Design **112** (1990), 59–68.

[13]   Wen–Tsün Wu: *Mechanical Theorem Proving in Geometries*, Texts and Monographs in Symbolic Computation, Springer, Wien–New York 1993.

Address:

HANS MICHAEL MÖLLER
Fachbereich Mathematik
Universität Dortmund
Vogelpothsweg 87
D–44221 Dortmund
Germany

An Orthonormal Bivariate Algebraic Polynomial Basis for $C(I^2)$ of Low Degree

J. Prestin and F. Sprengel

Abstract

For any fixed $\varepsilon > 0$, we construct an orthonormal Schauder basis $\{P_\mu\}_{\mu=0}^\infty$ for $C(I^2)$ consisting of bivariate algebraic polynomials P_μ with $(\deg_q P_\mu)^2 \le c_q(1+\varepsilon)\mu$ as a generalization of the corresponding univariate basis constructed in [1]. The orthogonality is with respect to the bivariate Chebyshev weight.

1 Introduction

The question of polynomial Schauder bases for spaces of continuous functions with the uniform norm has a long history. For a more detailed discussion, we refer to the monograph [2, Chap. 5] and to [1, 3] where some final results are proved. Here we follow the approach described in [1]. The main idea to construct an univariate basis $\{p_\mu\}$ consists of an appropriate frequency splitting. I.e., writing p_μ in terms of Chebyshev polynomials, all p_μ with $n_1 < \mu < n_2$ are built from Chebyshev polynomials T_k with $n_1(1 - \varepsilon) < k < n_2(1 + \varepsilon)$ (cf. (2.2) and Definition 2.1). This decomposition of the natural numbers in overlapping dyadic intervals will now be generalized in the bivariate case to overlapping rectangular blocks. This means that the bivariate polynomials are chosen as appropriate tensor products of the univariate polynomials (see Algorithm 3.1). The important properties of these polynomials will be summarized in our main Theorem 3.4. In this context, the basic question is how to measure the degree of the bivariate polynomials. For this reason, we consider ℓ^q-(quasi) norms with $0 < q \le \infty$ of the corresponding directional degrees (see (3.1)). Particular cases are $q = 1$, the (total) degree and $q = \infty$, the tensor product degree. Our construction could be easily applied also for $q = 0$. Namely, if one takes the limit $q \to 0$ in (3.1) one obtains $|(k, \ell)|_0 = \sqrt{k\ell}$, which yields polynomials corresponding to so-called hyperbolic crosses. However, this would not result in a basis for $C(I^2)$. In particular, the orthogonal projection operator is not bounded in this case.

Multivariate Approximation: Recent Trends and Results; W. Haußmann, K. Jetter and M. Reimer (eds.)
Mathematical Research, Vol. 101, pp. 177–188, ISBN 3–05–501770–6
© Akademie–Verlag, Berlin 1997

2   The Univariate Basis

We recall the construction of an univariate polynomial Schauder basis for $C(I)$ with $I := [-1, 1]$ where its elements are orthogonal with respect to the weighted inner product

$$\langle f, g \rangle_I \;=\; \frac{2}{\pi} \int\limits_{-1}^{1} f(x)g(x)\frac{dx}{\sqrt{1-x^2}} \;.$$

Then, the basis polynomials were given in terms of Chebyshev polynomials $T_n(x) = \cos n \arccos x$ $(n \in \mathbb{N}_0)$ which fulfil the orthogonality relations

$$\langle T_n, T_m \rangle_I \;=\; \begin{cases} 2 & \text{for } n = m = 0, \\ 1 & \text{for } n = m > 0, \\ 0 & \text{otherwise.} \end{cases}$$

For arbitrary $\delta > 0$, a polynomial sequence $\{p_\mu\}_{\mu \in \mathbb{N}_0}$ was defined such that the degree $\deg p_\mu \le \mu(1 + \delta)$. To this end, we introduce a natural number η satisfying

$$\begin{aligned} \eta &= 3, & \text{for } \tfrac{1}{2} &\le \delta, \\ \tfrac{3}{2^\eta - 2} \le \delta &< \tfrac{3}{2^{\eta-1} - 2}, & \text{for } 0 &< \delta < \tfrac{1}{2}. \end{aligned} \qquad (2.1)$$

For fixed δ, these conditions uniquely determine η, with $\eta \ge 3$. Then every index $\mu \in \mathbb{N}$, $\mu > 2^{\eta+1}$ determines uniquely the triplet of integers j, λ, s such that

$$\begin{aligned} \mu &= 2^j + (\lambda - 1)2^{j-\eta+1} + s \,, \\ \text{with} \quad j &\ge \eta+1, \quad 0 \le \lambda \le 2^{\eta-1} - 2, \quad 1 \le s \le 2^{j-\eta+1} \,. \end{aligned} \qquad (2.2)$$

Furthermore, the two functions,

$$g_1(x) \;:=\; \begin{cases} \dfrac{2+x}{\sqrt{2+2(x+1)^2}} & \text{for } -2 \le x < 0, \\[3mm] \dfrac{2-x}{\sqrt{2+2(x-1)^2}} & \text{for } \;\;0 \le x \le 2, \end{cases}$$

and

$$g_0(x) \;:=\; \begin{cases} 0 & \text{for } -2 \le x < -\tfrac{3}{2}, \\[3mm] \dfrac{x+\frac{3}{2}}{\sqrt{\frac{1}{2}+2(x+1)^2}} & \text{for } -\tfrac{3}{2} \le x < -\tfrac{1}{2}, \\[3mm] 1 & \text{for } -\tfrac{1}{2} \le x < 0, \\[3mm] \dfrac{2-x}{\sqrt{2+2(x-1)^2}} & \text{for } \;\;0 \le x \le 2, \end{cases}$$

were needed, the sampling of which gives polynomial coefficients.

Definition 2.1. (cf. [1])   *Let $\delta > 0$. Then with η as in (2.1), the polynomials p_μ are given by*

$$p_0 \;:=\; \frac{1}{\sqrt{2}}\,, \qquad p_k := T_k\,, \quad for\ k = 1,\ldots,2^{\eta+1} - 2\,,$$

$$p_{2^{\eta+1}-1} \;:=\; T_{2^{\eta+1}}\,, \qquad p_{2^{\eta+1}} := \frac{3}{\sqrt{10}}T_{2^{\eta+1}-1} + \frac{1}{\sqrt{10}}T_{2^{\eta+1}+1}\,,$$

and for $\mu > 2^{\eta+1}$ with j,λ,s as in (2.2), by

$$p_\mu \;:=\; 2^{(\eta-j)/2} \sum_{k=-3\cdot 2^{j-\eta}+1}^{2^{j-\eta}-1} g_1\left(1 + \frac{k}{2^{j-\eta}}\right) \sin\left(\frac{k(2s-1)\pi}{2^{j-\eta+2}}\right) T_{2^j+\lambda 2^{j-\eta+1}+k}\,,$$

if λ is odd, and by

$$p_\mu \;:=\; 2^{(\eta-j)/2} \sum_{k=-3\cdot 2^{j-\eta}+1}^{2^{j-\eta}-1} g_\lambda\left(1 + \frac{k}{2^{j-\eta}}\right) \cos\left(\frac{k(2s-1)\pi}{2^{j-\eta+2}}\right) T_{2^j+\lambda 2^{j-\eta+1}+k}\,,$$

if λ is even, with $g_\lambda := g_1$ for all $\lambda > 0$.

Theorem 2.2.   (cf. [1, 4])   *Let $\delta > 0$ be given. Then $\{p_\mu\}_{\mu\in\mathbb{N}_0}$ is an orthonormal polynomial Schauder basis of optimal degree for $C(I)$. I.e., we have for all $\mu,\nu \in \mathbb{N}_0$*

$$\deg\, p_\mu \;\le\; \mu(1+\delta)\,,$$
$$\langle p_\mu\,,\, p_\nu\rangle_I \;=\; \delta_{\mu,\nu}\,,$$

and for all $f \in C(I)$

$$\left\| f - \sum_{s=0}^{\mu}\langle f\,,\, p_s\rangle_I\, p_s \right\|_{C(I)} \;\le\; c\; E_{\lfloor\mu(1-\delta)\rfloor}(f, C(I))\,.$$

Here, $E_n(f, C(I))$ means the best approximation of f in the uniform norm on I by algebraic polynomials of degree n. In [4], the constant $c = c(\delta)$ was estimated as

$$c \;\le\; \frac{4}{\pi^2}\, \log\left(\frac{1}{\delta}\right) + 2700\,.$$

3   The Construction of the Bivariate Basis

Now, we want to construct a bivariate orthonormal algebraic polynomial Schauder basis $\{P_\mu\}_{\mu\in\mathbb{N}_0}$ similar to the univariate basis in the section before, i.e., we want orthonormality with respect to the bivariate Chebyshev weight

$$\langle f,g\rangle_{I^2} := \frac{4}{\pi^2} \iint_{I^2} f(x,y)g(x,y) \frac{dx\,dy}{\sqrt{1-x^2}\sqrt{1-y^2}},$$

reproduction of certain polynomials and low increasing of q-degree $\deg_q P_\mu$ for a fixed $q > 0$. For nonnegative k, ℓ, we define the ℓ^q-(quasi) norm as

$$|(k,\ell)|_q := \begin{cases} (k^q + \ell^q)^{1/q} & \text{if } 0 < q < \infty, \\ \max\{k,\ell\} & \text{if } q = \infty. \end{cases} \tag{3.1}$$

Then, we introduce the q-degree of a bivariate polynomial P as

$$\deg_q P = \min \left\{ \mu \in \mathbb{N}_0 \; : \; P(x,y) = \sum_{|(k,\ell)|_q \leq \mu} c_{k,\ell}\, T_k(x)\, T_\ell(y) \right\}.$$

Given $\varepsilon > 0$, we want our basis polynomials to satisfy

$$\left(\deg_q P_\mu\right)^2 \leq (1+\varepsilon)\, \frac{\mu}{A(q)},$$

where analogously to the univariate case the additional factor on the right-hand side is the reciprocal of the area

$$\begin{aligned} A(q) \; &:= \; \left| \left\{(x,y) \in \mathbb{R}_+^2 \; ; \; |(x,y)|_q \leq 1\right\} \right| \\ &= \; \begin{cases} \dfrac{\Gamma^2(\frac{1}{q})}{2q\,\Gamma(\frac{2}{q})} & \text{if } 0 < q < \infty, \\ 1 & \text{if } q = \infty. \end{cases} \end{aligned}$$

The construction of the bivariate basis is based on tensor products of polynomials p_μ from the univariate polynomial basis given in the previous section. Therefore, we choose

$$\delta := \min\left\{\frac{\varepsilon}{7}, \frac{1}{2}\right\} \tag{3.2}$$

and η and $\{p_\mu\}_{\mu\in\mathbb{N}_0}$ related to this δ by (2.1) and Definition 2.1. We want to give our basis in a constructive way. This can be described in two steps, the first is the initial step, the second is the general step. Because the curves $x^q + y^q = 1$ are convex for $0 < q < 1$ and concave for $q \geq 1$ we consider two different initialization steps. In the sequel, we set $1/q = 0$ for $q = \infty$.

Algorithm 3.1.   (Defining algorithm for the sequence $\{P_\mu\}_{\mu \in \mathbb{N}_0}$)

1.a) Initialization step for $q \geq 1$

1.1. $P_0 := p_0 \otimes p_0$, $\mu := 1$, $K_0 := 2^{1/q}\, 2^{\eta+1}$
1.2. for $i := 1, \ldots, 2^{\eta+1}$ do
      for $m := 1, \ldots, i$ do
         for $n := 1, \ldots, i$ do
            if $i - 1 < |(m,n)|_q \leq i$ then $P_\mu := p_m \otimes p_n$, $\mu := \mu + 1$
1.3. for $i := 2^{\eta+1} + 1, \ldots, \lceil K_0 \rceil$ do
      for $m := 1, \ldots, 2^{\eta+1}$ do
         for $n := 1, \ldots, 2^{\eta+1}$ do
            if $i - 1 < |(m,n)|_q \leq i$ then $P_\mu := p_m \otimes p_n$, $\mu := \mu + 1$
1.4. $j_0 := \eta + 1$, $\lambda_0 := 0$

1.b) Initialization step for $0 < q < 1$

1.1. $P_0 := p_0 \otimes p_0$, $\mu := 1$
1.2. find j_0, λ_0 such that
$$2\,(2^{j_0} + \lambda_0 2^{j_0 - \eta + 1})^q + 2^{q(\eta+1)} \leq 2\,(2^{j_0} + (\lambda_0 + 1)2^{j_0 - \eta + 1})^q$$
$$K_0 := 2^{1/q}\,(2^{j_0} + \lambda_0 2^{j_0 - \eta + 1})$$
$$j_1 := j_0 , \quad \lambda_1 := \lambda_0 + 1 , \quad K_1 := 2^{1/q}\,(2^{j_1} + \lambda_1 2^{j_0 - \eta + 1})$$
1.3. for $i := 1, \ldots, \lceil K_0 \rceil$ do
      for $m := 1, \ldots, i$ do
         for $n := 1, \ldots, i$ do
            if $i - 1 < |(m,n)|_q \leq i$ then $P_\mu := p_m \otimes p_n$, $\mu := \mu + 1$
1.4. for $j := 0, \ldots, j_1 + \lceil 1/q \rceil$ do
      for $k := 0, \ldots, j_1 + \lceil 1/q \rceil$ do
         for $\lambda := 0, \ldots, 2^{\eta-2} - 2$ do
            for $\nu := 0, \ldots, 2^{\eta-2} - 2$ do
               if $|(2^j + \lambda 2^{j-\eta+1}, 2^k + \nu 2^{k-\eta+1})|_q \leq K_1$
                  then for $r := 1, \ldots, 2^{j-\eta+1}$ do
                     $m := 2^j + (\lambda - 1)2^{j-\eta+1} + r$
                     for $s := 1, \ldots, 2^{k-\eta+1}$ do
                        $n := 2^k + (\nu - 1)2^{k-\eta+1} + s$
                        if $|(m,n)|_q > K_0$ then $P_\mu := p_m \otimes p_n$, $\mu := \mu + 1$
1.5. $j_0 := j_1$, $\lambda_0 := \lambda_1$, $K_0 := K_1$

2. General Step

2.1. j_0, λ_0, K_0, μ from the previous step

2.2. if $\lambda_0 < 2^{\eta-1}$

then $\lambda_1 = \lambda_0 + 1$, $j_1 := j_0$ else $\lambda_1 = 0$, $j_1 := j_0 + 1$

$K_1 := 2^{1/q} \left(2^{j_1} + \lambda_1 2^{j_1 - \eta + 1} \right)$

2.3 for $j := 0, \ldots, j_1 + \lceil 1/q \rceil$ do

for $k := 0, \ldots, j_1 + \lceil 1/q \rceil$ do

for $\lambda := 0, \ldots, 2^{\eta-2} - 2$ do

for $\nu := 0, \ldots, 2^{\eta-2} - 2$ do

if $K_0 < |(2^j + \lambda 2^{j-\eta+1}, 2^k + \nu 2^{k-\eta+1})|_q \leq K_1$

then for $r := 1, \ldots, 2^{j-\eta+1}$ do

$m := 2^j + (\lambda - 1) 2^{j-\eta+1} + r$

for $s := 1, \ldots, 2^{k-\eta+1}$ do

$n := 2^k + (\nu - 1) 2^{k-\eta+1} + s$

$P_\mu := p_m \otimes p_n$, $\mu := \mu + 1$

2.4. $j_0 := j_1$, $\lambda_0 := \lambda_1$, $K_0 := K_1$

2.5. goto 2.1.

Let us summarize properties of the P_μ defined in this way. By construction, we obtain orthonormality

$$\begin{aligned}
\langle P_\mu, P_\nu \rangle_{I^2} &= \langle p_{m_1} \otimes p_{n_1}, p_{m_2} \otimes p_{n_2} \rangle_{I^2} \\
&= \langle p_{m_1}, p_{m_2} \rangle_I \langle p_{n_1}, p_{n_2} \rangle_I = \delta_{m_1, m_2} \delta_{n_1, n_2} = \delta_{\mu, \nu} .
\end{aligned} \qquad (3.3)$$

Now, we fix a number μ and the values K_0, K_1 of the step (1b.1.4. or 2.3) of Algorithm 3.1 in which we counted μ. (The result (3.6) holds trivially for the μ counted earlier in steps 1a.1.2, 1a.1.3 or 1b.1.3.) By construction (see step 2.2 or 1b.1.2), we can estimate

$$K_0 \leq K_1 \leq K_0 (1 + 2^{-\eta}) \leq K_0 \left(1 + \frac{\delta}{3} \right) . \qquad (3.4)$$

Because of Definition 2.1 of the underlying univariate polynomial basis we obtain for the q-degree

$$\deg_q P_\mu \leq K_1 (1 + \delta) .$$

Furthermore, we use the geometry of the problem for estimating from below the number of polynomials already counted

$$\mu \geq A(q) K_0^2 (1 - 2^{-\eta})^2 \geq A(q) K_0^2 \left(1 - \frac{\delta}{3} \right)^2 . \qquad (3.5)$$

Hence,

$$\frac{\left(\deg_q P_\mu\right)^2}{\mu} \leq \frac{1}{A(q)} \frac{K_1^2 (1+\delta)^2}{K_0^2 \left(1-\frac{\delta}{3}\right)^2} \leq \frac{1}{A(q)} \frac{K_0^2 (1+\delta)^2 \left(1+\frac{\delta}{3}\right)^2}{K_0^2 \left(1-\frac{\delta}{3}\right)^2} .$$

Exploiting additionally the fact $\delta \leq 1/2$, we obtain after some simple calculations

$$\frac{(1+\delta)^2 \left(1+\frac{\delta}{3}\right)^2}{\left(1-\frac{\delta}{3}\right)^2} \leq (1+7\delta) .$$

With the definition (3.2) of δ, we obtain the desired result

$$\left(\deg_q P_\mu\right)^2 \leq (1+\varepsilon) \frac{\mu}{A(q)} \tag{3.6}$$

for the degrees of the polynomials in the sequence $\{P_\mu\}_{\mu \in \mathbb{N}_0}$. The way of ordering the polynomials will turn out to be an essential ingredient in proving the main result. As in [1], we collect the univariate polynomials in sample spaces

$$V = V_{2^j+(\lambda-1)2^{j-\eta+1}}^{2^{j-\eta}} = \text{span} \left\{ p_\mu \; ; \; \mu = 0,\ldots,2^j + (\lambda - 1)2^{j-\eta+1} \right\}$$

and wavelet packet spaces

$$W = W_{2^j,\lambda}^{2^{j-\eta}} = \text{span} \left\{ p_{2^j+(\lambda-1)2^{j-\eta+1}+s} \; ; \; s = 1,\ldots,2^{j-\eta+1} \right\} .$$

Observe that all polynomials in a wavelet packet space $W_{2^j,\lambda}^{2^{j-\eta}}$ are of the same degree. The ordering uses this structure. That means, the algorithm in the general step first collects all polynomials in one bivariate wavelet packet space $W_{2^j,\lambda}^{2^{j-\eta}} \otimes W_{2^k,\nu}^{2^{k-\eta}}$ before it proceeds to the next packet space.

Now we can prove some basic properties of the polynomial sequence $\{P_\mu\}_{\mu \in \mathbb{N}_0}$.

Lemma 3.2. *There exists an index μ_0, such that for all $\mu \geq \mu_0$ and every polynomial Q with*

$$\deg_\infty Q \leq 2^{-1/q} \sqrt{\frac{\mu}{A(q)}} (1 - \varepsilon),$$

it holds that

$$Q \in \text{span}\{P_k \; ; \; k = 0,\ldots,\mu\} .$$

Proof : The sequence $\{P_\mu\}_{\mu \in \mathbb{N}_0}$ was defined by Algorithm 3.1. Assume, we had already run Step 1 and once Step 2. We take the last counted parameter μ and denote it by μ_0. Now, we fix $\mu \geq \mu_0$. That μ was counted in a certain Step 2 with the parameters j_0, λ_0, K_0, and K_1. That means, with $N := 2^{j_0} + (\lambda_0 - 1)2^{j_0 - \eta + 1}$ and $M := 2^{j_0 - \eta}$, we have

$$V_N^M \otimes V_N^M \subset \mathrm{span}\left\{P_k \ ; \ k = 0, \ldots, \mu\right\}.$$

From the univariate construction, we can use the fact that

$$\Pi_{\lfloor N(1-\delta)\rfloor} \subset V_N^M$$

which gives

$$\Pi_{\lfloor N(1-\delta)\rfloor} \otimes \Pi_{\lfloor N(1-\delta)\rfloor} \subset \mathrm{span}\left\{P_k \ ; \ k = 0, \ldots, \mu\right\}. \tag{3.7}$$

With (3.4) and (3.5)

$$A(q)\, K_0^2 \left(1 - \frac{\delta}{3}\right)^2 \leq \mu \leq A(q)\, K_1^2 \leq A(q)\, K_0^2 \left(1 + \frac{\delta}{3}\right)^2,$$

we get the estimates

$$\sqrt{\mu} \ \leq \ \sqrt{A(q)}\, K_0 \left(1 + \frac{\delta}{3}\right),$$

$$\sqrt{\frac{\mu}{A(q)}}\, \frac{1}{1 + \frac{\delta}{3}} \ \leq \ K_0,$$

$$\sqrt{\frac{\mu}{A(q)}} \left(1 - \frac{\delta}{3}\right) \ \leq \ K_0 \ = \ 2^{1/q}\, N,$$

$$2^{-1/q}\, \sqrt{\frac{\mu}{A(q)}} \left(1 - \frac{\delta}{3}\right) (1 - \delta) \ \leq \ N\, (1 - \delta),$$

$$2^{-1/q}\, \sqrt{\frac{\mu}{A(q)}} \, (1 - \varepsilon) \ \leq \ N\, (1 - \delta).$$

By the assumption on Q, we conclude

$$Q \ \in \ \Pi_{\lfloor N(1-\delta)\rfloor} \otimes \Pi_{\lfloor N(1-\delta)\rfloor}$$

which together with (3.7) proves the lemma. $\qquad\square$

Let the partial sum operator for $f \in C(I^2)$ be defined by

$$S_\mu f := \sum_{k=0}^{\mu} \langle f, P_k \rangle_{I^2} \, P_k \, .$$

Lemma 3.3.   *The norm of the partial sum operator S_μ is bounded*

$$\|S_\mu\|_{C(I^2) \to C(I^2)} \leq C(q, \varepsilon) , \tag{3.8}$$

independently of μ.

Proof: Choose μ_0 as in the proof of Lemma 3.2. For $\mu < \mu_0$ it holds that

$$\|S_\mu\|_{C(I^2) \to C(I^2)} \leq \|S_{\mu_0}\|_{C(I^2) \to C(I^2)} \, .$$

Now, let $\mu \geq \mu_0$ where μ was counted in a certain Step 2 with j_0, λ_0, K_0 and K_1. Denote $f_s := \langle f, P_s \rangle_{I^2}$. Because Algorithm 3.1 first collects all polynomials in one bivariate wavelet packet space $W_{2^j,\lambda}^{2^{j-\eta}} \otimes W_{2^k,\nu}^{2^{k-\eta}}$ before it defines polynomials P_μ from the next packet space we can split the partial sum in the following way

$$\left\| \sum_{s=0}^{\mu} f_s P_s \right\|_{C(I^2)} \leq \left\| \sum_{P_s \in VV} f_s P_s \right\|_{C(I^2)} + \sum_{VW \subset VW} \left\| \sum_{P_s \in VW} f_s P_s \right\|_{C(I^2)}$$

$$+ \sum_{WV \subset WV} \left\| \sum_{P_s \in WV} f_s P_s \right\|_{C(I^2)} + \left\| \sum_{\substack{P_s \in WW, \\ s \leq \mu}} f_s P_s \right\|_{C(I^2)} \, .$$

With the notations VV, VW, WV, WW, we abbreviate the following spaces

$$VV \quad := \quad V_{2^{j_0}+(\lambda_0-1)2^{j_0-\eta+1}}^{2^{j_0}-\eta} \otimes V_{2^{j_0}+(\lambda_0-1)2^{j_0-\eta+1}}^{2^{j_0}-\eta} \, ,$$

$$VW \quad := \quad \bigcup V_{2^j+(\lambda-1)2^{j-\eta+1}}^{2^j-\eta} \otimes W_{2^k,\nu}^{2^k-\eta} \, ,$$

$$\text{where } 2^{j_0} + (\lambda_0 - 1)2^{j_0-\eta+1} \leq 2^k + (\nu - 1)2^{k-\eta+1} \, ,$$

$$\text{all polynomials in } VW := V_{2^j+(\lambda-1)2^{j-\eta+1}}^{2^j-\eta} \otimes W_{2^k,\nu}^{2^k-\eta}$$

$$\text{were already counted before } \mu \, ,$$

$$j, \lambda \text{ are chosen as big as possible} \, ,$$

$$WV \quad := \quad \bigcup W_{2^j,\lambda}^{2^j-\eta} \otimes V_{2^k+(\nu-1)2^{k-\eta+1}}^{2^k-\eta} \, ,$$

$$\text{where } 2^{j_0} + (\lambda_0 - 1)2^{j_0-\eta+1} \leq 2^j + (\nu - 1)2^{j-\eta+1} \, ,$$

$$\text{all polynomials in } WV := W^{2^{j-\eta}}_{2^j,\lambda} \otimes V^{2^{k-\eta}}_{2^k+(\nu-1)2^{k-\eta+1}}$$

$$\text{were already counted before } \mu \, ,$$

$$k, \nu \text{ are chosen as big as possible} \, ,$$

$$WW \; := \; W^{2^{j-\eta}}_{2^j,\nu} \otimes W^{2^{k-\eta}}_{2^k,\nu} \, ,$$

$$\text{where } P_\mu \in WW \, .$$

Using the bounds for the univariate projection norms (cf. [4])

$$\left\| \sum_{p_s \in V} \langle 1, |p_s| \rangle_I \, p_s \right\|_{C(I)} =: \; C_V \leq \frac{4}{\pi^2} \log\left(\frac{1}{\delta}\right) + 35 \, ,$$

$$\left\| \sum_{p_s \in W} \langle 1, |p_s| \rangle_I \, p_s \right\|_{C(I)} =: \; C_W \leq 2625 \, ,$$

we estimate

$$\left\| \sum_{P_s \in VV} f_s P_s \right\|_{C(I^2)} \leq \; \|f\|_{C(I^2)} \, C_V^2 \, ,$$

$$\left\| \sum_{P_s \in VW(WV)} f_s P_s \right\|_{C(I^2)} \leq \; \|f\|_{C(I^2)} \, C_V \, C_W \, ,$$

$$\left\| \sum_{P_s \in WW} f_s P_s \right\|_{C(I^2)} \leq \; \|f\|_{C(I^2)} \, C_W^2 \, .$$

Hence, we obtain

$$\|S_\mu f\|_{C(I^2)} \leq \|f_s\|_{C(I^2)} \left(C_V^2 + \sum_{VW \subset \mathcal{V}\mathcal{W}} C_V \, C_W + \sum_{WV \subset \mathcal{W}\mathcal{V}} C_W \, C_V + C_W^2 \right) .$$

It remains to estimate the number n_{VW} of the spaces VW in $\mathcal{V}\mathcal{W}$ and WV in $\mathcal{W}\mathcal{V}$. Assume $\lambda_0 < 2^{\eta-1} - 2$ (the estimation for $\lambda_0 = 2^{\eta-1} - 2$ can be done in a similar way). Then,

$$n_{VW} \leq \; 2^{-j_0+\eta} \left(2^{1/q} \left(2^{j_0} + (\lambda_0 + 1) \, 2^{j_0-\eta+1} \right) - \left(2^{j_0} + \lambda_0 \, 2^{j_0-\eta+1} \right) \right)$$

$$= \; 2 \left((2^{1/q} - 1) \, (2^{\eta-1} + \lambda_0) + 2^{1/q} \right)$$

$$\leq \; 2 \left((2^{1/q} - 1) \, (2^{\eta-1} + 2^{\eta-1} - 2) + 2^{1/q} \right)$$

$$n_{VW} \;\leq\; 2\left((2^{1/q} - 1)\,(2^\eta - 2) + 2^{1/q} \right)$$

$$\leq\; 2\left(2\,(2^{1/q} - 1)\left(\frac{3}{\delta} + 1 \right) + 2^{1/q} \right)$$

leads to

$$\|S_\mu\|_{C(I^2) \to C(I^2)} \;\leq\; C_V^2 + C_W^2 + 2\left(2(2^{1/q} - 1)\left(\frac{3}{\delta} + 1 \right) + 2^{1/q} \right) C_V\, C_W$$

$$=:\; C(q,\varepsilon)\,,$$

which proves the lemma. $\qquad\qquad\qquad\qquad\qquad\qquad\qquad\qquad\square$

Now we are able to state the main result of this paper.

Theorem 3.4. *Let $\varepsilon > 0$ be given. Then $\{P_\mu\}_{\mu \in \mathbb{N}_0}$ is an orthornormal polynomial Schauder basis of low degree for $C(I^2)$. I.e., the polynomial sequence $\{P_\mu\}_{\mu \in \mathbb{N}_0}$ satisfies for all $\mu, \nu \in \mathbb{N}_0$ the following properties*

$$\left(\deg_q P_\mu \right)^2 \;\leq\; \frac{\mu}{A(q)}\,(1 + \varepsilon)\,, \tag{3.9}$$

$$\langle P_\mu, P_\nu \rangle_{I^2} \;=\; \delta_{\mu,\nu} \tag{3.10}$$

and furthermore there exists $\mu_0 \in \mathbb{N}$ such that for all $\mu \geq \mu_0$ and all $f \in C(I^2)$

$$\|f - S_\mu f\|_{C(I^2)} \;\leq\; (1 + C(q,\varepsilon))\, E_{\mathcal{M}}(f, C(I^2))\,, \tag{3.11}$$

where

$$\mathcal{M} := \left\lfloor 2^{-1/q}\, \sqrt{\frac{\mu}{A(q)}}\,(1 - \varepsilon) \right\rfloor.$$

Here, $E_n(f, C(I^2))$ denotes the best approximation of f in the uniform norm on I^2 by algebraic polynomials from $\Pi_n \otimes \Pi_n$.

From the construction, we have for the constants

$$C(q,\varepsilon) \;\longrightarrow\; \infty \qquad \text{for } q \longrightarrow 0 \quad \text{or } \varepsilon \longrightarrow 0\,.$$

Let us mention here that an estimate of the form (3.11) is sufficient to state the Schauder basis property. However, replacing the inequality in Lemma 3.2 by a statement for $\deg_r Q$ would yield more general estimates than (3.11) with best approximation from different kinds of polynomial spaces.

Proof: The properties (3.9), (3.10) follow directly from the defining Algorithm 3.1 (see Equations (3.6), (3.3)). The approximation property (3.11) can be proved easily using the previous lemmata. With the same μ_0 as in the proof of Lemma 3.2, we estimate for arbitrary $\mu \geq \mu_0$ and $Q \in \Pi_{\mathcal{M}} \otimes \Pi_{\mathcal{M}}$

$$\begin{aligned}
\|f - S_\mu f\|_{C(I^2)} &= \|f - Q + S_\mu(Q - f)\|_{C(I^2)} \\
&\leq (1 + C(q,\varepsilon)) \, \|f - Q\|_{C(I^2)} \, .
\end{aligned}$$

Choosing Q as the polynomial of best approximation proves the theorem. $\square$

References

[1]	T. Kilgore, J. Prestin, K. Selig: *Orthogonal algebraic polynomial schauder bases of optimal degree*, J. Fourier Anal. Appl. **2** (1996), 597–610.

[2]	G. G. Lorentz, M. v. Golitschek, Y. Makovoz: *Constructive Approximation, Advanced Problems*, Grundlehren der mathematischen Wissenschaften **304**, Springer, Berlin 1996.

[3]	R. A. Lorentz, A. A. Sahakian: *Orthogonal trigonometric Schauder bases of optimal degree for $C(0, 2\pi)$*, J. Fourier Anal. Appl. **1** (1994), 103–112.

[4]	H. Voigt: *Algebraische Orthogonalpolynom–Basen*, Diploma thesis, University of Rostock, 1996.

Address:

Jürgen Prestin, Frauke Sprengel
Department of Mathematics
University of Rostock
D–18051 Rostock
Germany

Tensor Product Splines on Refined Grids in S-Convex Interpolation

J. W. Schmidt and M. Walther

Abstract

This paper is concerned with S–convexity preserving interpolation of data sets given on rectangular grids using biquadratic C^1 splines on refined grids. Utilizing the univariate results in nonnegative, monotone, and convex interpolation with quadratic C^1 splines on refined grids as well as the tensor product structure, sufficient conditions for the S-convexity are derived. Furthermore, if the additional knots are suitably chosen, the solvability of the arising set of inequalities can always be assured provided the data are in strictly S–convex position.

1 Introduction

Because of their practical applications as well as their theoretical attractiveness, restricted interpolations have received great attention in the past. Depending on the background of the interpolation problem, the preservation, e.g., of the convexity may be essential. For recent reviews on numerical methods in convex or other types of restricted interpolation we refer to [2, 5, 13, 19].

To attack convex interpolation successfully, it is important to have a discrete characterization for the given data set to be in convex position. The general definition, namely that at least one continuous convex interpolant exists should be made more manageable. In the univariate case, for a data set

$$(x_i, z_i), \quad i = 0, \ldots, n, \tag{1.1}$$

on the grid

$$\Delta : x_0 < x_1 < \cdots < x_n \tag{1.2}$$

a desirable characterization of the strict convexity reads

$$\tau_1 < \tau_2 < \cdots < \tau_n \tag{1.3}$$

Multivariate Approximation: Recent Trends and Results; W. Haußmann, K. Jetter and M. Reimer (eds.)
Mathematical Research, Vol. 101, pp. 189–202, ISBN 3–05–501770–6
© Akademie–Verlag, Berlin 1997

where $\tau_i = (z_i - z_{i-1})/h_i$ are the slopes and $h_i = x_i - x_{i-1}$ the step sizes. A convexity definition of this simplicity is not known for multivariate data sets. However, for bivariate data sets

$$(x_i, y_j, z_{i,j}), \quad i = 0,\ldots,n, \; j = 0,\ldots,m, \tag{1.4}$$

on the rectangular grid

$$\Delta \times \Sigma = \{x_0 < x_1 < \cdots < x_n\} \times \{y_0 < y_1 < \cdots < y_m\} \tag{1.5}$$

the related strict S–convexity can be directly characterized as follows [11],

$$\begin{aligned}
\tau_{1,j} &< \tau_{2,j} < \cdots < \tau_{n,j}, & j &= 0,\ldots,m, \\
\sigma_{i,1} &< \sigma_{i,2} < \cdots < \sigma_{i,m}, & i &= 0,\ldots,n, \\
\rho_{i,j} &> 0, & i &= 1,\ldots,n, \; j = 1,\ldots,m,
\end{aligned} \tag{1.6}$$

where in addition $k_j = y_j y_{j-1}$, and

$$\tau_{i,j} = (z_{i,j} - z_{i-1,j})/h_i, \quad \sigma_{i,j} = (z_{i,j} - z_{i,j-1})/k_j \tag{1.7}$$

are the partial slopes, while

$$\rho_{i,j} = (z_{i,j} - z_{i-1,j} - z_{i,j-1} + z_{i-1,j-1})/(h_i k_j) \tag{1.8}$$

is an approximation of the mixed partial derivative. An extension of this definition to three–dimensional gridded data is given in [12].

For functions the property of S–convexity has turned out to be useful in enclosing methods for solving nonlinear equations in partially ordered spaces; see e.g. [15]. The notation of S–convexity was proposed in the paper [10]. Bivariate C^2 functions s are S–convex if the partial derivatives satisfy the inequalities

$$\partial_1^2 s(x,y) \geq 0, \; \partial_1 \partial_2 s(x,y) \geq 0, \; \partial_2^2 s(x,y) \geq 0, \quad \forall(x,y). \tag{1.9}$$

Thus it is seen that convexity and S–convexity are not comparable, and that the biconvexity used in [3] is a consequence of the S–convexity. For a definition of the S–convexity of C^0 functions we refer again to [10].

Further we point out the rather well known negative result that convex interpolation is not always successful if the interpolating space is a *linear finite dimensional* subspace of $C^1[x_0, x_n]$; see [13] for a direct proof and [7] for a deductive proof in an abstract setting. Thus, in convex interpolation one should consider nonlinear spaces such as exponential splines, rational splines, or splines on refined grids with variable additional knots.

In the preceding papers [11, 16] special rational splines are applied in S–convex interpolation. There it is shown that the S–convexity can be preserved for sufficiently large rationality parameters, and we are restricted to heuristic search procedures in determining these parameters suitably.

In the present paper we attack the problem of S–convex interpolation using biquadratic C^1 splines on refined grids. Each subrectangle of $\Delta \times \Sigma$ is divided into four microrectangles by means of a split point. The main result is that in the case of strictly S–convex data we can always offer placements of the split points such that S–convex interpolation is successful. The placements are described by explicit formulae. A heuristic procedure is now not needed.

Finally, we mention the papers [9] and [17] where the above biquadratic splines are seen to be useful in range restricted interpolation as well as in interpolation subject to strip conditions on the first order derivatives.

2 Univariate Quadratic Splines on Refined Grids

Here we recall the properties of univariate quadratic C^1 splines needed in S–convex interpolation with tensor products of these splines. The refinement Δ_1 of the grid Δ is built by adding one knot

$$\xi_i = \beta_i x_{i-1} + \alpha_i x_i\,, \quad \alpha_i\,, \ \beta_i > 0,\ \alpha_i + \beta_i = 1\,, \tag{2.1}$$

on each subinterval $[x_{i-1}, x_i]$, $i = 1,\ldots,n$. As usual, the space of quadratic C^1 splines on Δ_1 is denoted by $S_2^1(\Delta_1)$. In convex interpolation these splines are used in [18].

Proposition 2.1. *For given z_i, p_i, $i = 0,\ldots,n$, the Hermite interpolation problem*

$$s(x_i) = z_i\,, \ s'(x_i) = p_i\,, \quad i = 0,\ldots,n\,, \tag{2.2}$$

is uniquely solvable for $s \in S_2^1(\Delta_1)$.

Further, there is an explicit representation of s, e.g., in Bernstein–Bézier form,

$$
\begin{aligned}
s(x) &= z_{i-1}v^2 + 2\left(z_{i-1} + \tfrac{1}{2}\alpha_i h_i p_{i-1}\right)uv + \zeta_i u^2 && \text{for } x \in [x_{i-1}, \xi_i]\,,\\
s(x) &= \zeta_i v^2 + 2\left(z_i - \tfrac{1}{2}\beta_i h_i p_i\right)uv + z_i u^2 && \text{for } x \in [\xi_i, x_i]\,,
\end{aligned}
\tag{2.3}
$$

with

$$\zeta_i = \alpha_i\left(z_i - \tfrac{1}{2}\beta_i h_i p_i\right) + \beta_i\left(z_{i-1} + \tfrac{1}{2}\alpha_i h_i p_{i-1}\right)\,. \tag{2.4}$$

Further, u and v are the barycentric coordinates with respect to the considered subintervals, namely

$$u = (x - x_{i-1})/(\alpha_i h_i), \; v = (\xi_i - x)/(\alpha_i h_i) \text{ for } x \in [x_{i-1}, \xi_i],$$
$$u = (x - \xi_i)/(\beta_i h_i), \; v = (x_i - x)/(\beta_i h_i) \quad \text{ for } x \in [\xi_i, x_i]. \tag{2.5}$$

Proposition 2.2. *In $S_2^1(\Delta_1)$ we have the following nonnegativity, monotonicity, and convexity criteria, respectively.*

(i) $s(x) \geq 0$ *for* $x \in [x_{i-1}, x_i]$ *if* $\lambda_k(s) \geq 0$ *for* $k = 1, 2, 3, 4$, *where*

$$\lambda_1(s) = \lambda_{1,i}(s) = s(x_{i-1}), \; \lambda_2(s) = \lambda_{2,i}(s) = s(x_i),$$
$$\lambda_3(s) = \lambda_{3,i}(s) = s(x_{i-1}) + \tfrac{1}{2}\alpha_i h_i s'(x_{i-1}),$$
$$\lambda_4(s) = \lambda_{4,i}(s) = s(x_i) - \tfrac{1}{2}\beta_i h_i s'(x_i).$$

(ii) $s'(x) \geq 0$ *for* $x \in [x_{i-1}, x_i]$ *if* $\mu_k(s) \geq 0$ *for* $k = 1, 2, 3$, *where*

$$\mu_1(s) = \mu_{1,i}(s) = s'(x_{i-1}), \; \mu_2(s) = \mu_{2,i}(s) = s'(x_i),$$
$$\mu_3(s) = \mu_{3,i}(s) = \frac{2}{h_i}\left(s(x_i) - s(x_{i-1})\right) - \alpha_i s'(x_{i-1}) - \beta_i s'(x_i).$$

(iii) $s''(x) \geq 0$ *for* $x \in (x_{i-1}, \xi_i) \cup (\xi_i, x_i)$ *if* $\nu_k(s) \geq 0$ *for* $k = 1, 2$, *where*

$$\nu_1(s) = \nu_{1,i}(s) = \frac{2}{h_i}\left(s(x_i) - s(x_{i-1})\right) - (1 + \alpha_i)s'(x_{i-1}) - \beta_i s'(x_i),$$
$$\nu_2(s) = \nu_{2,i}(s) = \alpha_i s'(x_{i-1}) + (1 + \beta_i)s'(x_i) - \frac{2}{h_i}\left(s(x_i) - s(x_{i-1})\right).$$

The proof is easily done since the functionals λ_k, μ_k, and ν_k being linear in s provide the B–ordinates in the Bernstein–Bézier representation of s, s', s'', respectively.

In view of the C^1 continuity, the convexity of s on $[x_0, x_n]$ is obtained if the condition (iii) is satisfied for $i = 1, \ldots, n$. In the case of the nonnegativity of s on $[x_0, x_n]$ some of the assumptions (i) written down for $i = 1, \ldots, n$ turn out to be superfluous, namely $s(x_1) \geq 0, \ldots, s(x_{n-1}) \geq 0$.

The grid Σ may be refined by the additional knots

$$\eta_j = \delta_j y_{j-1} + \gamma_j y_j, \quad \gamma_j, \delta_j > 0, \; \gamma_j + \delta_j = 1, \; j = 1, \ldots, m; \tag{2.6}$$

the result is denoted by Σ_1. The Proposition 2.2 holds also in the space $S_2^1(\Sigma_1)$ but in the functionals λ_k, μ_k, and ν_k we now have to substitute

$$\gamma_j \text{ for } \alpha_i, \; \delta_j \text{ for } \beta_i, \; k_j \text{ for } h_i, \text{ and } y_j \text{ for } x_i. \tag{2.7}$$

3   Univariate Convex Interpolation

In view of Proposition 2.2, part (iii), a spline $s \in S_2^1(\Delta_1)$ interpolating in the sense of Lagrange is convex on $[x_0, x_n]$ if (and only if) the parameters $p_0, p_1, \ldots, p_n$ fulfil the system

$$(1 + \alpha_i)p_{i-1} + (1 - \alpha_i)p_i \leq 2\tau_i \leq \alpha_i p_{i-1} + (2 - \alpha_i)p_i\,, \quad i = 1, \ldots, n\,. \quad (3.1)$$

Several proposals for explicit solutions of (3.1) are known, but none of these is suitable for an extension to our tensor product problem. We cannot use the proposal in [18] because there the ratios are determined to be fixed values. In [13] the α_i have to lie in intervals $(0, \kappa_i)$ but the p_i are computed recursively. The choices $\alpha_i \in (\omega_i, \kappa_i)$ given in [4], see also [13], have the drawback that both bounds ω_i and κ_i depend on i. Therefore, we are invited to develop a more suitable solution of the system (1.3).

Proposition 3.1.   *Let the given data set be in strictly convex position* (1.3) *and choose a number τ_0 such that $\tau_0 < \tau_1$. Further set*

$$\varepsilon = \tfrac{1}{2} \min_{i=1,\ldots,n} (\tau_i - \tau_{i-1}) > 0\,. \quad (3.2)$$

Then the system of inequalities (3.1) *characterizing convex interpolants $s \in S_2^1(\Delta_1)$ is solvable for ratios $\alpha_i \in (0, 1)$ with*

$$\alpha_i \leq \frac{2\varepsilon}{\tau_i - \tau_{i-1}}\,, \qquad i = 1, \ldots, n\,. \quad (3.3)$$

One solution reads

$$p_i = \tau_i + \varepsilon\,, \qquad i = 0, \ldots, n\,. \quad (3.4)$$

Proof : For the values (3.4) the system (3.1) reduces to

$$\alpha_i(\tau_i - \tau_{i-1}) \leq 2\varepsilon \leq (1 + \alpha_i)(\tau_i - \tau_{i-1})$$

being satisfied for the choices (3.2), (3.3). $\square$

4   Nonnegativity Lemma for Tensor Products

In this section sufficient nonnegativity conditions for tensor product splines $s \in S \otimes T$ defined on the rectangle $[a, b] \times [c, d]$ are derived. The univariate spline spaces S and T are assumed to be finite dimensional and linear. Further we suppose $S \subset C^0[a,b]$ and $T \subset C^0[c,d]$.

The justification of the following tensor product constructions is explained, e.g., in [1]. The space $S \otimes T$ is formed by all finite linear combinations of functions described by

$$(s \otimes t)(x,y) = s(x) \cdot t(y), \quad (x,y) \in [a,b] \times [c,d], \ s \in S, \ t \in T. \qquad (4.1)$$

The tensor product functional $\lambda \otimes \mu$ on $S \otimes T$ of two linear functionals λ on S and μ on T is defined by linear extension from the rule

$$(\lambda \otimes \mu)(s \otimes t) = \lambda(s) \cdot \mu(t), \quad s \in S, \ t \in T. \qquad (4.2)$$

The following lemma allows the construction of sufficient nonnegativity conditions in the tensor product space $S \otimes T$ by means of such conditions in S and T.

Lemma 4.1. *Let λ_i, $i = 1, \ldots, N$, be linear functionals on S such that for $s \in S$*

$$\lambda_i(s) \geq 0, \ i = 1, \ldots, N \implies s(x) \geq 0 \ \text{ for all } x \in [a,b] \qquad (4.3)$$

and μ_j, $j = 1, \ldots, M$, be linear functionals on T such that for $s \in T$

$$\mu_j(s) \geq 0, \ j = 1, \ldots, M \implies s(y) \geq 0 \ \text{ for all } y \in [c,d]. \qquad (4.4)$$

Then for $s \in S \otimes T$ the conditions

$$(\lambda_i \otimes \mu_j)(s) \geq 0, \ i = 1, \ldots, N, \ j = 1, \ldots, M, \qquad (4.5)$$

are sufficient for

$$s(x,y) \geq 0 \quad \text{ for all } (x,y) \in [a,b] \times [c,d].$$

It is easily seen that $N \geq \dim S$ and $M \geq \dim T$. Under the special assumption $N = \dim S$ and $M = \dim T$, a proof of Lemma 4.1 is given in [8] while the general case is treated in the paper [6].

Of course, Lemma 4.1 also applies if S and T are spaces of derivatives of the considered splines. Therefore the lemma is suitable for deriving sufficient conditions for the requirements (1.9) of S–convexity, too.

5 Existence and Construction of S–Convex Interpolants

This section is the main part of the present paper. It will be shown that the problem of S–convex C^1 interpolation is always solvable in the tensor product space $S_2^1(\Delta_1) \otimes S_2^1(\Sigma_1)$. The split points in the refined grid $\Delta_1 \times \Sigma_1$ have to be variable, and these points can be placed such that the S–convexity is always retained by the interpolants.

At first we recall that the interpolation problem

$$
\begin{aligned}
s(x_i, y_j) &= z_{i,j}\,, \\
\partial_1 s(x_i, y_j) &= p_{i,j}\,, \\
\partial_2 s(x_i, y_j) &= q_{i,j}\,, \\
\partial_1 \partial_2 s(x_i, y_j) &= r_{i,j}\,, \quad i = 0,\ldots,n\,,\ j = 0,\ldots,m\,,
\end{aligned}
\tag{5.1}
$$

is uniquely solvable for $s \in S_2^1(\Delta_1) \otimes S_2^1(\Sigma_1)$ if the values $z_{i,j}$, $p_{i,j}$, $q_{i,j}$, and $r_{i,j}$ are given; see [1]. While the $z_{i,j}$ are prescribed by the data set (1.4), the quantities $p_{i,j}$, $q_{i,j}$, and $r_{i,j}$ are used as parameters for satisfying the S–convexity conditions.

Next, applying Proposition 2.2 and Lemma 4.1 we obtain

Proposition 5.1. *The system of inequalities*

$$
(\nu_{k,i} \otimes \lambda_{l,j})(s) \geq 0\,, \ k = 1,2\,,\ l = 1,2,3,4\,,
\tag{5.2}
$$

$$
(\mu_{k,i} \otimes \mu_{l,j})(s) \geq 0\,, \ k = 1,2,3\,,\ l = 1,2,3\,,
\tag{5.3}
$$

$$
(\lambda_{k,i} \otimes \nu_{l,j})(s) \geq 0\,, \ k = 1,2,3,4\,,\ l = 1,2\,,
\tag{5.4}
$$

with $i = 1,\ldots,n$, $j = 1,\ldots,m$, *implies the S–convexity of an interpolant* $s \in S_2^1(\Delta_1) \otimes S_2^1(\Sigma_1)$.

In Proposition 5.1 it is essential that, for C^1 functions, the S–convexity on each microrectangle implies this property on the whole rectangle.

We write the conditions (5.2) which are sufficient for $\partial_1^2 s(x,y) \geq 0$ in detail, but without repetitions:

$$
0 \ \leq \ 2\tau_{i,j} - (1 + \alpha_i)p_{i-1,j} - (1 - \alpha_i)p_{i,j}\,,
\tag{5.5}
$$

$$
\begin{aligned}
0 \ \leq \ & 2\tau_{i,j-1} - (1 + \alpha_i)p_{i-1,j-1} - (1 - \alpha_i)p_{i,j-1} + \frac{\gamma_j k_j}{h_i}\left(q_{i,j-1} - q_{i-1,j-1}\right) \\
& - \tfrac{1}{2}\gamma_j k_j\left((1 + \alpha_i)r_{i-1,j-1} + (1 - \alpha_i)r_{i,j-1}\right)\,,
\end{aligned}
\tag{5.6}
$$

$$
\begin{aligned}
0 \;\leq\; & 2\tau_{i,j} - (1+\alpha_i)p_{i-1,j} - (1-\alpha_i)p_{i,j} - \frac{\delta_j k_j}{h_i}\left(q_{i,j} - q_{i-1,j}\right) \\
& + \tfrac{1}{2}\delta_j k_j \left((1+\alpha_i)r_{i-1,j} + (1-\alpha_i)r_{i,j}\right), & (5.7)
\end{aligned}
$$

$$
0 \;\leq\; -2\tau_{i,j} + \alpha_i p_{i-1,j} + (2-\alpha_i)p_{i,j}, \qquad\qquad\qquad (5.8)
$$

$$
\begin{aligned}
0 \;\leq\; & -2\tau_{i,j-1} + \alpha_i p_{i-1,j-1} + (2-\alpha_i)p_{i,j-1} - \frac{\gamma_j k_j}{h_i}\left(q_{i,j-1} - q_{i-1,j-1}\right) \\
& + \tfrac{1}{2}\gamma_j k_j \left(\alpha_i r_{i-1,j-1} + (2-\alpha_i)r_{i,j-1}\right), & (5.9)
\end{aligned}
$$

$$
\begin{aligned}
0 \;\leq\; & -2\tau_{i,j} + \alpha_i p_{i-1,j} + (2-\alpha_i)p_{i,j} + \frac{\delta_j k_j}{h_i}\left(q_{i,j} - q_{i-1,j}\right) \\
& - \tfrac{1}{2}\delta_j k_j \left(\alpha_i r_{i-1,j} + (2-\alpha_i)r_{i,j}\right). & (5.10)
\end{aligned}
$$

Here the indices run as $i = 1, \ldots, n$, $j = 1, \ldots, m$, with the exception of (5.5) and (5.8), where $i = 1, \ldots, n$, $j = 0, \ldots, m$.

The sufficient conditions (5.3) for $\partial_1\partial_2 s(x, y) \geq 0$ read in detail

$$
0 \;\leq\; r_{i,j}, \quad i = 0, \ldots, n, \; j = 0, \ldots, m, \qquad\qquad (5.11)
$$

$$
\begin{aligned}
0 \;\leq\; & 4\rho_{i,j} - \frac{2}{k_j}\left(\alpha_i\left(p_{i-1,j} - p_{i-1,j-1}\right) + \beta_i\left(p_{i,j} - p_{i,j-1}\right)\right) \\
& - \frac{2}{h_i}\left(\gamma_j\left(q_{i,j-1} - q_{i-1,j-1}\right) + \delta_j\left(q_{i,j} - q_{i-1,j}\right)\right) \\
& + \alpha_i\gamma_j r_{i-1,j-1} + \alpha_i\delta_j r_{i-1,j} + \beta_i\gamma_j r_{i,j-1} + \beta_i\delta_j r_{i,j}, \\
& i = 1, \ldots, n, \; j = 1, \ldots, m, & (5.12)
\end{aligned}
$$

$$
\begin{aligned}
0 \;\leq\; & \frac{2}{h_i}\left(q_{i,j} - q_{i-1,j}\right) - \alpha_i r_{i-1,j} - \beta_i r_{i,j}, \\
& i = 1, \ldots, n, \; j = 0, \ldots, m, & (5.13)
\end{aligned}
$$

$$
\begin{aligned}
0 \;\leq\; & \frac{2}{k_j}\left(p_{i,j} - p_{i,j-1}\right) - \gamma_j r_{i,j-1} - \delta_j r_{i,j}, \\
& i = 0, \ldots, n, \; j = 1, \ldots, m. & (5.14)
\end{aligned}
$$

The inequalities (5.4) assuring $\partial_2^2 s(x, y) \geq 0$ are analogous to (5.5)–(5.10).

The detailed representation (5.5)–(5.14) of the S–convexity conditions (5.2)–(5.4) shows that these inequalities depend quadratically on the ratios α_i, $\beta_i = 1 - \alpha_i$, γ_j, $\delta_j = 1 - \gamma_j$ of the split points. We are in the position to give explicit formulae for suitable choices α_i, γ_j. However, we believe that the obtained values can be improved, possibly by a numerical procedure. The aim should be ratios as near as possible to $1/2$, the middle of the feasible interval $(0, 1)$.

In order to find a solution of the system (5.2)–(5.4) we set

$$
\begin{aligned}
p_{i,j} &= \tau_{i,j} + \varepsilon\,, \\
q_{i,j} &= \sigma_{i,j} + \kappa\,, \\
r_{i,j} &= \rho_{i,j}\,, \quad i = 0,\ldots,n\,,\ j = 0,\ldots,m\,,
\end{aligned}
\tag{5.15}
$$

with the values $\tau_{i,j}$, $\sigma_{i,j}$, and $\rho_{i,j}$ defined by (1.7), (1.8). In addition, we choose $\tau_{0,j}$ and $\sigma_{i,0}$ such that

$$
\tau_{0,j} < \tau_{1,j}\,,\ \sigma_{i,0} < \sigma_{i,1}\,,
\tag{5.16}
$$

and

$$
\rho_{0,0} = \rho_{1,1}\,,\ \rho_{0,j} = \rho_{1,j}\,,\ \rho_{i,0} = \rho_{i,1}\,.
\tag{5.17}
$$

We assume that ε and κ are real numbers with

$$
\begin{aligned}
0 < 2\varepsilon < \tau_{i,j} - \tau_{i-1,j}\,, \qquad & i = 1,\ldots,n\,,\ j = 0,\ldots,m\,, \\
0 < 2\kappa < \sigma_{i,j} - \sigma_{i,j-1}\,, \qquad & i = 0,\ldots,n\,,\ j = 1,\ldots,m\,,
\end{aligned}
\tag{5.18}
$$

which exist in view of the strict S–convexity of the data set.

We are led to the following announced result.

Theorem 5.2.　　*Let be given a strictly S–convex data set $(x_i,\,y_j,\,z_{i,j})$, $i = 0,\ldots,n$, $j = 1,\ldots,m$. If the ratios are chosen according to*

$$
\alpha_i \in (0,1)\,,\ \alpha_i \le 1/M_i\,,\quad i = 1,\ldots,n\,,
\tag{5.19}
$$

$$
\gamma_j \in (0,1)\,,\ \gamma_j \le 1/N_j\,,\quad j = 1,\ldots,m\,,
\tag{5.20}
$$

with

$$
M_i = \max\left\{\max_{j=0,\ldots,m}\left\{\frac{\tau_{i,j} - \tau_{i-1,j}}{\varepsilon}\right\}\,,\ X_i\right\}\,,
$$

$$
X_i = \max_{j=1,\ldots,m}\left\{\frac{2\rho_{i-1,j} - \rho_{i,j}}{\rho_{i,j}}\,,\ \frac{h_i}{2}\frac{\rho_{i-1,j-1} - \rho_{i-1,j}}{\sigma_{i-1,j} - \sigma_{i-1,j-1} - 2\kappa}\,,\ h_i\frac{\rho_{i-1,j} - \rho_{i-1,j-1}}{\sigma_{i-1,j} - \sigma_{i-1,j-1}}\right\}
\tag{5.21}
$$

and

$$
N_j = \max\left\{\max_{i=0,\ldots,n}\left\{\frac{\sigma_{i,j} - \sigma_{i,j-1}}{\kappa}\right\}\,,\ Y_j\right\}\,,
$$

$$
Y_j = \max_{i=1,\ldots,n}\left\{\frac{2\rho_{i,j-1} - \rho_{i,j}}{\rho_{i,j}}\,,\ \frac{k_j}{2}\frac{\rho_{i-1,j-1} - \rho_{i,j-1}}{\tau_{i,j-1} - \tau_{i-1,j-1} - 2\varepsilon}\,,\ k_j\frac{\rho_{i,j-1} - \rho_{i-1,j-1}}{\tau_{i,j-1} - \tau_{i-1,j-1}}\right\}\,,
\tag{5.22}
$$

then S-convex C^1 spline interpolants $s \in S_2^1(\Delta_1) \otimes S_2^1(\Sigma_1)$ always exist on the correspondingly refined grids $\Delta_1 \times \Sigma_1$.

One spline solution is given by (5.15)–(5.18). Improved splines can be obtained, in general, by applying optimization procedures. For proposals of objective functions we refer, e.g., to [8, 9, 17]. The feasible domain is now given by (5.2)–(5.4).

Proof : Utilizing (5.15), the system (5.5) reduces to

$$(1 + \alpha_i)(\tau_{i,j} - \tau_{i-1,j}) \geq 2\varepsilon,$$

being satisfied because of (5.18). The inequalities (5.6) read now

$$2\tau_{i,j-1} - (1 + \alpha_i)\tau_{i-1,j-1} - (1 - \alpha_i)\tau_{i,j-1} + \frac{\gamma_j k_j}{h_i}(\sigma_{i,j-1} - \sigma_{i-1,j-1})$$

$$-\tfrac{1}{2}\gamma_j k_j((1 + \alpha_i)\rho_{i-1,j-1} + (1 - \alpha_i)\rho_{i,j-1}) - 2\varepsilon \;\; \geq \;\; 0.$$

They are equivalent to

$$(1 + \alpha_i)(\tau_{i,j-1} - \tau_{i-1,j-1}) - 2\varepsilon \geq (1 + \alpha_i)\frac{\gamma_j k_j}{2}(\rho_{i-1,j-1} - \rho_{i,j-1}).$$

Thus, since

$$\frac{1}{\gamma_j} \geq \frac{k_j}{2}\frac{\rho_{i-1,j-1} - \rho_{i,j-1}}{\tau_{i,j-1} - \tau_{i-1,j-1} - 2\varepsilon} \tag{5.23}$$

is assumed, the inequalities (5.6) are satisfied. Applying (5.18), the system

$$(1 - \frac{\delta_j}{2})(\tau_{i,j} - \tau_{i-1,j}) + \frac{\delta_j}{2}(\tau_{i,j-1} - \tau_{i-1,j-1}) \geq 2\varepsilon ,$$

which is stronger than (5.7), is seen to be valid.

The conditions (5.8), now $\alpha_i(\tau_{i,j} - \tau_{i-1,j}) \leq 2\varepsilon$ are seen to be satisfied because of

$$\frac{1}{\alpha_i} \geq \frac{\tau_{i,j} - \tau_{i-1,j}}{2\varepsilon} \tag{5.24}$$

implied by (5.19). In order to fulfil the system (5.9) we have assumed

$$\frac{1}{\alpha_i} \geq \frac{\tau_{i,j} - \tau_{i-1,j}}{\varepsilon} , \qquad \frac{1}{\gamma_j} \geq k_j\frac{\rho_{i,j-1} - \rho_{i-1,j-1}}{\tau_{i,j-1} - \tau_{i-1,j-1}} ,$$

leading to

$$\tau_{i,j-1} - \tau_{i-1,j-1} \leq \frac{\varepsilon}{\alpha_i}$$

and

$$\gamma_j k_j \left(\rho_{i,j-1} - \rho_{i-1,j-1}\right) \le \tau_{i,j-1} - \tau_{i-1,j-1} < 2\varepsilon < \frac{2\varepsilon}{\alpha_i}\,.$$

The system (5.9) is obtained by combining these inequalities to

$$\tau_{i,j-1} - \tau_{i-1,j-1} + \frac{\gamma_j k_j}{2}\left(\rho_{i,j-1} - \rho_{i-1,j-1}\right) \le \frac{2\varepsilon}{\alpha_i}\,.$$

The inequalities (5.10) read

$$\tau_{i,j} - \tau_{i-1,j} - \frac{\delta_j k_j}{2}\left(\rho_{i,j} - \rho_{i-1,j}\right) \le \frac{2\varepsilon}{\alpha_i}\,.$$

These requirements can be rewritten as

$$\left(1 - \frac{\delta_j}{2}\right)\left(\tau_{i,j} - \tau_{i-1,j}\right) + \frac{\delta_j}{2}\left(\tau_{i,j-1} - \tau_{i-1,j-1}\right) \le \frac{2\varepsilon}{\alpha_i}\,,$$

which are seen to be valid using (5.24). Obviously, the conditions (5.11) hold true in view of (5.15). The inequalities (5.12) read

$$\begin{aligned} 0 \;\le\; & 4\rho_{i,j} - 2\left(\alpha_i \rho_{i-1,j} + \beta_i \rho_{i,j}\right) - 2\left(\gamma_j \rho_{i,j-1} + \delta_j \rho_{i,j}\right) + \alpha_i \gamma_j \rho_{i-1,j-1} \\ & + \alpha_i \delta_j \rho_{i-1,j} + \beta_i \gamma_j \rho_{i,j-1} + \beta_i \delta_j \rho_{i,j}\,. \end{aligned}$$

These are equivalent to

$$\begin{aligned} 0 \;\le\; & \alpha_i \gamma_j \rho_{i-1,j-1} + (1 + \alpha_i)\left[\tfrac{1}{2}(1 + \gamma_j)\rho_{i,j} - \gamma_j \rho_{i,j-1}\right] \\ & + (1 + \gamma_j)\left[\tfrac{1}{2}(1 + \alpha_i)\rho_{i,j} - \alpha_i \rho_{i-1,j}\right]\,. \end{aligned}$$

For $\rho_{i,j} > 0$ and $\alpha_i,\ \gamma_j \in (0,1)$ this system of inequalities is satisfied if the terms in the brackets are nonnegative. These requirements lead to the conditions

$$\frac{1}{\alpha_i} \ge \frac{2\rho_{i-1,j} - \rho_{i,j}}{\rho_{i,j}}\,,\quad j = 1,\dots,m\,, \tag{5.25}$$

and

$$\frac{1}{\gamma_j} \ge \frac{2\rho_{i,j-1} - \rho_{i,j}}{\rho_{i,j}}\,,\quad i = 1,\dots,n\,, \tag{5.26}$$

incorporated in (5.19) and (5.20). In the case of (5.15), the requirements (5.13) reduce to the inequalities

$$0 \le \rho_{i,j} + \alpha_i(\rho_{i,j} - \rho_{i-1,j})\,;$$

these hold true since the ratios in view of (5.25) and (5.17) satisfy

$$\frac{1}{\alpha_i} \geq \frac{\rho_{i-1,j} - \rho_{i,j}}{\rho_{i,j}} \,, \ j = 0, \ldots, m \,.$$

Finally, we obtain analogously that (5.14) is fulfilled under the assumptions (5.20). Thus, the proof of Theorem 5.2 is complete. $\qquad\square$

Let us remark that the preceding considerations immediately apply to the determination of so–called biconvex interpolants in the sense of [3]. Biconvexity is already assured by the inequalities (5.2) and (5.4), not needing (5.3). Thus we can use the formulae (5.15)–(5.22) also for constructing biconvex C^1 interpolants. In addition, it is possible to simplify the requirements on the ratios by substituting

$$X_i = \max_{j=1,\ldots,m} \left\{ \frac{h_i}{2} \frac{\rho_{i-1,j-1} - \rho_{i-1,j}}{\sigma_{i-1,j} - \sigma_{i-1,j-1} - 2\kappa} \,, h_i \frac{\rho_{i-1,j} - \rho_{i-1,j-1}}{\sigma_{i-1,j} - \sigma_{i-1,j-1}} \right\}$$

$$Y_j = \max_{i=1,\ldots,n} \left\{ \frac{k_j}{2} \frac{\rho_{i-1,j-1} - \rho_{i,j-1}}{\tau_{i,j-1} - \tau_{i-1,j-1} - 2\varepsilon} \,, k_j \frac{\rho_{i,j-1} - \rho_{i-1,j-1}}{\tau_{i,j-1} - \tau_{i-1,j-1}} \right\} \,,$$

in (5.21) and (5.22), i.e., the requirements (5.25) and (5.26) are superfluous in biconvexity. In contrast to [3], no search procedure is needed when using the presented splines.

6 S–Convex Interpolation of C^2 Continuity

We briefly refer to an extension of the above results to S–convex interpolation of the continuity C^2. In the univariate case, a quartic C^2 spline s on Δ_1, i.e., $s \in S_4^2(\Delta_1)$, is uniquely represented by the parameters

$$z_i = s(x_i) \,, \ p_i = s'(x_i) \,, \ P_i = s''(x_i) \,, \ i = 0, \ldots, n \,, \ \Pi_i = s''(\xi_i) \,, \ i = 1, \ldots, n \,.$$

We denote by $\tilde{S}^2(\Delta_1) \subset S_4^2(\Delta_1)$ the subspace of splines satisfying

$$P_i = 0 \,, \ i = 0, \ldots, n \,, \ \Pi_i = 0 \,, \ i = 1, \ldots, n \,.$$

Then, Proposition 2.2 holds with the same functionals λ_k, μ_k, and ν_k if the space $S_2^1(\Delta_1)$ is replaced by $\tilde{S}^2(\Delta_1)$. The proof of the parts (ii) and (iii) is straightforward; part (i) is verified in detail in [14]. Therefore, in $\tilde{S}^2(\Delta_1) \otimes \tilde{S}^2(\Sigma_1)$ we are led again to the sufficient conditions (5.2), (5.3), and (5.4) of the S–convexity; and the choices (5.19) and (5.20) of ratios are suitable in $\tilde{S}^2(\Delta_1) \otimes \tilde{S}^2(\Sigma_1)$, and hence also in $S_4^2(\Delta_1) \otimes S_4^2(\Sigma_1)$.

Acknowledgment. The work of the second author was supported by Deutsche Forschungsgemeinschaft under grant Schm 968/2–2.

References

[1] C. de Boor: *A Practical Guide to Splines*, Springer–Verlag, New York–Heidelberg–Berlin 1978.

[2] P. Costantini: *Shape-preserving interpolation with variable degree polynomial splines*, in: *Advanced Course on Fairshape*, J. Hoschek and P. Kaklis (eds.), 87–114, Teubner, Stuttgart 1996.

[3] P. Costantini and F. Fontanella: *Shape-preserving bivariate interpolation*, SIAM J. Numer. Anal. **27** (1990), 488–506.

[4] R. Delbourgo and J. A. Gregory: *Shape preserving piecewise rational interpolation*, SIAM J. Sci. Statist. Comput. **6** (1985), 967–976.

[5] H. Greiner: *A survey on univariate data interpolation and approximation by splines of given shape*, Math. Comput. Modelling **15** (1991), 97–106.

[6] B. Mulansky: *Tensor products of convex cones*, in: *Multivariate Approximation and Splines*, G. Nürnberger, J. W. Schmidt and G. Walz (eds.), *Intern. Series Numer. Math.*, Birkhäuser–Verlag, Basel–Boston–Stuttgart (submitted).

[7] B. Mulansky and M. Neamtu: *Interpolation and approximation from convex sets*, J. Approx. Theory (to appear).

[8] B. Mulansky and J. W. Schmidt: *Nonnegative interpolation by biquadratic splines on refined rectangular grids*, in: *Wavelets, Images and Surface Fitting*, P.–J. Laurent, A. Le Méhauté and L. L. Schumaker (eds.), 379–386, AK Peters, Wellesley 1994.

[9] B. Mulansky, J. W. Schmidt, and M. Walther: *Tensor product spline interpolation subject to piecewise bilinear lower and upper bounds*, in: *Advanced Course on Fairshape*, J. Hoschek and P. Kaklis (eds.), 201–216, Teubner, Stuttgart 1996.

[10] F.–A. Potra: *On superadditive rates of convergence*, Math. Modell. Numer. Anal. **85** (1985), 671–685.

[11] J. W. Schmidt: *Rational biquadratic C^1-splines in S-convex interpolation*, Computing **47** (1991), 87–96.

[12]   J. W. Schmidt: *Positive and S–convex C^1–interpolation of gridded three dimensional data.* Numer. Funct. Anal. Optim. **16** (1995), 233–246.

[13]   J. W. Schmidt: *Staircase algorithm and construction of convex spline interpolants up to the continuity C^3,* Comput. Math. Appl. **41** (1996), 67–79. *Numerical Methods,* Miskolc 1994 (P. Rózsa, J. W. Schmidt, B. A. Szabó, guest eds.).

[14]   J. W. Schmidt: *Strip interpolations using splines on refined grids,* in: *Proceed. Intern. Workshop on Recent Advances in Applied Mathematics,* A. Hamoui and C. Großmann (eds.), 463–474, Kuwait Univ., Kuwait, 1996.

[15]   J. W. Schmidt and H. Leonhardt: *Eingrenzung von Lösungen mit Hilfe der Regula falsi,* Computing **6** (1970), 318–328.

[16]   J. W. Schmidt and W. Heß: *S–convex, monotone, and positive interpolation with rational bicubic splines of C^2–continuity,* BIT **33** (1993), 496–511.

[17]   J. W. Schmidt and M. Walther: *Gridded data interpolation with restrictions on the first order derivatives,* in: *Multivariate Approximation and Splines,* G. Nürnberger, J. W. Schmidt and G. Walz (eds.), *Intern. Series Numer. Math.,* Birkhäuser–Verlag, Basel–Boston–Stuttgart (submitted).

[18]   L. L. Schumaker: *On shape preserving quadratic spline interpolation,* SIAM J. Numer. Anal. **20** (1983), 854–864.

[19]   H. Späth: *One Dimensional Spline Interpolation Algorithms,* AK Peters, Wellesley, Boston 1995.

Address:

JOCHEN W. SCHMIDT, MARION WALTHER
Institute of Numerical Mathematics
Technical University Dresden
Mommsenstr. 13
D–01062 Dresden
Germany

Quasi–Balayage and A Priori Estimates for the Laplace Operator I

H. S. Shapiro

1 Introduction

The notion of balayage, as embodied in Proposition 2.1 below, is very important in potential theory. In the present paper we present a generalization, called *quasi–balayage*; while formally this is straightforward, and perhaps not of great intrinsic interest, it enables one to prove new a priori estimates for the Laplace operator, and also to derive some known ones in a novel and simple manner.

The paper is organized as follows. In Section 2 we review the classical balayage concept. This material is well known, and substantially contained in, e.g., the book [L]; its inclusion here is for the purpose of motivating the notion of quasi–balayage, which is presented in Section 3. In Section 4 some estimates are given that are needed in applying quasi–balayage. Sections 5 and 6 present applications. These include: *characterization of distributions of finite order on domains in* $\mathbb{R}^n$, *that annihilate harmonic functions* (this is essential to the study of unbounded quadrature domains); *solvability with bounds of* $\Delta u = f$ *where f is a distribution on a domain in* $\mathbb{R}^n$; and *a Phragmén–Lindelöf principle for solutions of Cauchy's problem for the Laplace equation* (this has applications to, among other things, proving the regularity of free boundaries arising in Hele–Shaw flows and the obstacle problem). Finally, Section 7 describes briefly ongoing joint work with Walter Hayman and Lavi Karp which allows sharpening some results of the present paper.

The author has also written a sequel extending some aspects of the present paper (see: Quasi–Balayage and A Priori Estimates for the Laplace Operator II, these Proceedings, pp. 231–254).

Acknowledgement. I am indebted to Lavi Karp for many illuminating discussions concerning the matters treated in this paper.

Multivariate Approximation: Recent Trends and Results; W. Haußmann, K. Jetter and M. Reimer (eds.)
Mathematical Research, Vol. 101, pp. 203–230, ISBN 3-05-501770-6
© Akademie-Verlag, Berlin 1997

2   Classical Balayage

2.1. Notations. Throughout this paper, $\mathbb{R}^n$ denotes Euclidean n–space. For an open set $\Omega \subset \mathbb{R}^n$, $H(\Omega)$ denotes the set of real–valued harmonic functions on Ω. $L^p(\Omega)$ denotes the usual Lebesgue space on Ω with respect to Lebesgue measure dx, and $HL^p(\Omega) = H(\Omega) \cap L^p(\Omega)$. For a closed set $K \subset \mathbb{R}^n$, $M(K)$ denotes the Banach space of bounded real measures μ on K, with norm $\|\mu\|_{M(K)}$ equal to the total variation of μ. $C(K)$ denotes the space of real–valued continuous functions on K, and $C_0(K)$ its subspace consisting of functions f such that (in case K is unbounded) $f(x) \to 0$ as $x \in K$ tends to ∞. $M(K)$ is, in a natural manner, the dual of $C_0(K)$ (note that in this paper the subscript c, rather than 0, denotes compact support).

The *kernel of Newtonian potential theory* is

$$K_2(x) \quad = \quad c_2 \, \log |x|, \tag{2.1}$$

$$K_n(x) \quad = \quad c_n |x|^{2-n}, \quad n \geq 3, \tag{2.2}$$

where the constants c_j are such that

$$-\Delta K_n \quad = \quad \delta, \tag{2.3}$$

Δ denoting the Laplace operator acting on distributions in $\mathbb{R}^n$, and δ the Dirac functional (point evaluation) at $0 \in \mathbb{R}^n$. For $\mu \in M(K)$ where K is a compact set in $\mathbb{R}^n$, *the Newtonian potential of μ* is denoted by U^μ :

$$U^\mu \quad := \quad K_n * \mu, \tag{2.4}$$

$*$ denoting convolution (the same definition applies if μ is any distribution of compact support in $\mathbb{R}^n$).

2.2 The (classical) *balayage principle* is

Proposition 2.1. *Let Ω be a bounded open connected set in $\mathbb{R}^n$ with closure denoted $\overline{\Omega}$, whose boundary is regular for the Dirichlet problem, and $\mu \in M(\overline{\Omega})$. Then, there exists a unique measure $\nu \in M(\partial\Omega)$ – where $\partial\Omega$ denotes the boundary of Ω – satisfying*

$$(\mathrm{i}) \qquad \|\nu\| \quad \leq \quad \|\mu\|,$$

$$(\mathrm{ii}) \qquad \int u \, d\mu \quad = \quad \int u \, d\nu \qquad \textit{for every } u \in C(\overline{\Omega}) \cap H(\Omega).$$

Remarks. One calls ν the *balayage*, or *swept measure*, of μ onto $\partial\Omega$. Choosing in particular $u(x) = K_n(x - y)$ in (ii), where $y \in \mathbb{R}^n \setminus \overline{\Omega}$ we see: $U^\nu(y) = U^\mu(y)$ for $y \in \mathbb{R}^n \setminus \overline{\Omega}$.

Let us outline the (very well known) proof of Proposition 2.1 since it is a model for the proofs needed in Section 3. Define a map $\Lambda : C(\overline{\Omega}) \cap H(\Omega) \to \mathbb{R}$ by

$$\Lambda u \;=\; \int u \, d\mu, \qquad u \in C(\overline{\Omega}) \cap H(\Omega).$$

Clearly Λ is linear. Also

$$|\Lambda u| \;\leq\; \|\mu\| \cdot \sup_{x \in \overline{\Omega}} |u(x)| \;=\; \|\mu\| \cdot \sup_{x \in \partial\Omega} |u(x)|, \qquad (2.5)$$

the last equality being the expression of the maximum principle for harmonic functions. If we denote by X the subspace of $C(\partial\Omega)$ consisting of restrictions of elements of $C(\overline{\Omega}) \cap H(\Omega)$ to $\partial\Omega$, (2.5) expresses that Λ is a bounded linear functional on X with norm $\leq \|\mu\|$. By the Hahn–Banach Theorem it extends to $C(\partial\Omega)$ with norm $\leq \|\mu\|$, and in view of F. Riesz' representation theorem this implies the existence of $\nu \in M(\partial\Omega)$ satisfying (i) and (ii).

The uniqueness of ν follows because, if ν_1 and ν_2 were two solutions, we would have $\int f \, d\nu_1 \;=\; \int f \, d\nu_2$ for every $f \in C(\partial\Omega)$ – since, in view of the classical solvability of the Dirichlet problem in Ω, there is $u \in C(\overline{\Omega}) \cap H(\Omega)$ with $u = f$ on $\partial\Omega$ – and now (ii) implies the desired conclusion.

Thus, *the balayage principle is precisely the Hahn–Banach dual of the maximum principle.* Note that *existence of ν satisfying* (i) *and* (ii) *is based solely on the maximum principle; solvability of the Dirichlet problem was invoked only to prove uniqueness of ν.*

It is instructive (again, as motivation for further work) to see how *the balayage principle implies the solvability of Dirichlet's problem* (at least in its "weak" formulation), in domains with smooth boundary. Suppose $\partial\Omega$ is a non–singular C^2 hypersurface and $\varphi \in C^2(\overline{\Omega})$ satisfies

$$\varphi \;=\; 0 \qquad \text{on } \partial\Omega. \qquad (2.6)$$

Let now $y \in \Omega$, and let δ_y be the Dirac measure at y, and ν_y its balayage to $\partial\Omega$ (at any rate, *some* balayage of δ_y. We needn't assume uniqueness for what follows; of course, it is well known that ν_y is expressible in terms of the normal derivative of Green's function, but we don't require that information). We have

$$\varphi(y) \;=\; \int \varphi \, d\delta_y \;=\; \int \varphi \, d[\delta_y - \nu_y] \qquad \text{because of (2.6)}$$

$$=\; -\int \varphi \Delta U^{\delta_y - \nu_y} \;=\; -\int \Delta\varphi(x) U^{\delta_y - \nu_y}(x)\, dx. \qquad (2.7)$$

The last equality is justified because both φ and $U^{\delta_y - \nu_y}$ (which really is the Green function) vanish on $\partial\Omega$. Indeed, assuming sufficient regularity of $\partial\Omega$ it is not hard to show that

(i) $U^{\delta_y - \nu_y}(x) \to 0$ uniformly as $x \to \partial\Omega$, and

(ii) $U^{\delta_y - \nu_y}$ is in $L^p(\Omega)$ for every $1 \le p < \dfrac{n}{n-2}$, $n \ge 3$, and for such p

$$\| U^{\delta_y - \nu_y} \|_p \;\le\; C(p, \Omega). \qquad (2.8)$$

We leave these verifications to the reader. Hence, from (2.7) :

$$|\varphi(y)| \;\le\; C(p, \Omega) \cdot \| \Delta\varphi \|_{p'} \qquad (y \in \Omega), \qquad (2.9)$$

where $p' = \dfrac{p}{p-1} > \dfrac{n}{2}$. We have thus obtained the a priori inequality :

Proposition 2.2. *If $\Omega \subset \mathrm{I\!R}^n$ is a bounded domain whose boundary $\partial\Omega$ is smooth, then for every $r > \dfrac{n}{2}$ there is a constant C depending only on Ω and r such that*

$$\| \varphi \|_\infty \;\le\; C \cdot \| \Delta\varphi \|_r \qquad (2.10)$$

(the subscripts denoting $L^p(\Omega)$ norms for $p = \infty$, r respectively), for all $\varphi \in C^2(\overline{\Omega})$ satisfying (2.6).

This, by a standard argument, implies the following version of solvability of Dirichlet's Problem in Ω :

Proposition 2.3. *For every bounded measure $\mu \in M(\Omega)$ there is a measurable function u on Ω with $u \in L^p(\Omega)$ for every $p < \dfrac{n}{n-2}$ satisfying $\| u \|_p \le C(n, p) \cdot \| \mu \|_M$ and*

$$\int_\Omega u(\Delta\varphi)\, dx \;=\; \int \varphi\, d\mu \qquad (2.11)$$

for every $\varphi \in C^2(\overline{\Omega})$ satisfying (2.6).

Remark. In particular, (2.11) holds for all φ in $C_c^\infty(\Omega)$, the infinitely differentiable functions with compact support. Hence (2.11) implies

$$\Delta u \;=\; \mu \qquad\qquad \text{in } \Omega \tag{2.12}$$

in the sense of distributions. We then moreover have, from (2.11)

$$\int_\Omega u \Delta \varphi \, dx \;=\; \int_\Omega \varphi \Delta u \, dx \tag{2.13}$$

for all φ satisfying (2.6) which is the "weak" formulation of "u vanishes on $\partial\Omega$". More precisely: relation (2.13) for all smooth φ on $\overline{\Omega}$ vanishing on $\partial\Omega$ *together with their first partial derivatives* is just a tautology; if the test functions are only required to vanish on $\partial\Omega$ (condition (2.6)), (2.13) expresses "Dirichlet" vanishing of u on $\partial\Omega$; whereas if (2.13) is known to hold for all $\varphi \in C^2(\overline{\Omega})$ – with *no* boundary conditions imposed on φ – that is the weak formulation of u having Cauchy data zero on $\partial\Omega$, i.e. u and its first–order partial derivatives vanish there.

To deduce Proposition 2.3 from Proposition 2.2 we consider the linear functional

$$\Lambda \;\;:\;\; \Delta\varphi \mapsto \int \varphi \, d\mu$$

from X to $\mathbb{R}$ where X is the vector subspace of $L^r(\Omega)$ defined by:

$$X \;\;:=\;\; \{\, \Delta\varphi \,:\, \varphi \in C^2(\overline{\Omega}) \text{ and } \varphi \text{ satisfies (2.6)} \,\}.$$

Then for each $r > \dfrac{n}{2}$

$$|\Lambda(\Delta\varphi)| \;\;\leq\;\; \|\mu\|_M \cdot \|\varphi\|_\infty \;\;\leq\;\; C\,\|\mu\|_M \cdot \|\Delta\varphi\|_r$$

for $\Delta\varphi \in X$, by (2.10), so Λ extends to a continuous linear functional on $L^r(\Omega)$ with bound $\leq C \,\|\mu\|_M$. The Riesz representation theorem now gives us $u \in L^p$ with $p = \frac{r}{r-1}$ satisfying (2.11) with

$$\|u\|_p \;\;\leq\;\; C(p,\Omega) \cdot \|\mu\|_M. \tag{2.14}$$

We have gone through these standard arguments because similar arguments will be used throughout the rest of this paper, and henceforth will be presented only very briefly if at all. Moreover (although for reasons of space that must be postponed to a later publication) they can be extended to establish solution with bounds of the Dirichlet problem in *unbounded* domains.

3 Quasi–Balayage

3.1 The Basic Idea. In classical balayage of, say, a point mass δ_y, $y \in \Omega$, to a measure ν_y supported on $\partial\Omega$, the support of ν_y is *all* of $\partial\Omega$: a measure $\sigma \in M(\Gamma_0)$, where Γ_0 is a proper closed subset of $\partial\Omega =: \Gamma$ could never satisfy

$$u(y) \;=\; \int u\,d\sigma$$

for all $u \in C(\overline{\Omega}) \cap H(\Omega)$, because that would imply for such u the *false* maximum principle

$$|u(y)| \;\leq\; A \cdot \sup_{x \in \Gamma_0} |u(x)|, \tag{3.1}$$

where $A = \|\sigma\|_M$. For some purposes we would however like to "sweep" δ_y to a measure supported on a set (like Γ_0) that does not topologically surround y. The way out is to restrict the set of test functions u in the basic identity (equation (ii) in Proposition 2.1) to a much narrower family than $C(\overline{\Omega}) \cap H(\Omega)$; for example we could restrict u to *the class HP_m of harmonic polynomials of degree at most m*. For this class a "quasi–maximum principle" like (3.1) sometimes holds and this gives rise to a corresponding "quasi–balayage measure" ν_y which can still be useful in proving estimates involving the Laplace operator in Ω such as we did in the illustrative example in Section 2.

3.2 Quasi–Maximum Principle and Quasi–Balayage. This section is elementary functional analysis, and could be done in the rather abstract setting of the continuous functions on a compact Hausdorff space.

However, since we are only interested in applications it suffices to carry out the details with regard to the space $C(\overline{B})$ where *B is an open ball in Euclidean n–space*.

Let S be a vector subspace of $C(\overline{B})$ and *suppose we have a closed set $K \subset \overline{B}$ and a positive constant A such that*

$$\|u\|_\infty \;\leq\; A \cdot \max_{x \in K} |u(x)|, \qquad \text{for all } u \in S, \tag{3.2}$$

where $\|\cdot\|_\infty$ denotes the sup–norm on $C(\overline{B})$. We call (3.2) a *quasi–maximum principle* for the class S. Let us give some examples:

(a) S is P_m, the class of polynomials in $x = (x_1, \ldots, x_n)$ of degree $\leq m$, and K is any closed subset of $\overline{B}$ with positive n–dimensional Lebesgue measure. In this case, (3.2) was established by Johnson and Wallin [JW]. One can also give some information about how A in (3.2) depends on m, n, the measure of K, and the radius of B, but we shall not enter into this here.

(b) S is HP_m, the set of harmonic polynomials of degree $\leq m$, and K is any closed subset of ∂B having positive $(n-1)$–dimensional Hausdorff measure. The correctness of (3.2) in this case is easy to establish (D. Khavinson and H. S. Shapiro, unpublished) although I don't know a reference for it.

In a forthcoming paper by Hayman, Karp and the present author a general principle is presented from which (a), (b) and many other special cases follow. Now, by exactly the reasoning used in the proof of Proposition 2.1, one proves

Proposition 3.1. *If (3.2) holds, then to every $\mu \in M(\overline{B})$ there corresponds $\nu \in M(K)$ such that*

$$\text{(i)} \qquad \| \nu \| \;\; \leq \;\; A \cdot \| \mu \|,$$

$$\text{(ii)} \qquad \int u \, d\mu \;\; = \;\; \int u \, d\nu \qquad \text{for all } u \in S.$$

We call ν the *quasi–balayage measure* belonging to μ (relative to the data S, B, K). In general ν is not unique.

An important advantage of this generalized balayage notion is, one also can "sweep" measures to (Schwartz) *distributions* supported on K; this allows K to be a much "thinner" set than is required for (3.2).

Suppose, for example, S is a vector subspace of $C^1(\overline{B})$ and, in place of (3.2) we have for $u \in S$

$$\| u \|_\infty \;\; \leq \;\; A \cdot \left[\; \max_{x \in K} \; | u(x) | \;\; + \;\; \sum_{i=1}^{n} \max_{x \in K} \; | \partial_i u(x) | \; \right], \qquad (3.3)$$

where ∂_i denotes $\partial / \partial x_i$. Then, by the usual Hahn–Banach argument we deduce

Proposition 3.2. *If (3.3) holds, to every $\mu \in M(\overline{B})$ there correspond $\{\nu_i\} \subset M(K)$ for $i = 0, 1, \ldots, n$, such that*

$$\text{(i)} \qquad \sum_{i=0}^{n} \| \nu_i \|_{M(K)} \;\; \leq \;\; A \cdot \| \mu \|_{M(\overline{B})},$$

$$\text{(ii)} \qquad \int u \, d\mu \;\; = \;\; \int u \, d\nu_0 \;\; + \;\; \sum_{i=1}^{n} \int (\partial_i u) \, d\nu_i \qquad (u \in S).$$

A significant case where (3.3) holds is that where $S = HP_m$ and K has positive Newtonian capacity. A detailed treatment of this situation and its consequences will be given in [HKS].

4   Estimates for Potentials of Measures Annihilating HP_m

4.1. Basic Estimates. We shall need estimates for the Newtonian potential U^σ of a bounded measure σ with compact support in $\mathbb{R}^n$ such that

$$\int f \, d\sigma \;=\; 0, \qquad\qquad \text{for all } f \in HP_m. \qquad (4.0)$$

We will also express this by writing "$\sigma \, @ \, HP_m$" ("σ annihilates HP_m") or $\sigma \in (HP_m)^@$.

Proposition 4.1. *For* $\sigma \in M\big(\overline{B(0;1)}\big) \cap (HP_m)^@$ *we have* :

$$|U^\sigma(x)| \;\leq\; C \cdot |x|^{-n-m+1} \, \|\sigma\|, \qquad |x| \geq 3, \qquad (4.1)$$

where $C = C(m,n)$.

If moreover $m \geq 2$, *then*

$$\|U^\sigma\|_{L^1(\mathbb{R}^n)} \;\leq\; C(m,n) \, \|\sigma\|. \qquad (4.2)$$

Note. Here and in the following C will denote a generic constant (likewise C_1, $C_2, \ldots, c_1, c_2, \ldots$) not necessarily the same at each occurrence. The notation $C(m,n)$ denotes, as usual, a constant that depends only on m, n; etc.

Proof: Throughout the proof the open unit ball $B(0;1)$ is denoted by B. We have first

Lemma 4.2. There is a function $h_x(\cdot)$ which is (for fixed x) in HP_m and satisfies

$$\Big| \, |x - y|^{2-n} - h_x(y) \, \Big| \;\leq\; C(m,n) \cdot |x|^{-n-m+1} \qquad (4.3)$$

for all x, y with $|x| \geq 3$, $|y| \leq 1$.

Proof of Lemma : It is enough to prove (4.3) with h_x replaced by p_x, which is for fixed x a polynomial (not necessarily harmonic!) of degree $\leq m$ with respect to y. Indeed, it is well known that to each $p \in P_m$ there is $h \in HP_m$ with $h = p$ on ∂B; so, if (4.3) holds with h_x replaced by p_x, it also holds when p_x is replaced by this interpolating h_x, initially for $|y| = 1$ and hence (in view of the maximum principle for harmonic functions) for $|y| \leq 1$. Now,

$$|x - y|^{2-n} \;=\; \left(|x - y|^2\right)^{1-\frac{n}{2}} \;=\; \left(|x|^2 - 2\,x \cdot y + |y|^2\right)^{1-\frac{n}{2}}$$

(where $x \cdot y$ denotes the scalar product of x and y)

$$= |x|^{2-n}\left[1 + \frac{|y|^2 - 2\,x \cdot y}{|x|^2}\right]^{1-\frac{n}{2}}.$$

Moreover,

$$\left|\,|y|^2 - 2\,x \cdot y\,\right| \;\leq\; 1 + 2|x| \;\leq\; (7/9)\cdot|x|^2 \quad \text{for } y \in \overline{B},\; |x| \geq 3. \tag{4.4}$$

Hence

$$|x - y|^{2-n} \;=\; |x|^{2-n}\sum_{k=0}^{\infty}\binom{1 - \frac{n}{2}}{k}\left(\frac{|y|^2 - 2\,x \cdot y}{|x|^2}\right)^k$$

with absolute convergence when $|y| \leq 1$, $|x| \geq 3$.

Now define

$$p_x(y) \;=\; |x|^{2-n}\sum_{k=0}^{m}\binom{1 - \frac{n}{2}}{k}\left(\frac{1 - 2\,x \cdot y}{|x|^2}\right)^k.$$

Note that p_x is a polynomial in y of degree $\leq m$ and, for $|y| = 1$,

$$\left|\,|x - y|^{2-n} - p_x(y)\,\right| \;=\; |x|^{2-n}\left|\sum_{k=m+1}^{\infty}\binom{1 - \frac{n}{2}}{k}\left(\frac{1 - 2\,x \cdot y}{|x|^2}\right)^k\right|$$

$$\leq C(m, n)\cdot|x|^{2-n-(m+1)}$$

for $|x| \geq 3$, and (4.3) is established. $\qquad\square$

Proof of Proposition 4.1 : Apart from a constant of normalization,

$$U^{\sigma}(x) \;=\; \int |x - y|^{2-n}\,d\sigma(y) \;=\; \int \left(|x - y|^{2-n} - h_x(y)\right)d\sigma(y)$$

where h_x is as in the lemma. Applying the lemma now yields (4.1). To get (4.2) observe first

$$\int\limits_{|x|\leq 3} |U^\sigma(x)|\, dx \;=\; C\cdot \int\limits_{|x|\leq 3}\left|\int |x-y|^{2-n}\, d\sigma(y)\right|\, dx$$

$$\leq C\cdot \int\limits_{|x|\leq 3}\left(\int |x-y|^{2-n}\, d\sigma^{\#}(y)\right)\, dx$$

(where $\sigma^{\#}$ is the "total variation" measure associated to σ)

$$= C\cdot \int\left(\int\limits_{|x|\leq 3} |x-y|^{2-n}\, dx\right)\, d\sigma^{\#}(y)$$

$$\leq C\cdot \left(\int\limits_{|z|\leq 4} |z|^{2-n}\, dz\right)\cdot \|\sigma\| \;=\; C_1\cdot\|\sigma\|.$$

The corresponding estimate for $\int_{|x|\geq 3} |U^\sigma(x)|\, dx$ follows from (4.1), and we get (4.2). $\qquad\qquad\square$

By very similar estimates (which are left to the reader) one obtains

Proposition 4.3. *Under the same assumptions as in Proposition* 4.1, *we have for $j = 1, 2,\ldots, n$:*

$$|\partial_j U^\sigma(x)| \;\leq\; C(m,n)\cdot\|\sigma\|\cdot|x|^{-m-n}, \qquad |x|\geq 3. \tag{4.5}$$

Moreover, if $m \geq 1$,

$$\|\partial_j U^\sigma\|_{L^1(\mathbb{R}^n)} \;\leq\; C(m,n)\cdot\|\sigma\|. \tag{4.6}$$

Remark. Corresponding estimates to (4.6) for second and higher derivatives of U^σ *do not* follow in this way, since second derivatives of the Newton kernel are not locally integrable.

4.2. Scaling. In applications we have to use estimates like (4.2) scaled to a ball $B(0; R)$ for $R > 0$. It is useful to introduce a notation for *scaled* (dilated) *measures*.

Definition 4.4. For $\sigma \in M(\mathbb{R}^n)$ and $a > 0$ let $\sigma_{(a)}$ be the element of $M(\mathbb{R}^n)$ defined by the relation

$$\int \varphi \, d\sigma_{(a)} \;=\; \int \varphi(ax) \, d\sigma \tag{4.7}$$

required to hold for all $\varphi \in C_0(\mathbb{R}^n)$. Equivalently,

$$\sigma_{(a)}(E) \;=\; \sigma(a^{-1}E) \qquad \text{for all Borel sets } E, \tag{4.8}$$

or

$$\widehat{\sigma}_{(a)}(\xi) \;=\; \widehat{\sigma}(a\xi) \qquad \text{where } \widehat{} \text{ denotes Fourier transformation.} \tag{4.9}$$

Note the important relation $\|\sigma_{(a)}\|_M = \|\sigma\|_M$.

Remark. In case σ is absolutely continuous with respect to Lebesgue measure, so that $d\sigma = f \, dx$ for some $f \in L^1(\mathbb{R}^n)$, it is easy to see that $d\sigma_{(a)} = f_{(a)} dx$ where

$$f_{(a)}(x) \;:=\; a^{-n} f(a^{-1}x). \tag{4.10}$$

Proposition 4.5. *If $\sigma \in M(\mathbb{R}^n)$ has compact support and U, V denote respectively the Newtonian potentials of σ, $\sigma_{(a)}$, then*

$$V(x) \;=\; a^{2-n} U(a^{-1}x) \tag{4.11}$$

Proof: $\quad \Delta U = \sigma \quad$ is equivalent to $\quad -|\xi|^2 \widehat{U}(\xi) = \widehat{\sigma}(\xi), \quad$ and

$$\Delta V = \sigma_{(a)} \quad \text{is equivalent to} \quad -|\xi|^2 \widehat{V}(\xi) = \widehat{\sigma}_{(a)}(\xi) = \widehat{\sigma}(a\xi)$$

by Fourier transformation. Replacing ξ by $a\xi$ in the first relation and comparing with the second yields

$$a^2 \, \widehat{U}(a\xi) \;=\; \widehat{V}(\xi)$$

which is equivalent to (4.11). $\qquad\qquad\qquad\qquad\qquad\qquad\qquad\qquad$ $\square$

5   A Priori Inequalities

5.1. Generalities. In this section we present some preliminary results that are well known in principle, although perhaps not in the form given here. We restrict ourselves to estimates in (weighted) L^1, L^∞ and measure norms (only for the Laplace operator) as well as the (dual) existence theorems they imply.

Let Ω denote an open connected set in $\mathbb{R}^n$. By $C_c^\infty(\Omega)$ we denote the (L. Schwartz) class of test functions, infinitely differentiable and with compact support in Ω. Let a denote any positive function in $C(\Omega)$. We shall work with the following spaces:

By $L^\infty(\Omega; a)$ we denote the Banach space of measurable functions f on Ω for which

$$\| f \|_{\infty;a} \quad := \quad \operatorname*{ess\,sup}_{x \in \Omega} \left[a(x)\,|f(x)| \right] \quad = \quad \| af \|_\infty \tag{5.1}$$

is finite. By $L^1(\Omega; a)$ we denote the Banach space of measurable functions g on Ω for which the norm

$$\| g \|_{1;a} \quad := \quad \int_\Omega a(x) \cdot |g(x)|\, dx \quad = \quad \| a \cdot g \|_1 \tag{5.2}$$

is finite; and by $M(\Omega; a)$ we denote the Banach space of measures μ on Ω for which the norm

$$\| \mu \|_{M;a} \quad := \quad \int_\Omega a(x)\, d\mu^{\#}(x) \tag{5.3}$$

is finite, $\mu^{\#}$ *denoting the "total variation measure" associated to* μ.

In a natural way, $L^\infty(\Omega, a^{-1})$ is the dual space of $L^1(\Omega; a)$. The duality is given by the bilinear form

$$\langle f, g \rangle \mapsto \int_\Omega f(x) g(x)\, dx; \qquad f \in L^\infty(\Omega; a^{-1}),\ g \in L^1(\Omega; a). \tag{5.4}$$

Clearly

$$| \langle f, g \rangle | \quad \le \quad \| f \|_{\infty;a^{-1}} \cdot \| g \|_{1;a}. \tag{5.5}$$

We also denote by $C_0(\Omega; a)$ the closure of the continuous compactly supported functions in $L^\infty(\Omega; a)$. Its dual is $M(\Omega; a^{-1})$.

Proposition 5.1. *Let a, b denote positive continuous functions on Ω, and assume*

$$\| \varphi \|_{\infty;a} \quad \le \quad \| \Delta\varphi \|_{\infty;b} \qquad\qquad \textit{for all } \varphi \in C_c^\infty(\Omega) \tag{5.6}$$

holds. Then, for each measure $\mu \in M(\Omega; a^{-1})$ *there exists a measure (not unique)* $\nu \in M(\Omega; b^{-1})$ *such that*

$$\int (\Delta\varphi)\, d\nu \;\; = \;\; \int \varphi\, d\mu, \qquad\qquad \textit{for all } \varphi \in C_c^\infty(\Omega). \qquad (5.7)$$

Moreover,

$$\| \nu \|_{M; b^{-1}} \;\; \leq \;\; \| \mu \|_{M; a^{-1}} \;\; . \qquad (5.8)$$

Remark. As a distribution ν in Ω, ν satisfies (in view of (5.7)) $\Delta\nu = \mu$ from which it easily follows that locally ν is absolutely continuous with respect to Lebesgue measure, and $d\nu/dx$ is in L_{loc}^p for every $1 \leq p < \dfrac{n}{n-2}$, $n \geq 3$.

Proof: We think of $V := \{\, \Delta\varphi \;:\; \varphi \in C_c^\infty(\Omega)\,\}$ as a vector subspace of $C_0(\Omega; b)$ and look at the map $\Lambda \;:\; V \to \mathbb{R}$ defined by

$$\Lambda(\Delta\varphi) \;\; = \;\; \int \varphi\, d\mu, \qquad\qquad \Delta\varphi \in V. \qquad (5.9)$$

Λ is well defined since if φ_1 and φ_2 are in $C_c^\infty(\Omega)$ with $\Delta\varphi_1 = \Delta\varphi_2$, $\varphi_1 - \varphi_2$ is in $H(\Omega)$ with compact support, and hence 0. From (5.9)

$$| \Lambda(\Delta\varphi) | \;\; = \;\; \left| \int (a\varphi)\, (a^{-1}\, d\mu) \right| \;\; \leq \;\; \| a\varphi \|_\infty \int a^{-1}\, d\mu^{\#}$$

$$\leq \;\; \| b\Delta\varphi \|_\infty \cdot \int a^{-1}\, d\mu^{\#}$$

in view of (5.6). By the Hahn–Banach Theorem Λ extends with preservation of norm from V to $C_0(\Omega; b)$ and by "the usual argument" there exists a representing measure ν for the extended functional, which then satisfies (5.7) and (5.8). $\qquad\qquad\square$

In case $\Omega = \mathbb{R}^n$ we can assert much more.

Proposition 5.2. *Let* a, b *denote positive continuous functions on* $\mathbb{R}^n$, *with* b *bounded. Suppose*

$$\| \varphi \|_{\infty; a} \;\; \leq \;\; \| \Delta\varphi \|_{\infty; b} \qquad\qquad \textit{for all } \varphi \in C_c^\infty(\mathbb{R}^n). \qquad (5.10)$$

Then, to every measurable function f *on* $\mathbb{R}^n$ *with* bf *essentially bounded there is a solution* u *to* $\Delta u = f$ *on* $\mathbb{R}^n$, *satisfying*

$$\| u \|_{\infty; a} \;\; \leq \;\; \| f \|_{\infty; b}. \qquad (5.11)$$

For the proof we require first a lemma.

Lemma 5.3. *Every function in $C_0(\mathbb{R}^n)$, i.e. continuous and tending to zero at ∞, is the uniform limit (on $\mathbb{R}^n$) of a sequence $\{\Delta\varphi_j\}$ where $\{\varphi_j\} \subset C_c^\infty(\mathbb{R}^n)$.*

Proof : It suffices to show that if $\mu \in M(\mathbb{R}^n)$ satisfies

$$\int (\Delta\varphi)\, d\mu \;\; = \;\; 0 \qquad\qquad \text{for all } \varphi \in C_c^\infty(\mathbb{R}^n) \qquad\qquad (5.12)$$

then $\mu = 0$. Now, by "Weyl's Lemma", (5.12) expresses that μ is absolutely continuous with respect to Lebesgue measure dx and $d\mu/dx =: h$ is harmonic on $\mathbb{R}^n$. Since $\| \mu \|_M = \int | h | \, dx$, h is integrable on $\mathbb{R}^n$ and hence identically 0. $\qquad\square$

Remark. Although we shall not need it in this paper we include here for its intrinsic interest the sharper result:

Lemma 5.4. *Every function in $C_0(\mathbb{R}^n)$ vanishing at 0 is the uniform limit (on $\mathbb{R}^n$) of a sequence $\{\Delta\varphi_j\}$ where $\{\varphi_j\} \subset C_c^\infty(\mathbb{R}^n \setminus \{0\})$.*

Proof : It suffices to show, if h is harmonic and integrable on $\mathbb{R}^n \setminus \{0\}$, then $h = 0$, since then any bounded measure on $\mathbb{R}^n$ that annihilates all $\Delta\varphi_j$ must reduce to a point mass at the origin. Let π denote any hyperplane through 0. Since each half–space of π is a null quadrature domain for harmonic functions (see e.g. [KM]), we see that $\int_{\mathbb{R}^n} h \, dx = 0$. Let now y be an arbitrary point of $\mathbb{R}^n \setminus \{0\}$ and let B denote the ball centered at y with radius $| y |$. Since $\mathbb{R}^n \setminus \overline{B}$ is a null quadrature domain, and h is harmonic and integrable over $\mathbb{R}^n \setminus \overline{B}$, the integral of h over $\mathbb{R}^n \setminus \overline{B}$, and consequently $\int_B h \, dx$, is zero. Hence $h(y) = 0$. $\square$

Remark. It is easy to adapt the above argument to show: if $h \in HL^1(\mathbb{R}^n \setminus E)$ where E consists of at most two points, then $h = 0$. This conclusion is no longer true if E consists of three points! On the other hand, $\mathbb{R}^n \setminus E$ is a null quadrature domain if E consists of at most n points.

Proof of Proposition 5.2 : Suppose first f is continuous on $\mathbb{R}^n$ and $b| f |$ is bounded; without loss of generality we may assume

$$b(x)| f(x) | \;\; \leq \;\; 1, \qquad\qquad x \in \mathbb{R}^n. \qquad\qquad (5.13)$$

Now fix $R > 0$ and let ψ_R denote a smooth function on $\mathbb{R}^n$ satisfying $0 \leq \psi_R(x) \leq 1$, $\psi_R(x) = 1$ for $|x| \leq R$ and $\psi_R(x) = 0$ for $|x| \geq 2R$.

By Lemma 5.3 there exists a sequence $\{\varphi_j\} \subset C_c^\infty(\mathbb{R}^n)$ satisfying

$$| \Delta\varphi_j(x) - \psi_R(x)f(x) | \ \leq \ \frac{1}{j} \qquad \text{for } x \in \mathbb{R}^n, \ j = 1, 2, \ldots \ . \quad (5.14)$$

In particular,

$$| \Delta\varphi_j(x) | \ \leq \ b(x)^{-1} + \frac{1}{j} \ \leq \ \left(1 + \frac{B}{j}\right) b(x)^{-1}, \quad (5.15)$$

where $B := \sup_x b(x) < \infty$ by hypothesis. Using assumption (5.10) we deduce

$$\| \varphi_j \|_{\infty;a} \ \leq \ 1 + \frac{B}{j},$$

i.e. $|\varphi_j(x)| \leq \left(1 + \dfrac{B}{j}\right) a(x)^{-1}$. It follows easily that there is a subsequence of $\{\varphi_j\}$ (still denoted $\{\varphi_j\}$) that converges distributionally to a measurable function u_R satisfying $|u_R(x)| \leq a(x)^{-1}$ a.e. on $\mathbb{R}^n$. In view of (5.14) we get (in the distributional sense)

$$\Delta u_R \ = \ \psi_R f.$$

Letting $R \to \infty$ we can extract a subsequence from $\{u_R\}$ which converges distributionally to a measurable function u on $\mathbb{R}^n$ satisfying $|u(x)| \leq a(x)^{-1}$ a.e. and $\Delta u = f$. This concludes the proof when f is continuous.

For the general case when f is merely measurable and satisfies (5.13) we have only to approximate f in a suitable manner by continuous functions, and apply the preceding. The slightly tedious, but straightforward details are omitted. $\square$

5.2. Some a Priori Estimates. We sketch first a classical estimate, not based on quasi–balayage.

Proposition 5.5. *For $\Omega = \mathbb{R}^n$, $n > 2$ and $1 < \lambda < n/2$, (5.6) holds with $a = (1+|x|^2)^{\lambda-1}$, $b = C \cdot (1+|x|^2)^\lambda$ where C is a positive constant depending only on n and λ.*

We only indicate the proof. For $\varphi \in C_c^\infty(\mathbb{R}^n)$, and $y \in \mathbb{R}^n$

$$\varphi(y) \ = \ -c_n \int \Delta\varphi(x) |x - y|^{2-n} dx$$

$$= -c_n \int \left(1 + |x|^2\right)^{\lambda} \Delta\varphi(x) \cdot \left(1 + |x|^2\right)^{-\lambda} |x - y|^{2-n} dx$$

holds true. Hence,

$$|\varphi(y)| \leq \| \left(1 + |x|^2\right)^{\lambda} \Delta\varphi \|_{\infty} \cdot c_n \int \left(1 + |x|^2\right)^{-\lambda} |x - y|^{2-n} dx. \qquad (5.16)$$

The last integral converges since $\lambda > 1$, and is the Newtonian potential of $(1 + |x|^2)^{-\lambda}$. It is fairly straightforward to show this potential is, for large $|y|$, asymptotically a constant times $(1 + |y|^2)^{1-\lambda}$. Substituting this in (5.16) we get

$$|\varphi(y)| \cdot (1 + |y|^2)^{\lambda-1} \leq C \cdot \| \left(1 + |x|^2\right)^{\lambda} \Delta\varphi \|_{\infty}, \qquad y \in \mathbb{R}^n$$

which implies the desired result. $\qquad\qquad\qquad\qquad\qquad\qquad\qquad\qquad\qquad\quad \square$

Remark. The restriction to $\lambda < n/2$ is essential. This was kindly pointed out to the author by Joaquim Ortega–Cerdà, thereby correcting an inaccuracy in an earlier version of the manuscript. There is a "cutoff" phenomenon, the potential of $(1 + |x|^2)^{-\lambda}$ remains of the order $|x|^{2-n}$ when $|x|$ is large, for all $\lambda > n/2$, and is of order $|x|^{2-n} \log |x|$ when $\lambda = n/2$.

This brings us naturally to our first application of quasi–balayage. In the above estimation we needed $\lambda > 1$ to ensure convergence of the integral in (5.16). But, there is great interest in estimates of the same type when $\lambda \leq 1$, indeed especially for λ large *negative*, when the above method fails completely, essentially because "the Newton kernel does not decay fast enough at ∞". We shall now obtain such an estimate using quasi–balayage. The estimate we get is not sharp, but suffices for significant applications.

Proposition 5.6. *For* $\Omega = \{ x \in \mathbb{R}^n : |x| > 1 \}$ *we have, for every* $s > 0$, *and all* $\varphi \in C_c^{\infty}(\Omega)$:

$$\left\| |x|^{-2s-4} \varphi \right\|_{\infty} \leq C(n,s) \cdot \left\| |x|^{-s} \Delta\varphi \right\|_{\infty} .$$

Proof : Let $y \in \Omega$ and write $R = |y|$. Then

$$\varphi(y) = \int \varphi \, d\delta_y = \int \varphi \, d[\delta_y - \nu_y] \qquad (5.17)$$

where ν_y is any measure supported in $\overline{B(0,1)}$. In fact, we let ν_y be a quasi-balayage measure of δ_y to $\overline{B(0,1)}$ relative to HP_m where $m > s + 1$. For

$p \in HP_m$, with $|p(x)| \leq 1$ on $\overline{B(0,1)}$, we have $|p(y)| \leq T_m(R)$ where T_m is the Chebyshev polynomial (of first kind) of degree m, so $|p(y)| \leq cR^m$. Indeed, restricting p to the line joining the origin to y, this is a consequence of a classical extremal proprety of Chebyshev polynomials, see [R, p. 108].

Therefore we can choose ν_y supported in $\overline{B(0,1)}$ (indeed, on the diameter passing through y) with $\|\nu_y\| \leq cR^m$ and $\sigma_y := \delta_y - \nu_y @ HP_m$. Now, (5.17) yields

$$\varphi(y) = \int \varphi \, d\sigma_y = -C \int \varphi \Delta U^{\sigma_y} = -C \int \Delta\varphi(x) \, U^{\sigma_y}(x) dx \qquad (5.18)$$

and defining τ_y by $\sigma_y = (\tau_y)_{(R)}$ so that

$$\|\tau_y\|_M \quad \leq \quad c_1 R^m \qquad\qquad (5.19)$$

we get from (5.18) and (4.11)

$$\varphi(y) \quad = \quad -C \int \Delta\varphi(x) \, U^{(\tau_y)_{(R)}}(x) \, dx \quad = \quad -C \int \Delta\varphi(x) \, R^{2-n} \, U^{\tau_y}(R^{-1}x) \, dx$$

whence, for every $s > 0$ and $m > s+1$

$$|\varphi(y)| \quad \leq \quad C \int_{|x|>1} |\Delta\varphi(x)| \cdot |x|^{-s} \left(|x|^s R^{2-n} \left| U^{\tau_y}(R^{-1}x) \right| \right) dx.$$

Thus,

$$R^{-s-2} |\varphi(y)| \quad \leq \quad C \left\| \, |x|^{-s} \Delta\varphi \right\|_\infty \int_{\mathbb{R}^n} |z|^s \, |U^{\tau_y}(z)| \, dz. \qquad (5.20)$$

Now, τ_y is supported in $\overline{B(0,1)}$ and satisfies (5.19), hence using the estimates in Proposition 4.1 we see that the last integral does not exceed

$$3^s \int_{|z|\leq 3} |U^{\tau_y}(z)| \, dz \quad + \quad C(m,n) \left(\int_{|z|\geq 3} |z|^s |z|^{-n-m+1} dz \right) \|\tau_y\|$$

$$\leq \quad C(m,n,s) \cdot \|\tau_y\| \quad \leq \quad C(m,n,s) \cdot R^m$$

and (5.20) yields

$$\|\,|x|^{-(m+s+2)}\varphi\,\|_\infty \;\leq\; C\cdot\|\,|x|^{-s}\Delta\varphi\,\|_\infty$$

with $C = C(m,n,s)$. This holds for all $s > 0$, and integer $m > s+1$ so we have, finally: for $\varphi \in C_c^\infty(\,\{\,|x| > 1\}\,)$

$$\|\,|x|^{-2s-4}\varphi\,\|_\infty \;\leq\; C(n,s)\cdot\|\,|x|^{-s}\Delta\varphi\,\|_\infty \qquad (5.21)$$

proving Proposition 5.6. $\square$

Remark. There is a vast literature on estimates like (5.21) with various parameters, weight functions etc. A recent paper which indicates the "state of the art" is [AGG]. We emphasize that we claim novelty only for the method used.

6 A Phragmén–Lindelöf Principle for Solutions to Cauchy's Problem

6.1. The "quadratic growth" estimate. We come now to the main results of the paper. In this section Ω denotes an open (usually unbounded) connected subset of $\mathbb{R}^n$, which shall be subjected to a geometric condition:

Definition 6.1. The domain Ω has *rich complement* (denoted $\Omega \in [RC]$) if

$$\liminf_{R\to\infty}\;\frac{|\,B(0;R)\setminus\Omega\,|}{|\,B(0;R)\,|} \;>\; 0. \qquad (6.1)$$

For example, if the complement of Ω contains a cone with interior points (even a "twisted" one) then $\Omega \in [RC]$.

Theorem 6.2. *Suppose we have a distribution u on $\mathbb{R}^n$, and a domain $\Omega \in [RC]$ such that the following hold :*

$$\text{(i)} \qquad \Delta u =: f \in L^\infty(\mathbb{R}^n) \qquad\qquad (\textit{this implies that } u \in C^1(\mathbb{R}^n)\,),$$

$$\text{(ii)} \qquad u(x) = 0, \qquad\qquad x \in \mathbb{R}^n \setminus \overline{\Omega},$$

$$\text{(iii)} \qquad |\,u(x)\,| \leq C(1+|x|\,)^N \qquad\qquad \textit{for } x \in \mathbb{R}^n \textit{ and positive } C \textit{ and } N.$$

Then

$$| u(x) | \leq C\,(1 + | x |)^2 \cdot \| \Delta u \|_\infty \,, \tag{6.2}$$

$$|(\partial_j u)(x) | \leq C\,(1 + | x |) \cdot \| \Delta u \|_\infty \qquad for\ j = 1, 2, \ldots, n \tag{6.3}$$

hold for $x \in \mathbb{R}^n$ *and constants* C.

Remarks. This can be seen as a Phragmén–Lindelöf principle for Cauchy's problem. Indeed, suppose Γ is a smooth hypersurface in $\mathbb{R}^n$ separating it into two components, one of which is Ω, and $\Omega \in [RC]$. Suppose $g \in L^\infty(\Omega)$ and the Cauchy problem

$$\Delta v = g \qquad in\ \Omega, \tag{6.4}$$

$$v\ \text{ and }\ \mathrm{grad}\,v \qquad \text{vanish on } \Gamma \tag{6.5}$$

has a solution v in Ω and $v(x) = O(\,| x |^N\,)$ in Ω, for large $| x |$. Then defining f, u by

$$f = g \ \text{ on }\ \Omega, \quad f = 0 \ \text{ on }\ \mathbb{R}^n \setminus \Omega,$$

$$u = v \ \text{ on }\ \Omega, \quad u = 0 \ \text{ on }\ \mathbb{R}^n \setminus \Omega$$

it is easy to check that u satisfies the hypotheses of the theorem (the "vanishing Cauchy data" condition (6.5) is crucial, since that is what guarantees that the distribution Δu is an element of $L^\infty(\mathbb{R}^n)$; mere vanishing of v on Γ would not suffice). Of course, the existence of a *global* solution v to the system (6.4), (6.5) even without any growth requirement is an exceptional situation, but one which is encountered in many problems (for one such, see Section 6.2 below). We also remark that neither condition (iii), nor the hypothesis $\Omega \in [RC]$ may be omitted in Theorem 6.2 (see $[S3]$ for some remarks in this direction).

However, the hypothesis $[RC]$ may be weakened, see Section 7 below.

In order to prove the theorem, we require a lemma to justify "integration by parts" on $\mathbb{R}^n$:

Lemma 6.3. *Suppose* $\Delta u \in L^\infty(\mathbb{R}^n)$, *and* $\Delta v =: \sigma$ *is a bounded measure with compact support. Suppose further that* uv, $v\Delta u$ *and* $u\,(\partial_j v)$ *are in* $L^1(\mathbb{R}^n)$ *for* $j = 1, 2, \ldots, n$. *Then*

$$\int u\,d\sigma \ - \ \int_{\mathbb{R}^n} u\Delta v\,dx \ = \ \int_{\mathbb{R}^n} v\Delta u\,dx. \tag{6.6}$$

Proof: Let $\theta \in C^\infty(\mathbb{R}^+)$ satisfy

$$\theta(t) = 1, \qquad\qquad 0 \le t \le 1,$$
$$\theta(t) = 0, \qquad\qquad t \ge 2,$$
$$0 \le \theta(t) \le 1, \qquad\qquad \text{for all } t \in \mathbb{R}^n$$

and let ψ_R denote the "cutoff" function defined by $\psi_R(x) = \theta(R^{-1}|x|)$, where $x \in \mathbb{R}^n$ and $R > 0$. Then

$$\int (\Delta u) \cdot v \, dx \;=\; \lim_{R\to\infty} \int (\Delta u) \cdot \psi_R \, v \, dx \;=\; \lim_{R\to\infty} \int u \cdot \Delta(\psi_R \cdot v) \, dx \qquad (6.7)$$

Now,

$$\Delta(\psi_R \, v) = (\Delta \psi_R) \cdot v + 2\,(\text{grad } \psi_R) \cdot (\text{grad } v) + \psi_R \Delta v.$$

Substituting this into the last integral in (6.7), and noting that for large R

$$\int u \, \psi_R \Delta v \, dx \;=\; \int u \, d\sigma,$$

we see that to prove (6.6) it suffices to show

$$\lim_{R\to\infty} \int u \, (\text{grad } \psi_R) \cdot (\text{grad } v) \, dx \;=\; 0 \qquad\qquad (6.8)$$

and

$$\lim_{R\to\infty} \int u(\Delta \psi_R) \, v \, dx \;=\; 0. \qquad\qquad (6.9)$$

Since $|\text{grad } \psi_R| \le C R^{-1}$, $|\Delta \psi_R| \le C R^{-2}$ it is clear that (6.8) and (6.9) are implied by our hypotheses. $\qquad\square$

Proof of Theorem 6.2 : Fix $y \in \mathbb{R}^n$ and look at the ball $B(0; R)$ with $R = |y|$. If R is sufficiently large (which we may assume), then

$$\frac{|B(0; R) \setminus \Omega|}{|B(0; R)|} \;\ge\; c > 0.$$

Consequently, given m, there is a constant $C = C(m, n, c)$ such that every $p \in HP_m$ satisfies the estimate

$$\max_{x \in \overline{B(0;R)}} |p(x)| \;\le\; C \cdot \max_{x \in \overline{B(0;R)} \setminus \Omega} |p(x)|.$$

(It is important that C is independent of R.) Hence δ_y admits a quasi–balayage measure ν_y onto $\overline{B(0; R)} \setminus \Omega$, relative to HP_m, satisfying $\|\nu_y\| \le C$. We have

$$u(y) \;=\; \int u \, d\delta_y \;=\; \int u \,(d\delta_y - d\nu_y) \;=\; \int u \, d\sigma_y$$

where $\sigma_y := \delta_y - \nu_y @ H P_m$, and $\| \sigma_y \| \le 1 + C = C_1$.

Thus,

$$u(y) \;=\; - \int u \Delta U^{\sigma_y} \;=\; - \int (\Delta u)(x) U^{\sigma_y}(x)\, dx. \qquad (6.10)$$

The last equality holds in view of Lemma 6.3 (with $v = U^{\sigma_y}$) provided we take m sufficiently large so that (in view of hypothesis (iii) of the theorem, and Proposition 4.1) the hypotheses of the lemma hold.

From (6.10)

$$| u(y) | \;\le\; \| \Delta u \|_\infty \, \| U^{\sigma_y} \|_{L^1(\mathbb{R}^n)} . \qquad (6.11)$$

Now, define the measure τ_y by $(\tau_y)_{(R)} = \sigma_y$. Then $\operatorname{supp} \tau_y \subset \overline{B(0;1)}$ and $\| \tau_y \| \le C_1$. By Proposition 4.5,

$$U^{\sigma_y}(x) \;=\; R^{2-n} U^{\tau_y}(R^{-1} x),$$

so

$$\int | U^{\sigma_y}(x)| \, dx \;=\; \int R^{2-n} | U^{\tau_y}(R^{-1} x) | \, dx \;=\; R^2 \int | U^{\tau_y}(z) | \, dz$$

$$\le \; C(m,n) R^2 \| \tau_y \|$$

by (4.2) and, recalling that $R = | y |$, this implies (6.2).

The proof of (6.3) is along similar lines, based on Proposition 4.3, and we shall not present the details here.

Corollary 6.4. *If $\varphi \in C_c^\infty(\Omega)$ with $\Omega \in [RC]$, then for all x*

$$| \varphi(x) | \;\le\; C (1 + | x |)^2 \, \| \Delta\varphi \|_\infty, \qquad (6.12)$$

$$|(\partial_j \varphi)(x)| \;\le\; C (1 + | x |) \, \| \Delta\varphi \|_\infty . \qquad (6.13)$$

6.2. Analogous Estimate at a Finite Boundary Point. If x^0 is any finite boundary point of Ω one can define the notion "Ω has rich complement near x^0", denoted $\Omega \in [RC; x^0]$, analogous to that in Definition 6.1 by the requirement

$$\liminf_{r \to 0} \ \frac{\mid B(x^0; r) \setminus \Omega \mid}{\mid B(x^0; r) \mid} \ > \ 0. \tag{6.14}$$

If we assume this, and that u has the properties assumed in Theorem 6.2, then an analogous reasoning shows

$$\mid u(x) \mid \ \le \ C \mid x - x^0 \mid^2, \tag{6.15}$$

$$\mid \operatorname{grad} u(x) \mid \ \le \ C \mid x - x^0 \mid. \tag{6.16}$$

These estimates are useful in studying the local regularity of free boundaries, see [KM, KS].

6.3. An Approximation Problem.

In this section, we study the following problem: We are given a domain $\Omega \subset \mathbb{R}^n$ and $h \in HL^1(\Omega)$. We ask whether h can be approximated arbitrarily closely in the norm of $L^1(\Omega)$ by some $u \in HL^1(\Omega)$ which moreover tends rapidly to zero near some given boundary point $x^0 \in \partial\Omega$. This type of approximation problem has come up in the study of quadrature domains. Here we will only consider the problem under the assumptions: Ω *is connected*, $x^0 = \infty$ *belongs to* $\partial\Omega$ *and* $\partial\Omega$ *is, near each of its (finite) points, a smooth hypersurface.*

Theorem 6.5. *Suppose $\Omega \in [RC]$. Then, for every $k > 0$ the set*

$$H_k \ := \ \{ u \in HL^1(\Omega) \ : \ \mid u(x) \mid = O(\mid x \mid^{-k}) \quad as \mid x \mid \to \infty \}$$

is dense in $HL^1(\Omega)$.

Remark. The hypothesis $\Omega \in [RC]$ cannot be omitted. One can show that if Ω is the subset $\{x \in \mathbb{R}^3 \ : \ x_1^2 + x_2^2 > 1\}$ of $\mathbb{R}^3$ (complement of a cylinder) the conclusion of the theorem is false. Details of this will be given in [HKS].

Proof of Theorem 6.5 : It suffices to show that if $f \in L^\infty(\Omega)$ and

$$\int_\Omega u \, f \, dx \ = \ 0 \qquad \qquad \text{for all } u \in H_k \tag{6.17}$$

then

$$\int_\Omega h \, f \, dx \ = \ 0 \qquad \qquad \text{for all } h \in HL^1(\Omega). \tag{6.18}$$

So, suppose f satisfies (6.17). If σ denotes any measure with compact support in $\mathbb{R}^n \backslash \overline{\Omega}$ annihilating HP_m, where $m+n-1 \geq k$, then $U^\sigma \in H_k$, by Proposition 4.1, hence

$$\int_\Omega U^\sigma f \, dx \;=\; 0.$$

Let now $\widetilde{f}$ denote the function equal to f on Ω, and 0 elsewhere. There exists a solution v to $\Delta v = \widetilde{f}$ on $\mathbb{R}^n$ which satisfies $v(x) = O(\,|x|^r\,)$ at ∞ for some r (in fact, $r = 3$ suffices, by results of Karp [K]). Hence

$$0 \;=\; \int_{\mathbb{R}^n} U^\sigma(x)\,\widetilde{f}(x)\,dx \;=\; \int_{\mathbb{R}^n} U^\sigma(x)\,\Delta v(x)\,dx\,.$$

By choosing m large enough, say $m = m_0$, U^σ decays rapidly enough at ∞ to justify transforming the last integral into

$$\int_{\mathbb{R}^n} v(x)\,\Delta U^\sigma(x)\,dx \;=\; -\int v\,d\sigma\,.$$

We have thus: $\displaystyle\int v\,d\sigma = 0$ whenever σ is a compactly supported measure on $\mathbb{R}^n \setminus \overline{\Omega}$ annihilating HP_{m_0}.

From this, and linear algebra, it follows that: *There exists $p \in HP_{m_0}$ such that $v = p$ on $\mathbb{R}^n \setminus \overline{\Omega}$.* Hence $V := v - p$ satisfies $\Delta V = \widetilde{f}$. Now V and $\widetilde{f}$ satisfy the hypotheses imposed by Theorem 6.2 (where the analogous functions were called u and f). We conclude that

$$|V(x)| \;\leq\; C(1 + |x|)^2, \tag{6.19}$$

$$|\operatorname{grad} V(x)| \;\leq\; C(1 + |x|). \tag{6.20}$$

To complete the proof of the theorem, we have to prove (6.18), i.e. $\displaystyle\int_\Omega h\Delta V = 0$, where h is any function in $HL^1(\Omega)$. In other words, we have to justify the "partial integration" step

$$\int_\Omega h\Delta V \;=\; \int_\Omega V\Delta h\,.$$

This is of course not a tautology, but its proof is straightforward with the use of the standard cut–off function and the estimates (6.19), (6.20). Since this step resembles the proof we gave for Lemma 6.3 (and moreover is quite standard in the approximation literature) we omit the details. $\square$

6.4. Measures that Annihilate Harmonic Functions. Implicit in the preceding proof is the following result, which is useful in the study of unbounded quadrature domains (see $[S1]$ for background).

Theorem 6.6. *Let $\Omega \subset \mathbb{R}^n$ be an open connected set with rich complement such that $\mathbb{R}^n \setminus \overline{\Omega}$ is non–empty. Suppose μ is a measure on Ω such that*

$$\int (1 + |x|)^{-r} \, d\mu^{\#}(x) \quad < \quad \infty \tag{6.21}$$

for some r with $0 < r < n - 2$ and for some (arbitrarily large) $k > 0$

$$\int h \, d\mu \quad = \quad 0 \tag{6.22}$$

for all harmonic functions h on Ω that are $O(\,|x|^{-k}\,)$ as $|x| \to \infty$. Then there is a locally integrable function u on $\mathbb{R}^n$ such that $u(x)(1 + |x|)^{-r-2}$ is integrable, $\Delta u = \mu$ and $u(x) = 0$ on $\mathbb{R}^n \setminus \overline{\Omega}$.

Proof : By combining Propositions 5.5 and 5.1 we see that there exists a measure ν on Ω satisfying

$$\int (1 + |x|)^{-r-2} \, d\nu^{\#}(x) \quad < \quad \infty \tag{6.23}$$

and $\Delta \nu = \mu$. The last equality implies ν is absolutely continuous with respect to Lebesgue measure dx, so we can write $d\nu = u \, dx$, and then $u \in L^1_{\mathrm{loc}}(\mathbb{R}^n)$, $\Delta u = \mu$ and

$$\int (1 + |x|)^{-r-2} \, |u(x)| \, dx \quad < \quad \infty \, .$$

The remaining assertions follow exactly as in the preceding Theorem. $\square$

6.5. Another Kind of Phragmén–Lindelöf Theorem.

By techniques similar to those used in proving Theorem 6.2 we can prove

Theorem 6.7. *Suppose* $\Delta u \in L^\infty(\mathbb{R}^n)$, *moreover* u *is of polynomial growth* (*i.e.* $O(|x|^k)$ *at* ∞ *for some* k) *and* $u = 0$ *on each of two parallel hyperplanes* (*say* $\{x_n = 0\}$ *and* $\{x_n = 1\}$). *Then* u *is bounded on each "slab"* $\{|x_n| \le A\}$ *for* $A > 0$.

Remarks. The bound depends on A in a manner which can be specified, but we won't pursue that here. We will do so in [HKS] where also generalizations will be given of Theorem 6.7 to cases where the pair of hyperplanes is replaced by other algebraic sets.

Proof of Theorem 6.7 : Let y be any point of the slab $S_A := \{|x_n| \le A\}$. Let K denote the intersection of the ball $B(y; 2A)$ with the union of the hyperplanes $\{x_n = 0\}$, $\{x_n = 1\}$. (We assume $A > 1$, so these intersections are nonempty.) Now, for each m there is a constant $C = C(m, n, A)$ such that every $p \in HP_m$ bounded by 1 on K is bounded by C on $B(y; 2A)$. (This follows from the fact that any $p \in HP_m$ vanishing on each of two parallel hyperplanes vanishes identically; for the detailed argument see [HKS]).

Consequently δ_y admits (with respect to HP_m) quasi–balayage measure $\nu_y \in M(K)$ with $\|\nu_y\| \le C$. Then (since u vanishes on supp ν_y)

$$u(y) \;=\; \int u\, d\delta_y \;=\; \int u\,(d\delta_y - d\nu_y) \;=\; \int u\, d\sigma_y \;=\; -\int u \Delta U^{\sigma_y}$$

where $\sigma_y = \delta_y - \nu_y \,@\, HP_m$. By an argument used earlier, U^{σ_y} decays so rapidly (if m is chosen large enough) that the last integral can be transformed to $\int U^{\sigma_y}(x)(\Delta u)(x)\, dx$ and hence

$$|u(y)| \;\le\; \|\Delta u\|_\infty \cdot \int_{\mathbb{R}^n} |U^{\sigma_y}(x)|\, dx.$$

The last integral is bounded by $C(m, n, A)$ and the proof is finished. $\square$

Remark. The hypothesis that u has polynomial growth is essential. In case $n = 2$, $u(x) := e^{x_1} \sin x_2$ satisfies $\Delta u = 0$ and vanishes on the hyperplanes $\{x_2 = 0\}$, $\{x_2 = \pi\}$ but $u(x_1, \frac{\pi}{2})$ is not bounded.

7 Extension of Some of the Preceding Results

In the preceding section, the concept of a domain having "rich complement" played an important role. Ultimately this was based on the fact that an element p of HP_m bounded by one on a subset K of the unit ball B having positive n–dimensional Lebesgue measure $\mid K \mid$, is bounded on B by a constant $C(m, n, \mid K \mid)$. But, it is not essential for such an estimate that $\mid K \mid$ be positive. For example, if $K \subset \partial B$ and its $(n - 1)$–dimensional measure is positive, we again get an estimate for p on B of the desired type. Thus, we could formulate a set of alternative theorems in which "rich complement" is measured with respect to *spheres* rather than balls, and $(n - 1)$–dimensional measure.

Continuing in this vein, one can (as remarked earlier) also get uniform estimates for $p \in HP_m$ on B if we have bounds for it *and its first derivatives* on a set $K \subset B$ of *positive Newtonian capacity.*

Since such sets may be essentially "thinner" than those with positive $(n - 1)$–dimensional measure, we get sharper results. The notion of "rich complement" gets replaced by the condition

$$\limsup_{R \to \infty} \frac{\operatorname{cap}\,(B(0; R) \setminus \Omega)}{\operatorname{cap}\, B(0; R)} > 0.$$

Interestingly, this condition has already appeared in several potential–theoretic studies with different goals than the present one (in some terminology, the complement of Ω would then be called, following [Le], "uniformly fat" at ∞; it turns out that this is an essentially stronger requirement than the celebrated Wiener criterion of regularity). This will be pursued further in [HKS].

References

[AGG]   C. Amrouche, V. Girault, J. Giroire: *Weighted Sobolev spaces for Laplace's equation in* $\mathbb{R}^n$, J. Math. Pures Appl. **73** (1994), 579–606.

[HKS]   W. K. Hayman, L. Karp, H. S. Shapiro: *Newtonian capacity and quasi–balayage (preliminary title)*, in preparation.

[JW]   A. Jonsson and H. Wallin: *Local polynomial approximation and Lipschitz type conditions on general closed sets*, Univ. of Umeå Math. Dept., Research Report # **1**, 1980.

[K]   L. Karp: *Generalized Newtonian potential and its applications*, J. Math. Anal. Appl. **174** (1993), 480–497.

[KM]   L. Karp, A. Margulis: *Newtonian potential theory for unbounded sources and applications to free boundary problems*, J. Analyse Math., to appear.

[KS]   L. Karp, H. Shahgholian: *Regularity of a free boundary problem*, Research Reports Math. # 2 (1996), Dept. of Math., Stockholm University.

[L]   N. S. Landkof: *Foundations of Modern Potential Theory*, Springer, Berlin–Heidelberg–New York 1972.

[Le]   J. L. Lewis: *Uniformly fat sets*, Trans. Amer. Math. Soc. **308** (1988), 177–196.

[R]   T. J. Rivlin: *Chebyshev Polynomials: From Approximation Theory to Algebra and Number Theory*, Wiley, New York–Chichester 1990.

[S1]   H. S. Shapiro: *Domains allowing exact quadrature identities for harmonic functions – an approach based on PDE*, in *Anniversary Volume on Approximation Theory and Functional Analysis*, P. L. Butzer et al. (eds.) Internat. Ser. Numer. Math. **65**, 335–354, Birkhäuser, Basel–Boston–Stuttgart 1984.

[S2]   H. S. Shapiro: *A weighted L^1 estimate for the Laplace operator*, in *Recent Advances in Fourier Analysis and its Applications*, J. S. Byrnes and J. F. Byrnes (eds.) NATO ASI Series **C 315**, 563–577, Kluwer, Dordrecht 1990.

[S3]   H. S. Shapiro: *Global geometric aspects of Cauchy's problem for the Laplace operator,* in *Geometrical and Algebraical Aspects in Several Complex Variables,* C. A. Berenstein and D. C. Struppa (eds.) Proc. Conf. Cetraro, Italy, 309–324, EditEl Publ. 1991.

Address:

HAROLD S. SHAPIRO
Mathematics Department
Royal Institute of Technology
S–100 44 Stockholm
Sweden

Quasi–Balayage and A Priori Estimates for the Laplace Operator II

H. S. Shapiro

The preceding article "Quasi–Balayage and A Priori Estimates for the Laplace Operator I", these Proceedings, pp. 203–230, is a (somewhat augmented) account of a talk presented by the author at the conference "Multivariate Approximation", at Haus Bommerholz, Dortmund in October 1996. The following sections extend these results. In § 8 we present a new proof, using quasi-balayage, of the "$|x|^2 \log |x|$ theorem" of L. Karp in [K]. This argument is in some points similar to that we gave in [S2] but is more elementary, avoiding the singular integral and embedding machinery used there. *En route*, some new results of independent interest are uncovered. In § 9 we include some remarks on a result of Karp and Margulis [KM] concerning the growth of the (generalized) Newtonian potential of a bounded function in a cylinder. Here quasi-balayage is not used, but we include this material since our treatment of it is in spirit close to techniques used in the present paper. We give also an application to a result of [GHR], in § 9. In § 10 we present an alternative approach to quasi-balayage based on Fourier analysis. This method is effective in dealing with generalizations such as balayage of a measure to a distribution of order one, which is important for applications.

8   A New Treatment of Karp's $|x|^2 \log |x|$ Theorem

8.1. The Theorem. In [K], Karp proved

Theorem 8.1. *For $f \in L^\infty(\mathbb{R}^n)$, there is a solution u to $\Delta u = f$ on $\mathbb{R}^n$ satisfying*

$$|u(x)| \le C\, W(x) \|f\|_\infty, \qquad\qquad x \in \mathbb{R}^n, \qquad\qquad (8.1)$$

where

$$W(x) := (1 + |x|)^2 \log(2 + |x|) \qquad\qquad (8.2)$$

and $C = C(n)$ is a constant.

The piquant logarithm in (8.2) is essential, indeed it is easy to check that if h is any nontrivial homogeneous harmonic polynomial in $x = (x_1, x_2, \ldots, x_n)$ of degree 2, then $v(x) := h(x) \log(2 + |x|^2)$ satisfies: Δv *is bounded in* $\mathbb{R}^n$, *but no solution u to $\Delta u = \Delta v$ is $o(W(x))$ as $|x| \to \infty$.*

Multivariate Approximation: Recent Trends and Results; W. Haußmann, K. Jetter and M. Reimer (eds.)
Mathematical Research, Vol. 101, pp. 231–254, ISBN 3–05–501770–6
© Akademie–Verlag, Berlin 1997

In fact, our methods yield a bit more than Theorem 8.1, namely

Theorem 8.2. *For $f \in L^\infty(\mathbb{R}^n)$ there is a solution u to $\Delta u = f$ on $\mathbb{R}^n$ satisfying (8.1) and also each of the following estimates:*

$$|\mathrm{grad}\, u(x)| \leq C(1+|x|)\log(2+|x|)\,\|f\|_\infty \quad , \quad x \in \mathbb{R}^n \tag{8.3}$$

$$|(\partial_j u)(x) - 2\partial_j u(x+y) + \partial_j u(x+2y)| \leq C\,|y|\,\|f\|_\infty \tag{8.4}$$

for $j = 1, 2, \ldots, n$ and all $x, y \in \mathbb{R}^n$.

$$\|\partial^\alpha u\|_{BMO(\mathbb{R}^n)} \leq C\,\|f\|_\infty \tag{8.5}$$

for all multi-indices α with $|\alpha| = 2$. In these estimates C denotes (various) constants depending only on n.

Remark. It is interesting to note that in (8.4) C is independent of y. For "small" y (8.4) expresses a uniform "Zygmund" smoothness condition and is well known. However, it is remarkable that (8.4) also holds for *large* $|y|$, and that is, in a sense the cause of the estimate (8.3). In [S2] we proved the key estimate (8.4) using Fourier analysis, here it will be based on Proposition 4.1.

8.2. Preparatory Results. The proof of Theorem 8.2 will be given after some preliminary results, which we think have some intrinsic interest.

Definition 8.3. *Given a vector space S of continuous functions on (say) $\mathbb{R}^n$, a* set of uniqueness *for S is a set $E \subset \mathbb{R}^n$ such that $f \in S$ and $f|_E = 0$ imply $f = 0$. We say E is minimal if no proper subset of E is a set of uniqueness.*

Lemma 8.4. *Let E be a finite subset of $\mathbb{R}^n$ which is a minimal set of uniqueness for S. Then E carries no nontrivial measure annihilating S.*

Proof: Denote $E = \{x^j\}_{j=1,2,\ldots,k}$ and suppose σ were a non-null measure carried by E and $\perp S$. Then $\sigma = \sum_{j=1}^{k} c_j\,\delta_{x^j}$ where, without loss of generality $c_k \neq 0$. For every $p \in S$ we have $\sum_{j=1}^{k} c_j\,p(x^j) = 0$. Hence, any $p \in S$ vanishing on $\{x^1, \ldots, x^{k-1}\}$ vanishes also on x^k and hence identically. This contradicts the supposed minimality of E. $\qquad\square$

Lemma 8.5. *For each m, there exist finite minimal sets of uniqueness for HP_m.*

Proof: It is obvious that there exist finite sets of uniqueness for HP_m (even for P_m). If $k = k(m)$ denotes the smallest cardinality of all such sets, then any uniqueness set of cardinality k is minimal. $\qquad\square$

Theorem 8.6. *Let σ denote a bounded measure with compact support on IR^n. If $U^\sigma \in L^1(\mathrm{IR}^n)$, then $\sigma @ HP_2$.*

Remark. This is a partial converse of Theorem 4.1. A full converse can be proved by the same method.

Proof : Suppose $U^\sigma \in L^1(\mathrm{IR}^n)$, and denote $u = cU^\sigma$ where c is chosen so that $-\Delta u = \sigma$. By Fourier transformation, $|\xi|^2 \widehat{u}(\xi) = \widehat{\sigma}(\xi)$. Now, $\widehat{\sigma}$ is (the restriction to IR^n of) an entire function with the Taylor expansion $\widehat{\sigma}(\xi) = \sum_{m=0}^\infty s_m(\xi)$, where

$$s_m(\xi) := \int \frac{[-i(x \cdot \xi)]^m}{m!} \, d\sigma(x)$$

is a homogeneous polynomial of degree m, $m = 0, 1, 2, \ldots$. (As usual the dot denotes the scalar product in IR^n). Now, since $\widehat{u} \in C(\mathrm{IR}^n)$, $|\xi|^{-2}\widehat{\sigma}(\xi)$ must be continuous at $\xi = 0$. For it even to be bounded we must have s_0 and s_1 vanishing identically. Hence $\sigma @$ harmonic polynomials of degree ≤ 1. Moreover since $|\xi|^{-2} s_m(\xi)$ vanishes at 0 for $m \geq 3$, *we must have* $|\xi|^{-2} s_2(\xi)$ *continuous at* $\xi = 0$. But, this function is constant on rays through 0, so to have continuity at 0, s_2 must be constant on $\{|\xi| = 1\}$, which implies

$$\int (x_1 \xi_1 + \ldots + x_n \xi_n)^2 \, d\sigma(x) = c \sum_1^n \xi_j^2, \qquad \xi \in \mathrm{IR}^n, \qquad (8.6)$$

for some constant c. Now, the identity (8.6) holds also for *complex* ξ and consequently we have shown: *σ annihilates all polynomials $(x_1 \xi_1 + \ldots + x_n \xi_n)^2$ whose ξ_j are complex numbers satisfying*

$$\xi_1^2 + \ldots + \xi_n^2 = 0. \qquad (8.7)$$

The proof is completed by observing that these polynomials (which obviously are harmonic) span all the homogeneous harmonic polynomials of degree 2. This is well known; we include the proof for completeness, in the following proposition.

Proposition 8.7 *Let $m \geq 0$ be an integer. The linear span of the polynomials $p_\xi := (\sum_1^n \xi_j x_j)^m$ whose ξ runs through all solutions of (8.7) in $\mathbf{C}^n$, is all the homogeneous harmonic polynomials of degree m.*

Proof · Consider the vector space V_m of homogeneous polynomials in $x_1, \ldots, x_n$ of degree m with complex coefficients (augmented by the element 0). This is a Hilbert space with respect to the *Fischer inner product* induced by the definition: $\langle x^\alpha, x^\beta \rangle$ is 0 for $\alpha \neq \beta$ and $\langle x^\alpha, x^\alpha \rangle = \alpha!$ for multi-indices α, β with

$|\alpha| = |\beta| = m$. Let $f \in V_m$ be orthogonal to all p_ξ with ξ satisfying (8.7). Since, as is easy to verify, $\langle f, p_\xi \rangle = f(\xi) \cdot m!$, f vanishes on the complex variety defined by (8.7), and since (we assume $n \geq 2$) this variety has no multiple irreducible components, $f = Pg$ where $P = \sum_1^n \xi_j^2$ and $g \in V_{m-2}$. But then, if h is any harmonic homogeneous polynomial of degree m, $\langle f, h \rangle = \langle Pg, h \rangle = \langle g, P(D)h \rangle = 0$ which completes the proof (for details on Fischer's inner product, see for example [S4]). $\qquad\square$

Proposition 8.8. *Let E denote any closed set of uniqueness for HP_2. For any s with $0 < s < 1$ we have the à priori inequality*

$$\left\| (1 + |x|)^{-2s-4} \, \varphi \right\|_\infty \leq C(n,s) \left\| (1 + |x|)^{-s} \Delta\varphi \right\|_\infty \tag{8.8}$$

for all $\varphi \in C_c^\infty(\mathbb{R}^n)$ which vanish on E.

Proof: This result is to be compared with Proposition 5.6. The proof is nearly identical except that now E plays the role there played by the unit ball. It is easy to check that for $p \in HP_2$, $\sup |p(x)|$ over $x \in \overline{B(0,1)}$ does not exceed a constant $C(E)$ times $\sup |p(x)|$ over $x \in E$ (there will be more discussion of this point in [HKS]; very briefly, the idea is that E being a set of uniqueness for HP_2, together with the finte dimensionality of this space, implies such a bound by a compactness argument). Therefore δ_y can be (quasi-)swept on to the set E, instead of $\overline{B(0;1)}$ with similar estimates as before. Since the rest of the proof is nearly a repetition of the earlier one, we omit details. $\qquad\square$

Proposition 8.9. *Let E denote any minimal (finite) set of uniqueness for HP_2. Then $V := \{\Delta\varphi : \varphi \in C_c^\infty(\mathbb{R}^n), \varphi|_E = 0\}$ is norm-dense in $C_0(\mathbb{R}^n)$.*

Proof : We have to show, if $\mu \in M(\mathbb{R}^n)$ annihilates V, $\mu = 0$. Now, $\mu \,@\, V$ means $\int \Delta\varphi \, d\mu = 0$ whenever $\varphi \in C_c^\infty(\mathbb{R}^n)$ satisfies $\int \Delta\varphi(x) |x-y|^{2-n} dx = 0$ for all $y \in E$. By linear algebra this gives (denoting $E = \{y^1, y^2, \ldots, y^k\}$) that there are constant $c_1, \ldots, c_k$ not all zero such that

$$\int (\Delta\varphi) \, d\mu = \int \left(\sum_{j=1}^k c_j \left| x - y^j \right|^{2-n} \right) \Delta\varphi(x) \, dx$$

holds for all $\varphi \in C_c^\infty(\mathbb{R}^n)$. This expresses the fact that $d\mu - \sum c_j \left| x - y^j \right|^{2-n}$ is identical with an everywhere harmonic function H on $\mathbb{R}^n$. In particular, $d\mu$ is absolutely continuous with respect to Lebesgue measure, say $d\mu = f(x) \, dx$ where $f \in L^1(\mathbb{R}^n)$, and

$$f(x) = \sum_{j=1}^k c_j \left| x - y^j \right|^{2-n} + H(x) \quad , \qquad x \in \mathbb{R}^n \setminus E. \tag{8.9}$$

Integrating (8.9) with respect to x in $B(z;1)$ shows that $H(z) \to 0$ as $|z| \to \infty$, hence $H = 0$ and consequently $\sum_{j=1}^{k} c_j |x - y^j|^{2-n}$ is in $L^1(\mathbb{R}^n)$. But, this is the potential of a measure σ carried by E, so by Theorem 8.6, $\sigma \,@\, HP_2$. But, this is in violation of Lemma 8.4, unless $\sigma = 0$, i.e. all the c_j (and hence f, and μ) vanish, which is what we had to show. $\qquad\square$

Proposition 8.10. *Let $0 < s < 1$, and suppose f is a measurable function on $\mathbb{R}^n$ with $|f(x)| \le (1 + |x|)^s$. Then, there is a solution to $\Delta u = f$ satisfying $|u(x)| \le C(n,s)(1 + |x|)^{4+2s}$.*

Proof: This follows by applying the reasoning used in proving Proposition 5.2, to the estimate (8.8). The only change is, instead of Lemma 5.3 we use the sharper Proposition 8.9. $\qquad\square$

Remark. Proposition 8.10 is a weak version of Karp's theorem, but we will use it as an "initialization" to get the sharp theorem.

Proposition 8.11. *Let $\varphi \in C_c^\infty(\mathbb{R}^n)$ vanish on a finite set E which is a uniqueness set for HP_4. Then for $x \in B(0;1)$ we have*

$$|\varphi(x)| \le C(E,n) \|\Delta\varphi\|_\infty. \tag{8.10}$$

Proof: It is no loss of generality to assume $\|\Delta\varphi\|_\infty = 1$. Let us apply Proposition 8.10 with $s = 1/4$ and $f = \Delta\varphi$. This yields a solution u to $\Delta u = \Delta\varphi$ with $|u(x)| \le C(n)(1 + |x|)^{9/2}$. Hence $h := u - \varphi$ is harmonic on $\mathbb{R}^n$ and $O\left(|x|^{9/2}\right)$ for large $|x|$, which implies $h \in HP_4$. Also, for $x \in E$ we have $|h(x)| = |u(x)| \le C(n)(1 + |x|)^{9/2}$ and hence $|h(x)| \le C_1(E,n)$ on $B(0;1)$. Thus, finally $|\varphi(x)| \le |h(x)| + |u(x)| \le C_2(E,n)$ for $x \in B(0;1)$. $\qquad\square$

8.3. Proof of Theorem 8.2. After these preliminaries, it is not too hard to prove Theorem 8.2. We shall present two variants of the proof. The germ of estimate (8.4) is contained in

Proposition 8.12. *For $\varphi \in C_c^\infty(\mathbb{R}^n)$, $x, y \in \mathbb{R}^n$ and $j \in \{1, 2, \dots, n\}$*

$$|(\partial_j \varphi)(x) - 2(\partial_j \varphi)(x + y) + (\partial_j \varphi)(x + 2y)| \le C\,|y|\,\|\Delta\varphi\|_\infty \tag{8.11}$$

where C depends only on n.

Proof: In this proof we denote $\partial_j \varphi$ by φ_j. We start from the identity $\varphi(x) = \int (\Delta\varphi)(z) K_n(x - z)\,dz$ where K_n is the Newton kernel. Then for $w \in \mathbb{R}^n$

$$\varphi(x + w) = \int \Delta\varphi(z) K_n(x + w - z)\,dz$$

and if σ is a measure with compact support

$$\int \varphi(x+w)\,d\sigma(w) = \int \Delta\varphi(z)\left[\int K_n(x+w-z)\,d\sigma(w)\right]dz$$

$$= \int \Delta\varphi(z)\,U^\sigma(z-x)\,dz.$$

Differentiating with respect to x_j,

$$\int \varphi_j(x+w)\,d\sigma(w) = \int \Delta\varphi(z)(\partial/\partial x_j)\,U^\sigma(z-x)\,dz,$$

hence

$$\left|\int \varphi_j(x+w)\,d\sigma(w)\right| \le \|\Delta\varphi\|_\infty\,\|(\partial/\partial x_j)\,U^\sigma\|_1. \tag{8.12}$$

We now choose the measure $\sigma = \sigma_y$ which places masses $1, -2, 1$ at $w = 0, y, 2y$ respectively. Note that this measure annihilates HP_1 so that, by Proposition 4.3 $(\partial/\partial x_j)U^\sigma$ is integrable. To estimate the integral observe that $\sigma = \tau_{(R)}$ where $R := 2|y|$ and $\|\tau\| = 4$, $\tau @ HP_1$ and τ has support in the unit ball. By Proposition 4.5, $U^\sigma(x) = R^{2-n}\,U^\tau(R^{-1}x)$, so

$$(\partial/\partial x_j)\,U^\sigma(x) = R^{1-n}\,(\partial_j\,U^\tau)\left(R^{-1}x\right)$$

and

$$\|(\partial/\partial x_j)\,U^\sigma\|_1 = R\,\|(\partial/\partial x_j)\,U^\tau\|_1 \le CR\,\|\tau\| = 8C\,|y|$$

by virtue of Proposition 4.3 where C depends only on n. Inserting this estimate in (8.12) yields (8.11). $\qquad\square$

Proposition 8.13. *Let $\varphi \in C_c^\infty(\mathbb{R}^n)$ vanish on the finite set E which is a uniqueness set for HP_4. Then for $x \in \mathbb{R}^n$*

$$|\varphi(x)| \le C\,(1+|x|)^2\,\log(2+x)\,\|\Delta\varphi\|_\infty \tag{8.13}$$

$$|(\partial_j\,\varphi)(x)| \le C\,(1+|x|)\,\log(2+x)\,\|\Delta\varphi\|_\infty \qquad (j = 1, 2, \ldots, n) \tag{8.14}$$

where C depend only on n, E.

Proof: By an elementary argument (given as Lemma 2 of [S2], and which we won't repeat here), (8.14) follows from (8.11) and (8.10). From (8.14) follows (8.13) by integration. $\qquad\square$

Proof of Theorem 8.2 : Given $f \in L^\infty(\mathbb{R}^n)$, with $\|f\|_\infty \le 1$ we wish first to solve $\Delta u = f$ with

$$|u(x)| \le C\,W(x), \tag{8.15}$$

W being given by (8.2). By a limiting argument we used earlier, using the fact that f is the pointwise a.e. limit of a sequence $\{f_j\} \subset C_0(\mathbb{R}^n)$ with $\|f_j\|_\infty \le 1$ we see that it is sufficient to conduct the proof for $f \in C_0(\mathbb{R}^n)$. Let now E denote a finite uniqueness set for HP_4, and let $S \subset C_0(\mathbb{R}^n)$ denote the closed subspace spanned by $\{\Delta\varphi : \varphi \in C_c^\infty(\mathrm{I\!R}^n), \varphi \text{ vanishes on } E\}$. In view of earlier discussion, S has finite codimension, and hence there are linearly independent functionals $\wedge_1, \ldots, \wedge_N$ on $C_0(\mathrm{I\!R}^n)$ (for some finite $N \geq 1$) such that

$$S = \{f \in C_0(\mathrm{I\!R}^n) : \wedge_1 f = \ldots = \wedge_N f = 0\}. \tag{8.16}$$

Clearly we can construct $\theta_1, \ldots, \theta_N$ in $C_c^\infty(\mathrm{I\!R}^n)$ with support in $B(0,1)$ which are biorthogonal to the $\wedge_j$, i.e.

$$\wedge_j \theta_k = \delta_{jk} \text{ (Kronecker symbol)}, \qquad\qquad 1 \leq j, k \leq N. \tag{8.17}$$

Let now f be given in $C_0(\mathrm{I\!R}^n)$ with $\|f\|_\infty \leq 1$, then

$$g := f - \sum_{k=1}^{N} (\wedge_k f)\,\theta_k \tag{8.18}$$

is in S, and estimate (8.13) implies easily that $\Delta v = g$ has a solution v satisfying

$$|v(x)| \leq C\,W(x)\,\|g\|_\infty \leq C_1\,W(x)\,\|f\|_\infty.$$

On the other hand, $\Delta w = \sum_{k=1}^{N} (\wedge_k f)\,\theta_k$ has a solution w bounded on $\mathrm{I\!R}^n$ by a constant times the sup norm of $h := \sum_{k=1}^{N} (\wedge_k f)\,\theta_k$ since h is supported in the unit ball (just take $w = U^h$). Hence $\Delta(v + w) = f$ so $u := v + w$ solves $\Delta u = f$ with the desired estimates. The verification of (8.4) presents no difficulty and is left to the reader. The BMO estimate (8.5) is not elementary, being based on the Fefferman estimate

$$\|\partial^\alpha \varphi\|_{BMO(\mathrm{I\!R}^n)} \leq C\,\|\Delta\varphi\|_\infty, \qquad\qquad |\alpha| = 2 \tag{8.19}$$

valid for $\varphi \in C_c^\infty(\mathrm{I\!R}^n)$. Using this we get a solution to $\Delta u = f$ satisfying (8.1) to (8.5), initially for $f \in S$, and then extend to all $f \in C_0(\mathrm{I\!R}^n)$, and finally $f \in L^\infty(\mathrm{I\!R}^n)$ by auxiliary arguments, as done earlier. We omit the details. $\quad\square$

Remark. There is another way to arrange the logic of the argument leading (as in [S2]) to the L^1 estimate which is the dual form of Theorem 8.1:

Theorem 8.14. *For $\varphi \in C_c^\infty(\mathrm{I\!R}^n)$,*

$$\int_{\mathrm{I\!R}^n} |\varphi(x)|\,dx \leq C \int_{\mathrm{I\!R}^n} W(x)\,|\Delta\varphi(x)|\,dx \tag{8.20}$$

where C, W are as in Theorem 8.1.

Proof :

$$\int |\varphi(x)|\,dx = \sup \left| \int \varphi(x)\,g(x)\,dx \right|$$

where the sup is over all $g \in C_0(\mathbb{R}^n)$ with $\|g\|_\infty = 1$. Hence, in view of Proposition 8.9, if E is any finite, minimal set of uniqueness for HP_2, and $\varepsilon > 0$ there is $\psi \in C_c^\infty(\mathbb{R}^n)$ vanishing on E such that $\|\Delta\psi\|_\infty \le 1$ and

$$\int |\varphi(x)|\,dx \le \left| \int \varphi(x)\,\Delta\psi(x)\,dx \right| + \varepsilon. \tag{8.21}$$

The first term on the right equals

$$\left| \int \psi(x)\,\Delta\varphi(x)\,dx \right| \le \int |\Delta\varphi(x)|\,(C\,W(x))\,dx$$

by estimate (8.13) applied to ψ. Since ε in (8.21) is arbitrary the proof is finished. $\square$

Remark. Theorem 8.14 implies Theorem 8.1 by a standard Hahn-Banach argument. Theorem 8.1 in turn immediately implies Theorem 8.14. The proof given above for Theorem 8.14 is new, however and simpler than that in [S2].

9   A Remark on Karp's "Generalized Potential"

9.1. For a "body" (which here shall mean an *open connected set* $\Omega \subset \mathbb{R}^n$) with *bounded density* (that is, a function $f \in L^\infty(\Omega)$) it is useful to associate a Newtonian potential even in cases where (because of unboundedness of Ω) the usual integral defining it is divergent. This is important e.g. in studying unbounded quadrature domains. Karp [K] introduced the fruitful idea (setting $\tilde{f}(x) = f(x)$ for $x \in \Omega$ and $\tilde{f}(x) = 0$ on $\mathbb{R}^n \setminus \Omega$) to adopt as the definition of Newtonian potential of the body (Ω, f) (always modulo a constant normalization factor) a solution u to $\Delta u = \tilde{f}$ on $\mathbb{R}^n$ that has (in a suitable sense) minimal growth at ∞. In view of Theorem 8.1 there is always a solution u that is $O(|x|^2 \log|x|)$ at ∞. This function is uniquely determined, modulo an additive element of HP_2. Of course, if Ω is "small" there may well be a solution to $\Delta u = \tilde{f}$ of smaller growth (e.g. if Ω is bounded the usual definition of Newtonian potential gives a *bounded* solution u, indeed even one which is $\sim c\,|x|^{2-n}$ at ∞ for $n \ge 3$). Karp has raised the interesting question of finding, for a given Ω the minimal rate of growth of solutions to $\Delta u = \tilde{f}$. In [KM] this problem was solved for circular cylinders. In particular they proved a result which, specialized to $n = 3$ asserts

Theorem 9.1. [KM] *Let $\Omega = \{x \in \mathbb{R}^3 : x_2^2 + x_3^2 < 1\}$. Given $f \in L^\infty(\mathbb{R}^3)$ vanishing outside Ω, there is a solution to $\Delta u = f$ on $\mathbb{R}^3$ satisfying*

$$|u(x)| \le C \log(2 + |x|) \|f\|_\infty, \qquad x \in \mathbb{R}^3. \tag{9.1}$$

It is easy to see this estimate is, on the plane $\{x_1 = 0\}$, of the right order of magnitude. Indeed, taking f_0 to be the characteristic function of Ω, we may choose the logarithmic potential of the characteristic function of the unit disk in the (x_2, x_3) plane. This $u = u(x_2, x_3)$, considered as a function on $\mathbb{R}^3$, satisfies $\Delta u = f_0$. Also, $u \sim c \log(x_2^2 + x_3^2)$ for large $x_2^2 + x_3^2$ and it is easy to see that no other solution to $\Delta u = f_0$ has a smaller order of magnitude.

The (rather modest) goal of this section is some remarks inspired by Theorem 9.1 (leading to a new proof of this result).

First of all, let us formulate the dual version of the generalized potential.

Proposition 9.2. *The following are equivalent, for any $\Omega \subset \mathbb{R}^n$ and positive continuous function A on $\mathbb{R}^n$:*

(a) *Given $f \in L^\infty(\mathbb{R}^n)$, with $f = 0$ outside Ω there is a solution to $\Delta u = f$ on $\mathbb{R}^n$ satisfying*

$$|u(x)| \le A(x)\|f\|_\infty, \qquad x \in \mathbb{R}^n. \tag{9.2}$$

(b) *For every $\varphi \in C_c^\infty(\mathbb{R}^n)$,*

$$\int_\Omega |\varphi(x)|\, dx \le \int_{\mathbb{R}^n} A(x)\,|\Delta\varphi(x)|\, dx. \tag{9.3}$$

Proof: The proof is similar to arguments given earlier, and left to the reader.

Our strategy in proving results of type (a) will therefore be to try to prove estimates of type (9.3). Theorem 8.1 is of course the case $\Omega = \mathbb{R}^n$ and has been thoroughly discussed.

Note first of all that since

$$\varphi(x) = -c_n \int |x - y|^{2-n}\,(\Delta\varphi)(y)\, dy \tag{9.4}$$

we have the estimate (9.3) with

$$A(x) = c_n \int_\Omega |x - y|^{2-n}\, dy, \tag{9.5}$$

but of course this yields nothing in the case (that here is of interest) where the right side of (9.5) is identically $+\infty$. To get a nontrivial estimate in such cases we can use the device employed by Karp and others, of subtracting off from the Newton kernel in (9.4) (considered as a function of y, with $x \neq 0$ as a parameter) one or more terms of its Taylor expansion about $y = 0$. This amounts to regrouping the expansion we used in § 4.1 in ascending powers of y. We have, up to quadratic terms in y,

$$|x - y|^{2-n} = |x|^{2-n} + (n - 2)|x|^{-n}(x \cdot y) \tag{9.6}$$

$$+ \left(\frac{n}{2} - 1\right)|x|^{-n-2}\left[n(x \cdot y)^2 - |x|^2|y|^2\right] + \dots .$$

Observe that the homogeneous (in y) polynomial terms of degree $0, 1, 2, \dots$ appearing on the right side are harmonic, as functions of y and therefore can be subtracted from $|x-y|^{2-n}$ in (9.4) without changing the value of the integral. For example, subtracting only the term of degree 0 gives, in place of (9.4)

$$\varphi(x) = c_n \int \left(|x - y|^{2-n} - |x|^{2-n}\right)(\Delta\varphi)(y)\,dy.$$

Hence

$$\int_\Omega |\varphi(x)|\,dx \leq c_n \int_{\mathbb{R}^n} |\Delta\varphi(y)| \left[\int_\Omega \left||x - y|^{2-n} - |x|^{2-n}\right|\,dx\right]dy \tag{9.7}$$

$$= \int B(y)\,|\Delta\varphi(y)|\,dy$$

where

$$B(y) := c_n \int_\Omega \left||x - y|^{2-n} - |x|^{2-n}\right|\,dx. \tag{9.8}$$

Since the integral in (9.8) falls off like $|x|^{1-n}$ at ∞ the integral may converge in cases where $\int_\Omega |x - y|^{2-n}\,dx$ does not. As we'll check shortly, this is the case with the cylinder in $\mathbb{R}^3$.

In like manner, we get the estimates

$$\int_\Omega |\varphi(x)|\,dx \leq \int C(y)\,|\Delta\varphi(y)|\,dy \tag{9.9}$$

where

$$C(y) := c_n \int_\Omega \left||x - y|^{2-n} - |x|^{2-n} - (n - 2)|x|^{-n}(x \cdot y)\right|\,dx \tag{9.10}$$

which gives a nontrivial estimate whenever $\int_\Omega |x|^{-n}\, dx < \infty$, and one final one (which we won't write out) that gives a nontrivial estimate whenever $\int_\Omega |x|^{-n-1}\, dx < \infty$, hence for every Ω whose complement contains a neighborhood of 0 (this path leads essentially to Karp's proof of Theorem 8.1).

After these preliminaries, let us turn to the proof of Theorem 9.1. First of all, from (9.7) and (9.8) we get

Proposition 9.3. *With Ω as in Theorem 9.1, we have for $\varphi \in C_c^\infty \left(\mathbb{R}^n\right)$,*

$$\int_\Omega |\varphi(x)|\, dx \leq \int_{\mathbb{R}^3} B(x)\, |\Delta\varphi(x)|\, dx \tag{9.11}$$

where

$$B(y) = c \int_\Omega \left| |x-y|^{-1} - |x|^{-1} \right|\, dx. \tag{9.12}$$

Corollary. *For every $f \in L^\infty \left(\mathbb{R}^3\right)$ vanishing outside Ω there is a solution to $\Delta u = f$ satisfying $|u(x)| \leq B(x)\, \|f\|_\infty$ for $x \in \mathbb{R}^3$.*

To proceed we need some estimates of the integral in (9.12). We suppress the constant factor and write

$$B_0(y) := \int_\Omega \left| |x-y|^{-1} - |x|^{-1} \right|\, dx. \tag{9.13}$$

First we make a slight digression and show that for some y (9.13) can be evaluated:

Lemma 9.4. *For y in the plane $\{y_1 = 0\}$, we have*

$$B_0(y) = 2\pi \left(1 + \log |y|\right) \quad , \quad |y| \geq 2. \tag{9.14}$$

Proof : For $y = (0, y_2, y_3)$ with $y_2^2 + y_3^2 \geq 4$ the plane perpendicularly bisecting the segment joining the origin to y does not meet Ω, and consequently, for such y

$$|x - y| \geq |x|, \qquad x \in \Omega,$$

so

$$B_0(y) = \int_\Omega \left(\frac{1}{|x|} - \frac{1}{|x-y|} \right) dx$$

$$= \lim_{a \to \infty} \int_D \left\{ \int_{-a}^{a} \left[\left(x_1^2 + x_2^2 + x_3^2 \right)^{-1/2} \right. \right.$$

$$\left. \left. - \left(x_1^2 + (x_2 - y_2)^2 + (x_3 - y_3)^2 \right)^{-1/2} \right] dx_1 \right\} dx_2 \, dx_3$$

where D denotes the disk $\{x_2^2 + x_3^2 < 1\}$ in $\mathbb{R}^2$. The inner integral equals (after some manipulation)

$$2 \int_{Qa}^{Pa} \left(1 + t^2 \right)^{-1/2} dt \tag{9.15}$$

where the limits of integration are abbreviated by writing

$$P = \left(x_2^2 + x_3^2 \right)^{-1/2}, \qquad Q = \left((x_2 - y_2)^2 + (x_3 - y_3)^2 \right)^{-1/2}.$$

Hence the limit as $a \to \infty$ of the integral in (9.15) is $\leq \lim_{a \to \infty} \int_{Qa}^{Pa} \frac{dt}{t} = \log(P/Q)$ and we get

$$B_0(y) = \int_D \log \left[\frac{(x_2 - y_2)^2 + (x_3 - y_3)^2}{x_2^2 + x_3^2} \right] dx_2 \, dx_3$$

$$= \pi \log \left(y_2^2 + y_3^2 \right) + 2\pi$$

as was to be shown. $\square$

Remark. We won't need this estimate in proving Theorem 9.1.

Our key estimate for proving Theorem 9.1 is

Lemma 9.5. *For $y \in \mathbb{R}^3$,*

$$B_0(y) \leq C \, |y| \left(1 + y_2^2 + y_3^2 \right)^{-1/2}. \tag{9.16}$$

Proof: Since

$$\left| |x - y|^{-1} - |x|^{-1} \right| \leq |y| \, |x|^{-1} \, |x - y|^{-1},$$

$$B_0(y) \leq |y| \int_\Omega |x|^{-1} |x - y|^{-1} \, dx$$

$$\leq |y| \left[\int_\Omega |x|^{-2} \, dx \right]^{1/2} \left[\int_\Omega |x - y|^{-2} \, dx \right]^{1/2}.$$

The first bracket is a constant, and

$$\int\limits_{\Omega} |x - y|^{-2}\, dx = \pi \int\limits_{D} \left[(x_2 - y_2)^2 + (x_3 - y_3)^2\right]^{-1/2} dx_2\, dx_3$$

from which (9.16) follows.   □

Lemma 9.6. *Let $E(x) := |x|\left(1 + x_2^2 + x_3^2\right)^{-1/2}$, $x \in \mathbb{R}^3$. For any $z \in \mathbb{R}^3$ the mean value of E over the ball*

$$B_z := \{x \in \mathbb{R}^3 : |x - z| < |z|\} \tag{9.17}$$

does not exceed an absolute constant.

Proof: Write $R = |z|$. Then

$$\int\limits_{B_z} E\, dx \leq \int\limits_{B(0;2R)} E\, dx \leq C\, R^3$$

by trivial estimates, which implies the assertion.   □

Lemma 9.7. *The mean value of*

$$G(x) := \log\left(2 + x_2^2 + x_3^2\right), \qquad x \in \mathbb{R}^3 \tag{9.18}$$

over B_z (as defined by (9.17)) does not exceed $C \log(2 + |z|)$ where C is an absolute constant.

Proof:
$$\int\limits_{B_z} G\, dx \leq \int\limits_{B(0;2R)} G\, dx \leq |B(0;2R)| \log\left(2 + 4R^2\right)$$

from which the assertion follows.   □

Proof of Theorem 9.1 : As already noted, there is a solution U to $\Delta U = F$, where F denoted the characteristic function of Ω, satisfying

$$|U(x)| \leq C \log\left(2 + x_2^2 + x_3^2\right), \qquad x \in \mathbb{R}^3. \tag{9.19}$$

We assume, as we may, $\|f\|_\infty = 1$ and denote by u the solution to $\Delta u = f$ satisfying (by virtue of the corollary to Proposition 9.3 and Lemma 9.5)

$$|u(x)| \leq C\, E(x) \tag{9.20}$$

where E is as in Lemma 9.6. Now, $U - u$ and $U + u$ are subharmonic on $\mathbb{R}^3$.

Hence

$$U(z) - u(z) \le |B_z|^{-1} \int_{B_z} (U - u)\, dx \le |B_z|^{-1} \int_{B_z} (|U| + |u|)\, dx$$

$$C_1 + C_2 \log(2 + |z|) \le C_3 \log(2 + |z|)$$

by Lemmas 9.6 and 9.7 (together with (9.19)).

Hence

$$-u(z) \le |U(z)| + C_3 \log(2 + |z|) \le C_4 \log(2 + |z|), \qquad (9.21)$$

again using (9.19). In turn, using subharmonicity of $U + u$ leads to the same estimate (9.21) with $-u(z)$ replaced by $u(z)$. This concludes the proof. $\square$

Remarks. 1. We have not been able to prove a sharper estimate like (9.19) for $|u(x)|$, and do not know if that is true.

2. The technique applied here applies also to higher-dimensional cylinders, as investigated in [KM]. The idea to compare the generalized potentials of f and F should be applicable also to other choices of Ω.

9.2. Finally, as an illustration of the application of generalized potentials, let us outline a new proof of (a generalized version of) a recent theorem [GHR] characterizing domains admitting the same quadrature identity as the cylinder. In this section, again Ω denotes the cylinder in $\mathbb{R}^3$ defined by $\{x_2^2 + x_3^2 < 1\}$ (we restrict ourselves to 3 dimensions only for purposes of simplifying the exposition, the method presented is quite general). It is well known, and easy to check that Ω admits the "quadrature identity"

$$\int_{\Omega} h\, dx = c \int_{-\infty}^{\infty} h(x_1, 0, 0)\, dx_1, \qquad h \in HL^1(\Omega), \qquad (9.22)$$

where c is a positive constant.

Theorem 9.8. ([GHR]) *Let $D \subset \mathbb{R}^3$ be a connected open set satisfying*

D is the interior of its closure, $\qquad\qquad\qquad\qquad\qquad$ (9.23)

D contains the x_1-axis, $\qquad\qquad\qquad\qquad\qquad$ (9.24)

for some $R > 0$, D is contained in the cylinder $\{x_2^2 + x_3^2 < R^2\}$, $\qquad$ (9.25)

and

$$\int_{D} h\, dx = c \int_{-\infty}^{\infty} h(x_1, 0, 0)\, dx, \qquad h \in HL^1(D) \qquad (9.26)$$

where c is the same as in (9.22). Then $D = \Omega$.

Remarks. 1. The topological hypothesis (9.23) can be weakened, at the cost of a more complicated formulation.

2. In the proof given below we only assume, instead of (9.25), the weaker condition:

$$For\ some\ R > 0,\ D \subset \{x \in \mathbb{R}^3 : |x_3| < R\}. \tag{9.25a}$$

Indeed, one can further weaker (9.25a) considerably, but we won't pursue this aspect here.

In the following proof we refer to [KM] for a few calculations which we shall call "well known"; see also [S5].

Lemma 9.9. *Let λ denote the measure on $\mathbb{R}^3$ defined by $\int \varphi \, d\lambda = \int_{-\infty}^{\infty} \varphi(x_1, 0, 0)\, dx_1$ for $\varphi \in C_c(\mathbb{R}^3)$, and X_Ω the characteristic function of Ω. Then*

$$\Delta u = X_\Omega - \lambda \tag{9.27}$$

has a solution u supported in $\overline{\Omega}$, and satisfying

$$u(x) \geq 0, \qquad\qquad x \in \mathbb{R}^3, \tag{9.28}$$

$$u(x) \leq C\left(1 + |x|^2\right), \qquad x \in \mathbb{R}^3. \tag{9.29}$$

Remark. Much stronger estimates than (9.29) are possible, but this suffices for our purposes.

Lemma 9.10. *Under the hypotheses of Theorem 9.8 there is a solution v to*

$$\Delta v = X_D - \lambda \tag{9.30}$$

supported in $\overline{D}$ and satisfying

$$|v(x)| \leq C\left(1 + |x|^2\right), \qquad\qquad x \in \mathbb{R}^3. \tag{9.31}$$

Proofs : All this is well known, see e.g. [KM].

Lemma 9.11. *Let S_R denote the "slab" $\{x \in \mathbb{R}^3 : |x_3| < R\}$. There is a smooth superharmonic function F on S_R satisfying*

$$F(x) > 0, \qquad\qquad x \in S_R \tag{9.32}$$

$$\lim_{x \in S_R,\, |x| \to \infty} \left(|x|^2 - \varepsilon F(x)\right) = -\infty, \qquad\qquad for\ each\ \varepsilon > 0. \tag{9.33}$$

Proof : Consider the polynomial

$$H\left(x_1, x_2, x_3\right) = Re\left[\left(x_1 + i\,x_3\right)^4 + \left(x_2 + i\,x_3\right)^4\right].$$

It is harmonic and, as a function of x_1, x_2 with x_3 "frozen" in the interval $[-R, R]$ has leading form $x_1^4 + x_2^4$, hence $\to +\infty$ as $|x| \to \infty$ with $x \in S_R$. Hence it attains a minimum value $m > -\infty$ on $\overline{S_R}$. Define $F = H - m$.

Thus, F is superharmonic (even harmonic!) and satisfies the requirements.
□

Proof of Theorem 9.8 : We assume $D \neq \Omega$ and shall derive a contradiction. We first show $D \setminus \overline{\Omega}$ *is not empty*. Indeed, were this false then $D \subset \Omega$ with $\Omega \setminus D$ not empty. From (9.22) and (9.26) we then get

$$\int_{\Omega \setminus D} h\, dx = 0, \qquad \text{for all } h \in HL^1(\Omega).$$

But this is false, since there are strictly positive functions h_0 in $HL^1(\Omega)$ (for example,

$$h_0\left(x\right) = \left|x - x^1\right|^{-1} - \left|x - x^2\right|^{-1}$$

where $x^1 = (0,0,2)$, $x^2 = (0,0,3)$ is such a function). Hence there exists a $y \in \partial D \setminus \overline{\Omega}$. Since v (defined in Lemma 9.10) is strictly subharmonic near y and vanishes at y, there is a point z near y $(z \notin \overline{\Omega})$ with $v(z) > 0$. Now, consider the function

$$w\left(x\right) := v\left(x\right) - u\left(x\right) - \varepsilon\, F\left(x\right), \qquad x \in S_R,$$

where $\varepsilon > 0$ is chosen so small that $w\left(z\right) = v\left(z\right) - \varepsilon\, F\left(z\right)$ is positive. Since, by virtue of (9.29), (9.31) and (9.33), $w\left(x\right)$ is negative for $x \in S_R$ and $|x|$ sufficiently large we conclude:

$$M := \max_{x \in S_R} w\left(x\right) \geq w\left(z\right) > 0. \tag{9.34}$$

This maximum is attained at some point x^0. Since u, F are everywhere nonnegative we must have $x^0 \in D$. But

$$\Delta w = \Delta v - \Delta u - \varepsilon \Delta F = X_D - X_\Omega \tag{9.35}$$

so w is subharmonic on a neighborhood of x^0. Indeed, (9.35) shows that w is subharmonic on all of D, hence $w\left(x\right) = M$, $x \in \overline{D}$. But at points $y \in \partial D$, $w\left(y\right) \leq 0$. This is a contradiction, and the theorem is proved. □

Remarks.   1. We constructed "by hand" a positive integrable harmonic function on a cylinder. This suggests an interesting question: *For which* $\Omega \subset \mathbb{R}^n$ *is there a positive function in* $HL^1(\Omega)$? When Ω is a slab in $\mathbb{R}^3$ we have not been able to decide this. Also, more generally, when is there a positive harmonic function on Ω admitting a given majorization (e.g. $O\left(|x|^{-\lambda}\right)$) at ∞ for some prescribed $\lambda > 0$? For example, for arbitrarily large λ there is a nontrivial cone in $\mathbb{R}^n$ for which this holds, but it seems the cone must be made narrower when λ increases. Note that there is never a positive integrable harmonic function on a halfspace, since that is a null quadrature domain.

2. It is also of interest to determine, for how "large" domains (in place of S_R in Lemma 9.11) there exists a positive superharmonic (or as the case may be) harmonic function satisfying (9.33), or the analogous property with $|x|^2$ replaced by some other comparison function. This kind of result is crucial for further generalization of results of type Theorem 9.8.

3. As a final question (going back to § 8): Can one characterize the minimal sets of uniqueness for HP_2 (or HP_m, for that matter)? For HP_1 of course such a set consists precisely of $n + 1$ points of $\mathbb{R}^n$ that lie on no hyperplane.

10   A Fourier Approach to Quasi–Balayage

The use of Fourier analysis is a good way to prove some of the estimates of § 4, as we now sketch briefly. We require the following (very well known) "embedding theorem" giving an estimate for the L^1 norm of a Fourier transform; we will give the proof for the reader's convenience.

Lemma 10.1. *Let* $f \in C^k(\mathbb{R}^n)$ *and suppose* $\partial^\alpha f \in L^p(\mathbb{R}^n)$, *for* $|\alpha| \le k$, *where* $1 \le p \le 2$. (*It suffices, more generally, that* $\partial^\alpha f$, *as a distribution, is in* $L^p(\mathbb{R}^n)$ *for all* α, ($|\alpha| \le k$)). *Then if* $k > \frac{n}{p}$, $\hat{f}$ *is integrable and*

$$\|\hat{f}\|_1 \le C(n,k,p)\left(\sum_{|\alpha|\le k} \|\partial^\alpha f\|_p \right). \tag{10.1}$$

Proof: We conduct the proof assuming $p > 1$, leaving to the reader the obvious modifications when $p = 1$. By the Hausdorff–Young inequality, for $|\alpha| \le k$

$$\int \left|(i\xi)^\alpha \hat{f}(\xi)\right|^r d\xi \le \left(\int |\partial^\alpha f(x)|^p dx \right)^{1/(p-1)},$$

where $r := p/(p-1)$ and all integrals are over $\mathbb{R}^n$. Hence

$$\int \left|\widehat{f}(\xi)\right|^r \left(\sum_{|\alpha|\leq k} |\xi^\alpha|^r\right) d\xi \leq \sum_{|\alpha|\leq k} \left(\int |\partial^\alpha f(x)|^p \, dx\right)^{1/(p-1)},$$

which implies

$$\int \left|\widehat{f}(\xi)\right|^r (1+|\xi|)^{kr} \, d\xi \leq C(n,k,p) \left(\sum_{|\alpha|\leq k} \|\partial^\alpha f\|_p\right)^r. \tag{10.2}$$

By Hölder's inequality

$$\int \left|\widehat{f}(\xi)\right| d\xi \leq \left[\int \left|\widehat{f}(\xi)\right|^r (1+|\xi|)^{kr}\right]^{1/r} \left[\int (1+|\xi|)^{-kp} \, d\xi\right]^{1/p}. \tag{10.3}$$

If $kp > n$ the last integral is finite and, combining with (10.2) we get (10.1).
□

As a first application, we give a new proof of the following estimate from § 4:

Proposition 10.2. *Let $\sigma \in M(\mathbb{R}^n)$ have support in $\overline{B(0,1)}$ and $\sigma @ HP_2$. Then U^σ is integrable over $\mathbb{R}^n$ and*

$$\|U^\sigma\|_{L^1(\mathbb{R}^n)} \leq C(n)\|\sigma\|_M. \tag{10.4}$$

The proof will be in two steps: first we'll show U^σ is integrable, and then deduce the estimate for $\|U^\sigma\|_1$. On the Fourier side, since

$$(U^\sigma)^\wedge(\xi) = |\xi|^{-2}\widehat{\sigma}(\xi) \tag{10.5}$$

we have to show the function on the right belongs to the Wiener algebra $A(\mathbb{R}^n)$ of Fourier transforms of functions in $L^1(\mathbb{R}^n)$. Since $A(\mathbb{R}^n)$ is a local algebra, it suffices to show the restrictions of this function to $B(0;2)$ and to $\{\xi : |\xi| > 1\}$ each extend to an element in $A(\mathbb{R}^n)$. To handle the second one we use the well known fact that there is an integrable function k_λ, for each $\lambda > 0$, such that $\widehat{k_\lambda}(\xi) = |\xi|^{-\lambda}$, for $|\xi| > 1$. (This depends only on the fact that $|\xi|^{-\lambda}$ is smooth on $\mathbb{R}^n \setminus \{0\}$ and homogeneous of negative degree; see e.g. [S6, Lemma 4.2].) So, on $\{|\xi| > 1\}$ we have

$$|\xi|^{-2}\widehat{\sigma}(\xi) = \widehat{k_2}(\xi)\widehat{\sigma}(\xi)$$

and the right hand member is the Fourier transform of the integrable function $k_2 * \sigma$, and hence in the Wiener algebra. To show that the restriction of $|\xi|^{-2}\widehat{\sigma}(\xi)$ to $B(0;2)$ is in (the restriction to this set of) $A(\mathbb{R}^n)$ we introduce a cutoff function $\psi \in C_c^\infty(\mathbb{R}^n)$ satisfying $\psi(\xi) = 1$ for $|\xi| \leq 2$, and will show that

$$F(\xi) := |\xi|^{-2}\,\hat{\sigma}(\xi)\,\psi(\xi) \tag{10.6}$$

is in $A(\mathbb{R}^n)$. For this it is enough, by Lemma 10.1 (note that the roles of x and ξ are reversed) to show for some $1 < p < 2$ that $\partial^\alpha F \in L^p(\mathbb{R}^n)$ for all $|\alpha| \le k$, where $k > \frac{n}{p}$. Now,

$$\hat{\sigma}(\xi) = \int \exp(-i(x\cdot\xi))\,d\sigma(x)$$

$$= \int \left\{ \sum_{m=0}^{\infty} \frac{[-i(x\cdot\xi)]^m}{m!} \right\} d\sigma(x) = \sum_{m=0}^{\infty} s_m(\xi)$$

where s_m is a homogeneous polynomial of degree m (or else identically zero). Since $\sigma \,@\, HP_2$ we see at once that $s_0 = s_1 = 0$, and by a "Fischer's theorem" argument used earlier s_2 is a constant multiple of $\sum_1^n \xi_i^2$. Hence, in checking $F \in A(\mathbb{R}^n)$ we may replace $\hat{\sigma}(\xi)$ in (10.6) by

$$S(\xi) := \hat{\sigma}(\xi) - s_2(\xi) = \sum_{m=3}^{\infty} s_m(\xi).$$

Now, in checking the integrability of $\left|\partial^\alpha\left(|\xi|^{-2}S(\xi)\psi(\xi)\right)\right|^p$ we have only to take account of the behaviour at $\xi = 0$. Since $|\xi|^{-2}S(\xi)$ is a sum of homogeneous functions each with degree of homogeneity ≥ 1, we see that

$$\partial^\alpha\left(|\xi|^{-2}S(\xi)\right) \in L^p \qquad for \qquad 1 < p < \frac{n}{n-1}$$

holds for all α, $|\alpha| \le n$. Since $\psi \in C_c^\infty(\mathbb{R}^n)$ we conclude that

$$\partial^\alpha F \in L^p(\mathbb{R}^n)\,; 1 < p < \frac{n}{n-1}, \qquad |\alpha| \le n. \tag{10.7}$$

So, we have shown $U^\sigma \in L^1(\mathbb{R}^n)$, but have still not a useable *estimate* for $\|U^\sigma\|_1$. For this step we require

Lemma 10.3. *Let $E \subset \mathbb{R}^n$ be a finite uniqueness set for HP_2. Then there is a constant $C = C(n, E)$ such that for $|x| \le 1$,*

$$|\varphi(x)| + \|\mathrm{grad}\,\varphi(x)\| \le C\,\|\Delta\varphi\|_\infty \tag{10.8}$$

holds for all $\varphi \in C_c^\infty(\mathbb{R}^n)$ vanishing on E.

Proof: This is shown by modification of an argument used earlier. By Karp's Theorem 8.1 we can solve $\Delta u = \Delta \varphi$ with

$$|u(x)| \le C_1 (1 + |x|)^2 \log(2 + |x|) \, \|\Delta\varphi\|_\infty \tag{10.9}$$

$$\|\operatorname{grad} u(x)\| \le C_2 (1 + |x|) \log(2 + |x|) \, \|\Delta\varphi\|_\infty. \tag{10.10}$$

Then $h := u - \varphi$ is harmonic on $\mathbb{R}^n$ and satisfies estimate (10.9) for large $|x|$, hence is in HP_2. Since $h = u$ on E, (10.9) gives a bound for $|h(x)|$ when $x \in E$ and hence $|h(x)| \le C_3 (1+|x|)^2$, $\|\operatorname{grad} h(x)\| \le C_4 (1+|x|)$ hold on $\mathbb{R}^n$. These estimates and (10.9), (10.10) imply (10.8). $\qquad\square$

Remark.   The present approach to quasi-balayage thus requires Karp's theorem (at least in a weak form, like a $o(|x|^3)$ estimate in (10.9)) as part of the machinery.

We can now easily finish the proof of Proposition 10.2.

Lemma 10.4. *Given* $g \in L^1(\mathbb{R}^n)$,

$$\|g\|_1 = \sup \left| \int g \, s \, dx \right| \tag{10.11}$$

where the supremum is over all $s \in C_0(\mathbb{R}^n)$ *with* $\|s\|_\infty \le 1$.

Proof: Obvious. Note, for later purposes, it suffices if in (10.11) we let s range over any set *dense in* the unit ball of $C_0(\mathbb{R}^n)$. $\qquad\square$

Remark. It is important to observe that (10.11) is only correct once we know *a priori* that $g \in L^1$. The mere finiteness of $\sup |\int g \, s \, dx|$ over, say, some collection of compactly supported s which is dense in the unit ball of $C_0(\mathbb{R}^n)$ does not imply $g \in L^1$! For example ($g = 1$) there is a dense subset s of $C_0(\mathbb{R}^n)$ such that $\int s \, dx = 0$.

Proof of Proposition 10.2, concluded : We have to estimate $\|U^\sigma\|_1$. To do so, let E be a minimal set of uniqueness for HP_2. By Proposition 8.9, $\{\Delta\varphi : \varphi \in C_c^\infty(\mathbb{R}^n), \varphi|_E = 0\} =: S$ is dense in $C_0(\mathbb{R}^n)$, so by Lemma 10.4,

$$\|U^\sigma\|_1 = \sup \left| \int U^\sigma (\Delta\varphi) \, dx \right|$$

over $\Delta\varphi \in S$, satisfying $\|\Delta\varphi\|_\infty \le 1$.

But

$$\int U^\sigma (\Delta\varphi) \, dx = - \int \varphi \, d\sigma$$

so

$$\|U^\sigma\|_1 \le \|\sigma\|_M \cdot \sup \{ |\varphi(x)| : |x| \le 1 \} \le C \|\sigma\|_M$$

by Lemma 10.3. $\qquad\square$

In like manner, we prove

Proposition 10.5. *Let $\sigma \in M(\mathbb{R}^n)$ have support in $\overline{B(0;1)}$, and $\sigma \,@\, HP_1$. Then for $j = 1, 2, \ldots, n$ we have $\partial_j U^\sigma \in L^1(\mathbb{R}^n)$ and $\|\partial_j U^\sigma\|_1 \le C(n)\|\sigma\|_M$.*

Proof : Again the proof divides into two parts: first to show the integrability of $\partial_j U^\sigma$ and then to estimate $\|\partial_j U^\sigma\|_1$. Denote $\partial_j U^\sigma$ by v. Then

$$\widehat{v}(\xi) = -i\,\xi_j\,|\xi|^{-2}\,\widehat{\sigma}(\xi).$$

The Taylor expansion of $\widehat{\sigma}$ now lacks terms of degree $0, 1$. So, near $\xi = 0$, $\widehat{v}(\xi)$ is a sum of smooth functions each with degree of homogeneity ≥ 1. Thus, by the same reasoning as above $\widehat{v}(\xi)$ restricted to $B(0;2)$ is the restriction of an element of $A(\mathbb{R}^n)$. The restriction to $\{|\xi| > 1\}$ is also handled as before, since $\xi_j\,|\xi|^{-2}$ is the restriction of an element of $A(\mathbb{R}^n)$. Hence $v \in L^1(\mathbb{R}^n)$.

To estimate $\|v\|_1$ we use the same test functions as in the preceding proof, and write

$$\|v\|_1 = \sup\left|\int v\Delta\varphi\,dx\right| \quad \text{over } \Delta\varphi \in S.$$

Now,

$$\left|\int v\,\Delta\varphi\,dx\right| = \left|\int (\partial_j U^\sigma)\,\Delta\varphi\,dx\right| = \left|\int (\partial_j\,\varphi)\,d\sigma\right|,$$

and now (10.8) shows this is $\le C\,\|\Delta\varphi\|_\infty\,\|\sigma\|_M$.   □

Remark. If σ is any distribution of compact support with $\sigma \,@\, HP_2$, then $(U^\sigma)^\wedge(\xi)$ restricted to a neighborhood of 0 belongs to $A(\mathbb{R}^n)$ restricted to that neighborhood (same argument as above applies). But, the restriction to a neighborhood of ∞ is in general no longer in the corresponding restriction algebra of $A(\mathbb{R}^n)$. A similar remark applies to $(\partial_j U^\sigma)^\wedge(\xi)$ when $\sigma \,@\, HP_1$.

Analogously, for $\sigma \in M(\mathbb{R}^n)$ the second derivatives $\partial^\alpha U^\sigma$ with $|\alpha| = 2$ have Fourier transforms which, on $\{|\xi| > 1\}$, are bounded but not in (the restriction of) the algebra $A(\mathbb{R}^n)$ nor even of the bigger algebra of Fourier transforms of bounded measures. Thus, analysis of estimates for second derivatives of potentials by Fourier methods is no longer elementary but involves one in Calderón-Zygmund singular integrals. We hope to return to this point in later work.

Finally, let us prove a quasi-balayage result which will be needed in future work. Let σ denote a *distribution of order one* on $\mathbb{R}^n$, with support in the unit ball. This means σ satisfies the following (equivalent) conditions. (Here B denotes the open unit ball.)

(D1) *σ is a linear map from $C^\infty\left(\mathbb{R}^n\right)$ to $\mathbb{R}$ (or $\mathbf{C}$, as the case may be) satisfying*

$$|\langle \varphi, \sigma \rangle| \leq C \, |||\varphi||| \tag{10.12}$$

where $\langle \varphi, \sigma \rangle$ denotes the action of σ on φ and

$$|||\varphi||| := \sup \left(|\varphi(x)| + \|\operatorname{grad} \varphi(x)\|\right) \ \ \text{over } x \in \overline{B}. \tag{10.13}$$

(D2) *There are measures μ_j $(j = 0, 1, \ldots, n)$ in $M\left(\mathbb{R}^n\right)$ supported in $\overline{B}$ with*

$$\sum_{j=0}^{n} \|\mu_j\|_M \leq C_1 \tag{10.14}$$

and

$$\langle \varphi, \sigma \rangle = \int \varphi \, d\mu_0 + \sum_{j=1}^{n} \int \left(\partial_j \varphi\right) d\mu_j \tag{10.15}$$

for all $\varphi \in C^\infty\left(\mathbb{R}^n\right)$.

Remarks. The constants C, C_1 in (10.12), (10.15) are comparable. The measures μ_i in (10.15) are of course not unique. Also, the test class $C^\infty\left(\mathbb{R}^n\right)$ could indifferently be replaced by $C^1\left(\mathbb{R}^n\right)$, $C_c^\infty\left(\mathbb{R}^n\right)$, $C^\infty\left(\overline{B}\right)$, $C^1\left(\overline{B}\right)$ without changing the class of σ so defined. Finally, these definitions are equivalent to each of the following:

(D3) *There are measures μ_j as in (D2) and satisfying (10.14), with*

$$\hat{\sigma}(\xi) = \hat{\mu}_0(\xi) - \sum_{j=1}^{n} i\,\xi_j \, \hat{\mu}_j(\xi). \tag{10.16}$$

(D4) *There are measures μ_j as in (D2) and satisfying (10.14) with*

$$\sigma = \mu_0 + \sum_{j=1}^{n} \partial_j \mu_j \tag{10.17}$$

in the sense of distributions.

The verification of these equivalences is elementary and left to the reader.

Proposition 10.6. *If σ is a distribution of order one with support in $\overline{B}$ and $\sigma @ HP_2$, then U^σ is integrable with $\|U^\sigma\| \leq C(n)\,\|\sigma\|_D$, where $\|\sigma\|_D$ denotes either of the constants C in (10.12), or C_1 in (10.14).*

Proof: Using (10.16) we have

$$\left(U^\sigma\right)^{\wedge}(\xi) = |\xi|^{-2}\left[\hat{\mu}_0(\xi) - \sum_{j=1}^{n}\left(i\,\xi_j\right)\hat{\mu}_j(\xi)\right].$$

From this, by reasoning used earlier, the restriction of $(U^\sigma)^\wedge(\xi)$ to $\{|\xi| > 1\}$ is in the restriction of $A(\mathbb{R}^n)$. As already remarked, the restriction to $B(0;2)$ is in the restriction of $A(\mathbb{R}^n)$ there, by virtue of $\sigma @ HP_2$ alone.

This shows $U^\sigma \in L^1(\mathbb{R}^n)$. Moreover, for $\Delta\varphi \in S$ as above

$$\int U^\sigma(\Delta\varphi)\,dx = -\langle\varphi,\sigma\rangle = -\left[\int \varphi\,d\mu_0 + \sum_{j=1}^{n}\int(\partial_j\varphi)d\mu_j\right].$$

Using (10.8) gives

$$\left|\int U^\sigma(\Delta\varphi)\,dx\right| \leq A\sum_{j=0}^{n}\|\mu_j\|_M$$

for a suitable constant A (depending only on n and the choice of the set E in defining the class S), and now an application of Lemma 10.4, with $g = U^\sigma$, finishes the proof. $\qquad\square$

In precisely the same way, one demonstrates

Proposition 10.7. *If σ is a distribution of order one with support in $\overline{B}$, and $\sigma @ HP_1$, then $\partial_j U^\sigma$ is integrable with $\|\partial_j U^\sigma\|_1 \leq C(n)\|\sigma\|_D$.*

References

[AGG] C. Amrouche, V. Girault and J. Giroire: *Weighted Sobolev spaces for Laplace's equation in* $\mathbb{R}^n$, J. Math. Pures Appl. **73** (1994), 579–606.

[GHR] M. Goldstein, W. Haussmann and L. Rogge: *A harmonic quadrature formula characterizing bi–infinite cylinders*, Michigan Math. J. **42** (1995), 175–191.

[HKS] W. K. Hayman, L. Karp and H. S. Shapiro: *Newtonian capacity and quasi-balayage* (preliminary title), in preparation.

[JW] A. Jonsson and H. Wallin: *Local polynomial approximation and Lipschitz type conditions on general closed sets*, Univ. of Umeå, Math. Dept., Research Report #1, 1980.

[K] L. Karp, *Generalized Newtonian potential and its applications*, J. Math. Anal. Appl. **174** (1993), 480–497.

[KM] L. Karp and A. Margulis, *Newtonian potential theory for unbounded sources and applications to free boundary problems*, J. Analyse Math., to appear.

[KS] L. Karp and H. Shahgholian, *Regularity of a free boundary problem,* Research Reports Math. #**2**, (1996), Dept. of Math., Stockholm University.

[L] N. S. Landkof, *Foundations of Modern Potential Theory,* Springer, Berlin–Heidelberg–New York 1972.

[Le] J. L. Lewis, *Uniformly fat sets,* Trans Amer. Math. Soc. **308** (1988), 177–196.

[S1] H. S. Shapiro, *Domains allowing exact quadrature identities for harmonic functions — an approach based on PDE,* in *Anniversary Volume on Approximation Theory and Functional Analysis,* P. L. Butzer et al. (eds.), Internat. Ser. Numer. Math. **65**, 335–354, Birkhäuser, Basel–Boston–Stuttgart 1984.

[S2] H. S. Shapiro, *A weighted L^1 estimate for the Laplace operator,* in *Recent Advances in Fourier Analysis and its Applications,* J. S. Byrnes and J. F. Byrnes (eds.), NATO ASI Series **C 315**, Kluwer, Dordrecht 1990.

[S3] H. S. Shapiro, *Global geometric aspects of Cauchy's problem for the Laplace operator,* in *Geometrical and Algebraical Aspects in Several Complex Variables,* C. A. Berenstein and D. C. Struppa (eds.), Proc. Conf. Cetraro, Italy, 309–324, EditEl Publ. 1991.

[S4] H. S. Shapiro, *An algebraic theorem of E. Fischer and the holomorphic Goursat problem,* Bull. London Math. Soc. **21** (1989), 513–537.

[S5] H. S. Shapiro, *The Schwarz Function and its Generalization to Higher Dimensions,* Wiley, New York 1992.

[S6] H. S. Shapiro, *Approximation by trigonometric polynomials to periodic functions of several variables,* in *Abstract Spaces and Approximation,* P.L. Butzer and B. Sz.-Nagy (eds.), Internat. Ser. Numer. Math. **10**, 203–217, Birkhäuser, Basel–Boston–Stuttgart 1969.

Address:

HAROLD S. SHAPIRO
Mathematics Department
Royal Institute of Technology
S–100 44 Stockholm
Sweden

Interpolation and Hyperinterpolation on the Sphere

I. H. Sloan

Abstract

This paper is concerned with interpolation and hyperinterpolation on the sphere $S^{r-1} \subseteq \mathbb{R}^r$ by polynomials of degree $\leq n$. (Hyperinterpolation was introduced in I.H. Sloan, J. Approx. Theory, **83** (1995), 238–254.) We study the interpolation operators Λ_n and hyperinterpolation operators L_n as maps from $C(S^{r-1})$ to $L^2(S^{r-1})$. In this setting we prove an apparently new result for interpolation, that for $r \geq 3$ and $n \geq 3$ every choice of the interpolation points $x_1, \ldots, x_d$ (where $d = d_n$ is the dimension of the space of polynomials of degree $\leq n$) leads to $\|\Lambda_n\| >| S^{r-1} |^{\frac{1}{2}}$. In contrast, for the hyperinterpolation operator it is known that $\|L_n\| =| S^{r-1} |^{\frac{1}{2}}$. The proof of the latter result is sketched, and the still unanswered questions for both methods are discussed.

1 Introduction

In this paper we consider interpolation and hyperinterpolation on the unit sphere $S^{r-1} \subseteq \mathbb{R}^r$ (where hyperinterpolation is a generalisation of interpolation to be explained later), in the setting in which the operators map from $C(S^{r-1})$ to $L^2(S^{r-1})$. The image space for the interpolation and hyperinterpolation operators is P_n, the set of polynomials of degree $\leq n$ restricted to S^{r-1}. Let $\Lambda_n = \Lambda_n(x)$ denote the interpolation operator for the fundamental system $x = \{x_1, x_2, \ldots, x_d\} \subseteq S^{r-1}$, where $d = d_n$ is the dimension of P_n. For $f \in C(S^{r-1})$ the interpolant $\Lambda_n f$ is defined by

$$\Lambda_n f \in C(S^{r-1}), \ \Lambda_n f(x_j) = f(x_j), \ j = 1, \ldots, d.$$

Let the norm of Λ_n be

$$\|\Lambda_n\| = \sup \left\{ \|\Lambda_n f\| : f \in C(S^{r-1}), \ \|f\|_\infty \leq 1 \right\}, \tag{1.1}$$

where

$$\|\Lambda_n f\| = \left(\int_{S^{r-1}} | f(x) |^2 \, dx \right)^{\frac{1}{2}}, \tag{1.2}$$

Multivariate Approximation: Recent Trends and Results; W. Haußmann, K. Jetter and M. Reimer (eds.)
Mathematical Research, Vol. 101, pp. 255–268, ISBN 3–05–501770–6
© Akademie–Verlag, Berlin 1997

with dx denoting surface measure for the sphere S^{r-1}, and

$$\|f\|_\infty = \sup_{x \in S^{r-1}} |\, f(x)\,|, \; f \in C(S^{r-1}).$$

How small can $\|\Lambda_n\|$ be? A trivial lower bound is

$$\|\Lambda_n\| \geq |\, S^{r-1}\,|^{\frac{1}{2}}, \tag{1.3}$$

where $|\, S^{r-1}\,|$ is the surface area of the unit sphere S^{r-1}. This follows from

$$\|\Lambda_n 1\|^2 = \int_{S^{r-1}} 1^2 dx = |\, S^{r-1}\,|,$$

where 1 is the polynomial with the constant value 1. In Section 2 we show that this lower bound is not attainable except in very special cases. Specifically, we prove there the following theorem:

Theorem 1.1. *Let $r \geq 3$ and $n \geq 3$. Then*

$$\|\Lambda_n\| > |\, S^{r-1}\,|^{\frac{1}{2}}\,.$$

As far as we are aware, this result is new.

In this paper we also review hyperinterpolation, introduced recently in [9]. The hyperinterpolation operator L_n, like the interpolation operator Λ_n, is a projection defined on $C(S^{r-1})$ with range P_n, but its norm is better behaved. A specific instance of hyperinterpolation is defined not by choosing a set of points, as in the interpolation case, but rather by choosing a quadrature rule Q,

$$Qf = \sum_{k=1}^{m} w_k g(x_k) \approx \int_{S^{r-1}} g(x)dx, \tag{1.4}$$

with the requirement that the weights $\{w_k\}$ are positive, the points $\{x_k\}$ belong to S^{r-1}, and, importantly, that the rule integrates exactly all polynomials of degree $\leq 2n$. It is known, as we discuss later, that such rules always exist.

Once the rule Q is chosen, we construct $L_n f$, for $f \in C(S^{r-1})$, by the following explicit formula:

$$L_n f = \sum_{j=1}^{d} (f, p_j)_m p_j, \tag{1.5}$$

where

$$(v, z)_m = \sum_{k=1}^{m} w_k v(x_k) z(x_k) = Q(vz), \tag{1.6}$$

and where $\{p_j\}_{j=1}^d$ is any orthonormal basis of P_n (for example, we may take $\{p_j\}$ to be the normalized spherical harmonics of degree $\leq n$). The quantity $(v, z)_m$ is therefore just a quadrature approximation to the exact inner product

$$(v, z) = \int_{S^{r-1}} v(x)z(x)\,dx. \tag{1.7}$$

Note that orthonormality of the set $\{p_j\}_{j=1}^d$ holds equally with respect to $(\cdot, \cdot)_m$ and $(\cdot, \cdot)$, because $p_j p_k$ is a polynomial of degree $\leq 2n$.

An alternative definition of $L_n f$ is

$$L_n f \in P_n, \ (L_n f, \chi)_m = (f, \chi)_m, \ \text{for all } \chi \in P_n, \ \text{and all } f \in C(S^{r-1}). \tag{1.8}$$

Thus $L_n f$ is a discrete approximation to the L^2 projection onto P_n, obtained by applying a quadrature approximation to the inner product. It is easily seen that L_n is a projection, since (1.8) is satisfied by $L_n f = f$ if $f \in P_n$.

A key property of the hyperinterpolation operator constructed in this way is the stability property stated in the following theorem. The result, which is in striking contrast to the result for interpolation stated in Theorem 1.1 above, is proved in [9]. The norm $\| \cdot \|$ is defined as in (1.1).

Theorem 1.2. *Under the condition that the rule Q is exact for all polynomials of degree $\leq 2n$,*

$$\|L_n\| = |\, S^{r-1}\, |^{\frac{1}{2}}\, .$$

It is well known that the norm of Λ_n or L_n plays an important role in the estimation of the error in $\Lambda_n f$ or $L_n f$, through

$$\|\Lambda_n f - f\| = \|\Lambda_n(f - \chi) - (f - \chi)\|,$$

for χ an arbitrary polynomial in P_n, from which follows

$$\|\Lambda_n f - f\| \leq \left(\|\Lambda_n\| + |\, S^{r-1}\, |^{\frac{1}{2}}\right) E_n(f), \tag{1.9}$$

where

$$E_n(f) = \inf\{\|f - \chi\|_\infty : \chi \in P_n\}, \tag{1.10}$$

the error of best uniform approximation of f by an element of P_n.

The proof of Theorem 1.1 is presented in the next section. Then in Section 3 we discuss hyperinterpolation, and sketch the proof of Theorem 1.2.

In Section 4 we consider the possible quadrature rules for hyperinterpolation. Finally, in Section 5 we compare the results for interpolation and hyperinterpolation, and draw attention to some still unanswered questions.

2   Proof of Theorem 1.1

We now prove Theorem 1.1. Let $l_1, l_2, \ldots, l_d$ be the Lagrange polynomials for the fundamental system $x = \{x_1, x_2, \ldots, x_d\}$, i.e. $l_j \in P_n, j = 1, \ldots, d$, and

$$l_j(x_k) = \delta_{j,k}, \quad j, k = 1, \ldots, d.$$

(The existence of the Lagrange polynomials follows from the circumstance that $\{x_1, x_2, \ldots, x_d\}$ is a fundamental system.) Then we may write, as usual,

$$\Lambda_n f = \sum_{j=1}^{d} f(x_j) l_j,$$

from which follows

$$\|\Lambda_n f\|^2 = (\Lambda_n f, \Lambda_n f) = \sum_{j=1}^{d} \sum_{k=1}^{d} f(x_j) L_{j,k} f(x_k), \tag{2.1}$$

where

$$L_{j,k} = (l_j, l_k). \tag{2.2}$$

The set $\{l_j\} \subseteq P_n$ is linearly independent because $\{x_1, x_2, \ldots, x_d\}$ is a fundamental system, thus the Gram matrix L is nonsingular, and L has d positive eigenvalues. Much is known about the matrix L; see, in particular [7]. It is closely related to the 'reproducing kernel' $G(x, y)$, defined by

$$G(x, y) = \sum_{j=1}^{d} p_j(x) p_j(y), \tag{2.3}$$

whose reproducing property in the space P_n is expressed by the easily verified property

$$(f, G(x, .)) = f(x), \forall f \in P_n, \forall x \in S^{r-1}. \tag{2.4}$$

Let $g_k, k = 1, \ldots, d$, be the polynomials in P_n defined by

$$g_k(x) = G(x_k, x), k = 1, \ldots, d.$$

The set $\{g_k\} \subseteq P_n$ is linearly independent, since

$$(l_j, g_k) = (l_j, G(x_k, \cdot)) = l_j(x_k) = \delta_{j,k}, \quad j, k = 1, \ldots, d. \tag{2.5}$$

Now define $G = \{G_{j,k}\}$ to be the matrix defined by

$$G_{j,k} = (g_j, g_k) = G(x_j, x_k), \; j, k = 1, \ldots, d.$$

Then it is known that G is the inverse of L,

$$L^{-1} = G.$$

This follows from (2.5), which implies (since both $\{l_j\}$ and $\{g_k\}$ are bases for P_n)

$$
\begin{aligned}
l_j &= \sum_{k=1}^{d} (l_j, l_k) \, g_k, \\
g_k &= \sum_{j=1}^{d} (g_k, g_j) \, l_j,
\end{aligned}
$$

so that

$$\delta_{j,k} = (l_j, g_k) = \sum_{m=1}^{d} (l_j, l_m)(g_m, g_k) = (LG)_{j,k} \, .$$

To prove Theorem 1.1 it is clearly enough to prove the existence of $f^\star \in C(S^{r-1})$ such that $\|f^\star\|_\infty = 1$, and

$$\|\Lambda_n f^\star\| > | S^{r-1} |^{\frac{1}{2}} \, . \tag{2.6}$$

The proof that follows is not constructive, in that we prove the existence of such a function $f^\star$ indirectly, by an averaging argument. We may view the expression (2.1) as a positive–definite quadratic form

$$\|\Lambda_n f\|^2 = \sum_{j=1}^{d} \sum_{k=1}^{d} \xi_j L_{j,k} \xi_k,$$

where we have replaced $f(x_j)$ by ξ_j.

This quadratic form is defined over the unit cube, since

$$| \xi_j | = | f(x_j) | \leq 1 \quad \text{for} \quad j = 1, \ldots, d,$$

thus $\|\Lambda_n f\|^2$ must take its maximum value at one of the vertices of the unit cube; that is, when f takes one or other of the extreme values ± 1 at each of the interpolation points x_j. We shall define 2^d such functions f, and average over them. Thus we define $f_i = f_{i_1, i_2, \ldots, i_d} \in C(S^{r-1})$, where each component of $i = (i_1, i_2, \ldots, i_d)$ takes the values ± 1 independently, and where

$$f_i(x_j) = i_j \text{ for } j = 1, \ldots, d, \ \|f_i\|_\infty = 1. \tag{2.7}$$

Now we compute

$$
\begin{aligned}
\operatorname*{average}_i \|\Lambda_n f_i\|^2 &= \operatorname*{average}_i \sum_{j=1}^{d} \sum_{k=1}^{d} i_j L_{j,k} i_k \\
&= \operatorname*{average}_i \left(i_1 L_{1,1} i_1 + i_1 L_{1,2} i_2 + \cdots \right).
\end{aligned}
$$

We see something interesting arising from averaging the components of $i = (i_1, i_2, \ldots, i_d)$ over ± 1: that the off–diagonal terms in this expression all average to zero, while the jth diagonal term averages to $L_{j,j}$ because $(i_j)^2 = 1$. Thus

$$\operatorname*{average}_i \|\Lambda_n f_i\|^2 = \sum_{j=1}^{d} L_{j,j}.$$

Since there must be at least one f_i for which the value of $\|\Lambda_n f_i\|$ is as big as the average value, we conclude from the definition of $\|\Lambda_n\|$ that

$$\|\Lambda_n\|^2 \geq \sum_{j=1}^{d} L_{j,j}. \tag{2.8}$$

Now let $\lambda_1, \lambda_2, \ldots, \lambda_d$ be the eigenvalues of $G = L^{-1}$. We know that $\lambda_j > 0$ for $j = 1, \ldots, d$. Thus we can write

$$\sum_{j=1}^{d} L_{j,j} = \operatorname{tr} L = \sum_{j=1}^{d} \frac{1}{\lambda_j}.$$

A well known theorem says that the arithmetic mean of a finite set $\{a_1, \ldots, a_n\}$ of positive numbers is greater than or equal to the harmonic mean of the set, that is

$$\frac{a_1 + \ldots + a_n}{n} \geq \frac{n}{a_1^{-1} + \ldots + a_n^{-1}}, \tag{2.9}$$

with equality if and only if $a_1 = a_2 = \ldots = a_n$. Applying this result to the reciprocal eigenvalues $\{\lambda_1^{-1}, \ldots, \lambda_d^{-1}\}$, we obtain

$$\sum_{j=1}^{d} L_{j,j} \geq \frac{d^2}{\sum_{j=1}^{d} \lambda_j},$$

with equality if and only if

$$\lambda_1 = \ldots = \lambda_d.$$

Now we use (see [7], p. 127)

$$\sum_{j=1}^{d} \lambda_j = \operatorname{tr} G = \sum_{k=1}^{d} G(x_k, x_k) = dG(x, x) = \frac{d^2}{\mid S^{r-1} \mid},$$

where we made use of the well known fact (see [7]) that $G(x, x)$ is independent of x. Thus we have established

$$\sum_{j=1}^{d} L_{j,j} \geq \mid S^{r-1} \mid,$$

with equality if and only if

$$\lambda_1 = \ldots = \lambda_d.$$

But it is known (again see [7], p. 127) that equality of the eigenvalues is impossible if $r \geq 3$ and $n \geq 3$. Briefly, the argument proceeds by showing that if the eigenvalues were equal then this would imply that the 'Lagrangian square sums' satisfy

$$\sum_{j=1}^{d} l_j(x)^2 = 1, \forall x \in S^{r-1}.$$

But from a result of Bos [3] it is known that the latter condition would imply that the points $\{x_1, x_2, \ldots, x_d\}$ form what is known as a spherical $2n$ design; that is to say, that an equal weight quadrature rule based on these points would integrate all polynomials of degree $\leq 2n$ exactly. The concept of a spherical t design (that is, a set of points for which the equal weight rule based on these points integrates all polynomials of degree $\leq t$ exactly) was introduced in [4], who also showed that every spherical $2n$ design necessarily has $m \geq d = d_n$ points. A design with exactly $m = d$ points is called a 'tight' spherical design. That is what we would have in the present case, because the number of points in the fundamental system $\{x_1, x_2, \ldots, x_d\}$ is d. But (and this is the crux of the argument) it is also known from [2] that there are *no* tight spherical $2n$ designs when $r \geq 3$ and $n \geq 3$.

Thus at the end of this chain of argument we conclude that for $r \geq 3$ and $n \geq 3$

$$\sum_{j=1}^{d} L_{j,j} > \mid S^{r-1} \mid,$$

where the inequality is strict. In turn it follows from (2.8) that

$$\|\Lambda_n\| >\mid S^{r-1}\mid^{\frac{1}{2}},$$

and Theorem 1.1 is proved.

3   Hyperinterpolation on the Sphere

Now we look in more detail at the hyperinterpolation approximation ([9])

$$L_n f = \sum_{j=1}^{d_n} (f, p_j)_m p_j,$$

where $(v, z)_m$ is the approximation to the exact inner product (v, z) obtained by application of the m–point positive–weight quadrature rule Q, see (1.4) and (1.6). The crucial assumption about the rule Q is that its degree of precision is at least $2n$, where n is the degree of the polynomial space P_n.

Our first task is to sketch the proof of Theorem 1.2 given in [9]; that is, we must show that $\|L_n\|$ has the exact value $\mid S^{r-1}\mid^{\frac{1}{2}}$ under the above assumption on the degree of precision of Q.

Suppose we are given $f \in C(S^{r-1})$. Because $(L_n f)^2$ is a polynomial of degree $\leq 2n$, the square of the L^2 norm of $(L_n f)^2$ is

$$\|L_n f\|^2 = (L_n f, L_n f) = (L_n f, L_n f)_m.$$

From the lemma below, we see that the right hand side of the last expression is bounded above by $(f, f)_m$, thus we can conclude

$$\|L_n f\|^2 \leq (f, f)_m = \sum_{k=1}^{m} w_k f(x_k)^2 \leq \sum_{k=1}^{m} w_k \|f\|_\infty^2 =\mid S^{r-1}\mid \|f\|_\infty^2,$$

where the last step follows because the rule Q is exact for the constant polynomial 1. It follows that $\|L_n\| \leq\mid S^{r-1}\mid^{\frac{1}{2}}$, and the proof of Theorem 1.2 is completed by use of the corresponding lower bound, which is proved in the same manner as (1.3).

Thus to complete the proof all that remains is to prove the following lemma:

Lemma 3.1. *For $f \in C(S^{r-1})$,*

$$(L_n f, L_n f)_m \leq (f, f)_m. \tag{3.1}$$

To prove this lemma we need only observe that from the alternative definition (1.8) we have

$$(f - L_n f, \chi)_m = 0, \forall \chi \in P_n,$$

from which follows immediately

$$(L_n f, L_n f)_m = (f, f)_m - (f - L_n f, f - L_n f)_m \le (f, f)_m,$$

completing the proof of the lemma, and in turn of Theorem 1.2.

4 Quadrature on the Sphere

Now that Theorem 1.2 is proved, it is natural that we ask some questions about the quadrature rule Q. We have seen that the hyperinterpolation approximation $L_n f$ achieves an ideally small norm when considered as a mapping from $C(S^{r-1})$ to $L^2(S^{r-1})$, a property denied to the interpolation approximation $\Lambda_n f$ except when r or n is small. But the construction of $L_n f$ requires that the quadrature rule Q be a positive weight rule with degree of precision $2n$. Do such rules always exist? And if they do exist, how many points do they need?

To help us in considering these questions, it is useful first to note the following simple results from [9].

Theorem 4.1. *Let L_n be the hyperinterpolation approximation ([9]) constructed with a quadrature rule Q that is exact for all polynomials of degree $\le 2n$. Then*

(a) $m \ge d$,

(b) $m = d$ *if and only if $L_n f$ is interpolatory,*

(c)   *If $m = d$ then $w_1 = \cdots = w_m = \dfrac{|S^{r-1}|}{m}$.*

These results all follow easily by observing the properties of the $d \times m$ matrix $P = \{p_{j,k}\}$, defined by

$$p_{j,k} = \sqrt{w_k} p_j(x_k), \; j = 1, \ldots, d, \; k = 1, \ldots, m.$$

The rows of this matrix are orthonormal, from which follows immediately $m \ge d$. If $m > d$ then it is of course impossible for the set $\{x_k\}_{k=1}^m$ to be a fundamental system for interpolation. On the other hand, if $m = d$ then the matrix P is an orthogonal matrix, and therefore has orthonormal columns, from which it follows that

$$\sum_{j=1}^{d} p_j(x_k)p_j(x_{k'}) = \frac{1}{w_k}\delta_{k,k'}, \quad k,k' = 1, \ldots, m = d, \tag{4.1}$$

and now it follows from the definition ([9]) of $L_n f$ that

$$L_n f(x_k) = f(x_k), \quad k = 1, \ldots, m = d,$$

which proves part (b). It also follows from (4.1) on setting $k = k'$ that

$$w_k = \left(\sum_{j=1}^{d} p_j(x_k)^2\right)^{-1} = G(x_k, x_k)^{-1}, \quad k = 1, \ldots, m = d.$$

As we have remarked before, $G(x, x)$ is independent of x, thus the weights w_k are all equal, and part (c) of the theorem now follows from the observation that

$$\sum_{k=1}^{m} w_k = \mid S^{r-1} \mid.$$

The last part of the last theorem has an important consequence. That result tells us that if $m = d$ then the weights in the quadrature rule Q on which the hyperinterpolation approximation is built must be equal. Since the rule must also integrate exactly all polynomials of degree $\leq 2n$, the rule Q must be a spherical $2n$ design (see Section 2); and moreover it must be a *tight* spherical $2n$ design, since $m = d$. But we may also recall from Section 2 that, as shown by [2], there are *no* tight spherical $2n$ designs if $r \geq 3$ and $n \geq 3$. Thus we have the following result.

Theorem 4.2. *Let $L_n f$ be the hyperinterpolation approximation (1.4), with an m-point quadrature rule Q that is exact for all polynomials of degree $\leq 2n$. If $r \geq 3$ and $n \geq 3$ then $m > d$.*

Note that this result does not tell us how many quadrature points are needed if $r \geq 3$ and $n \geq 3$, only that the number is certainly greater than d, so that the hyperinterpolation approximation cannot be interpolatory.

So now let us return to our question about quadrature rules Q that are suitable for hyperinterpolation. Do positive weight quadrature rules of precision $2n$ always exist? The answer is a decisive yes, because it is known that such rules exist even if we insist that all the weights are equal! A celebrated result by Seymour and Zaslavsky [8] says that a spherical t design on S^{r-1} exists for every value of $t \geq 0$ and every $r \geq 2$. It follows that an equal weight quadrature rule Q of precision $2n$ exists for every $r \geq 2$ and every $n \geq 0$. This result allows us to answer our first question in the affirmative. On the other hand, it does not say how large the number of points m needs to be, nor does it tell us how to construct efficient spherical designs.

We have remarked before that for every spherical $2n$ design on S^{r-1} we must have at least d points, that is $m \geq d$, and that spherical designs with $m = d$ cannot exist if $r \geq 3$ and $n \geq 3$. Little is known about practical spherical designs on the sphere. Concrete constructions of spherical t designs have been given for all values of r and t (see [1] and [6]), but it is thought that the numbers of points in these constructions are far from best possible. So what quadrature rules of precision $\geq 2n$ exist with a reasonable number of quadrature points m? And how large a number of points m do they need? Various positive weight rules of the appropriate precision are discussed in [9]. For the case of the sphere $S^2 \subseteq R^3$ the known rules with the smallest numbers of points m seem to be those proposed by Lebedev [5], which follow ideas of Sobolev [10]. For the Lebedev rules it is pointed out in [9] that

$$\frac{m}{d} \to \frac{4}{3} \text{ as } n \to \infty.$$

Whether better rules than this exist in the sense of having smaller values of m for the same precision is not currently known. Product Gauss rules, which are certainly easier to use, are about 50% more expensive again, having

$$\frac{m}{d} = 2.$$

For larger values of the dimension r even less is known.

It should also be said that there have been as yet very few implementations of the hyperinterpolation construction. One such, however, is given by Wienert [11], who used what we here call hyperinterpolation with product Gauss rules as a practical tool for approximation of smooth functions over the sphere S^2.

5   Discussion

In this paper we have compared the properties of the interpolation approximation $\Lambda_n f$ and the hyperinterpolation approximation $L_n f$. The former is defined by choosing a fundamental system of points $\{x_1, \ldots, x_d\}$ on the sphere, the latter by choosing an m–point quadrature rule which integrates exactly all polynomials of degree $\leq 2n$.

The interesting cases to consider are those in which both $r \geq 3$ and $n \geq 3$. In this situation we know from the discussion in Section 4 that $m > d$. In other words, for these nontrivial values of r and n hyperinterpolation requires more information, in the sense of more function values, than interpolation.

But then we know also, from Theorems 1.1. and 1.2, that when $r \geq 3$ and $n \geq 3$

$$\|\Lambda_n\| > \|L_n\| = |\, S^{r-1}\, |^{\frac{1}{2}},$$

suggesting that the quality of the hyperinterpolation approximation is generally better than that of interpolation, no matter how the interpolation points may be chosen.

Can we quantify the difference in quality between the interpolation and hyperinterpolation approximations? One measure would be the size of

$$\Lambda(n) = \inf \left\{ \|\Lambda_n\| : (x_1, x_2, \ldots, x_d) \in (S^{r-1})^d \right\},$$

compared to $\|L_n\| = |\, S^{r-1}\, |^{\frac{1}{2}}$. At the moment, however, there seems to be no information on this, not even empirical information. Is $\Lambda(n)$ bounded or unbounded as $n \to \infty$ for fixed r? Is the ratio of $\Lambda(n)/\|L_n\|$ bounded or unbounded as $r \to \infty$ for fixed n? There are also analogous questions for the hyperinterpolation approximations: if $m(n)$ denotes the minimum number of points for quadrature rules that are exact for all polynomials of degree $\leq 2n$, how does the ratio $m(n)/d_n$ behave as $n \to \infty$ for fixed r, and as $r \to \infty$ for fixed n?

Clearly, much more work is needed before we can claim to really understand interpolation and hyperinterpolation on the sphere in the setting $C \to L^2$.

6 Acknowledgments

The support of the Australian Research Council is gratefully acknowledged. The author is also grateful to Dr Manfred Reimer and the organisers and participants of the 1996 Multivariate Approximation Conference at Haus Bommerholz, at which a version of this work was presented, and during which Theorem 1.1 was proved.

References

[1]   B. Bajnok: *Chebyshev quadrature formulas on the sphere*, Congr. Numer. **85** (1991), 214–218.

[2]   E. Bannai, E. M. Damerell: *Tight spherical designs I*, J. Math. Soc. Japan **31** (1979), 199–207.

[3]   L. Bos: *Some remarks on the Fejér problem for Lagrange interpolation in several variables*, J. Approx. Theory **60** (1990), 133–140.

[4]   P. Delsarte, J. M. Goethals, J .J. Seidel: *Spherical codes and designs*, Geom. Dedicata **6** (1977), 363–388.

[5]   V. I. Lebedev: *Quadrature on a sphere*, Comput. Math. Math. Phys. **16**, no. 2, (1976), 10–24.

[6]   P. Rabau, B. Bajnok, *Bounds for the number of nodes in Chebshev type quadrature formulas*, J. Approx. Theory **67** (1991), 199–214.

[7]   M. Reimer: *Constructive Theory of Multivariate Functions*, B.I. Wissenschaftsverlag, Mannheim–Wien–Zürich 1990.

[8]   P. D. Seymour, T. Zaslavsky: *Averaging sets: a generalisation of mean values and spherical designs*, Adv. Math. **52** (1984), 213–240.

[9]   I. H. Sloan: *Polynomial interpolation and hyperinterpolation over general regions*, J. Approx. Theory **83** (1995), 238–254.

[10]   S. L. Sobolev: *Cubature formulae on the sphere invariant under finite groups of rotations*, Soviet Math. Dokl. **3** (1962), 1307–1310.

[11]   L. Wienert: *Die numerische Approximation von Randintegraloperatoren für die Helmholtzgleichung in* $\mathbb{R}^3$, Dissertation, Göttingen 1990.

Address:

Ian H. Sloan
School of Mathematics
University of New South Wales
Sydney 2052
Australia

Interpolation and Wavelets on Sparse Gauß–Chebyshev Grids

F. Sprengel

Abstract

Nested spaces of multivariate functions on the square forming a non-stationary multiresolution analysis are investigated. The scaling functions of these spaces are fundamental Lagrange interpolants on a sparse Gauß–Chebyshev grid. The approach based on Boolean sums leads to sample spaces of significantly lower dimension. The algorithms for complete decomposition and reconstruction are of simple structure and low complexity.

1 Introduction

We are interested here in wavelet methods on a bounded domain, in our case on the square $[-1, 1]^2$, which are based on the construction of univariate wavelets on an interval. Therefore, we consider special bivariate sample and wavelet spaces of functions on the square. As our starting point we take univariate interpolatory scaling functions and wavelets similar to those introduced by K-ilgore and Prestin [11], which were recently further developed and generalized by Plonka, Selig, and Tasche [12, 17]. This approach works for the weighted Hilbert space $L^2_w([-1, 1])$ and with generalized shifts for functions on the interval instead of special scaling functions and wavelets at the boundary as in e.g. [4, 5, 13]. We want to introduce here new bivariate interpolatory s-caling functions and wavelets by investigating Boolean sums of interpolation operators. In this way, we overcome the problem of a very fast growth of the dimensions that we have in the ordinary tensor product case. That means, we obtain sample spaces with dimension $\mathcal{O}(j\, 2^j)$ while tensor products yield dimension $\mathcal{O}(2^{2j})$ for the level j. However, because of the simple structure of the spaces one can take use of fast algorithms provided in [12] for the univari-ate case. The concept of Boolean sums and sparse grids is often applied in interpolation and approximation (see e.g. [1, 6]) as well as in finite element methods (see e.g. [9, 10, 20]). It uses essentially the interpolatory properties of

Multivariate Approximation: Recent Trends and Results; W. Haußmann, K. Jetter and M. Reimer (eds.)
Mathematical Research, Vol. 101, pp. 269–286, ISBN 3–05–501770–6
© Akademie–Verlag, Berlin 1997

the underlying univariate scaling functions and wavelets. Similar approaches are the approximation of periodic functions by trigonometric polynomials with frequencies from the hyperbolic cross (see e.g. [8, 18, 19]) and the investigation of hyperbolic wavelets [7] in the cardinal case.

Applying the Boolean sum method, we are able to construct multivariate scaling functions and wavelets from univariate interpolatory scaling functions and wavelets. We obtain the bivariate sample spaces

$$V_j^B = \sum_{r=0}^{j} V_r \otimes V_{j-r},$$

and the bivariate wavelet spaces

$$W_{j+1}^B = \bigoplus_{r=0}^{j+1}{}^{\perp} W_r \otimes W_{j+1-r},$$

from the underlying univariate sample spaces V_r and wavelet spaces W_r, respectively. The univariate sample spaces V_r are images of interpolation operators L_r. Analogously, the bivariate sample spaces V_j^B are images of blending interpolation operators B_j defined by j-th order Boolean sums.

We can give the algorithms for complete decomposition and reconstruction which are based on the algorithms for the univariate case. They have simple structure and low complexity of $\mathcal{O}(j^2\,2^j)$ essential operations for the level j.

2   Univariate Interpolation and Sample and Wavelet Spaces on the Interval

Let $d \in \mathbb{N}$ be a fixed parameter. Put $d_j := d2^j, j \in \mathbb{N}_0$. We denote our reference interval by $I := [-1,1]$ and the Chebyshev weight by $w(x) := (1-x^2)^{-1/2}\,(x \in (-1,1))$. Let $L_w^2(I)$ be the weighted Hilbert space of all measurable functions $f : I \longrightarrow \mathbb{R}$ with

$$\int_I f(x)^2 w(x)\,dx < \infty.$$

For $f, g \in L_w^2(I)$, the corresponding inner product is given by

$$\langle f, g \rangle_I := \frac{2}{\pi} \int_I f(x)g(x)w(x)\,dx\,.$$

By Π_n we denote the set of all real valued polynomials of degree at most n restricted on I. Furthermore let $T_n \in \Pi_n$ be the Chebyshev polynomials $T_n(x) := \cos(n \arccos x)$ and $\mathcal{G}_j$ the grid of the Gauß–Chebyshev nodes

$$\mathcal{G}_j := \{g_{j,k} := \cos k\tfrac{\pi}{2^j}\ ;\ k = 0, \ldots, d_j\}.$$

The Chebyshev shift (cf. [3]) $s_h f$ of a function f is defined as

$$(s_h f)(x) := \tfrac{1}{2} f(xh - \tfrac{1}{w(x)w(h)}) + \tfrac{1}{2} f(xh + \tfrac{1}{w(x)w(h)}), \qquad x \in I.$$

For every level j, we need the shifts $\sigma_{j,k} := s_{g_{j,k}}$, $k = 0, \ldots, d_j$. We choose scaling functions ϕ_j as modified Langrange interpolants on the Gauß–Chebyshev grid, i.e.,

$$(\sigma_{j,k}\phi_j)(g_{j,\ell}) = \frac{1}{\varepsilon_{j,k}}\, \delta_{k,l}, \tag{2.1}$$

where here and in the following

$$\varepsilon_{j,k} := \begin{cases} \tfrac{1}{2} & \text{if}\quad k = 0 \text{ or } k = d_j, \\ 1 & \text{if}\quad k = 1, 2, \ldots, d_j - 1. \end{cases}$$

We define the sample spaces V_j by

$$V_j := \operatorname{span}\{\phi_{j,k} := \sigma_{j,k}\phi_j\ ;\ k = 0, \ldots, d_j\} \tag{2.2}$$

with the dimension $\dim V_j = d_j + 1$. We assume the spaces V_j to form a non–stationary multiresolution of the weighted Hilbert space $L_2(I)$, i.e.,

$$V_j \subset V_{j+1},\ j \in \mathbb{N}_0 \quad \text{and} \quad \operatorname{clos}_{L^2_w(I)} \bigcup_{j \in \mathbb{N}_0} V_j = L^2_w(I).$$

The interpolation operator L_j projecting onto $\operatorname{Im} L_j = V_j$ with the interpolation nodes $\mathcal{G}_j$ is given as

$$L_j f := \sum_{k=0}^{d_j} \varepsilon_{j,k}\, f(g_{j,k})\, \phi_{j,k}.$$

Now let us consider the univariate wavelet spaces W_{j+1} defined by

$$V_{j+1} = V_j \oplus^\perp W_{j+1},$$

where $\oplus^\perp$ denotes the orthogonal sum with respect to the inner product in $L^2_w(I)$. Assume the wavelet spaces W_{j+1} to be generated by the shifts of a single function ψ_{j+1}

$$W_{j+1} := \operatorname{span}\{\psi_{j+1,2k+1} := \sigma_{j+1,2k+1}\psi_{j+1}\ ;\ k = 0, 1, \ldots, d_j - 1\} \tag{2.3}$$

with wavelet functions $\psi_{j+1,2k+1}$ satisfying also interpolatory conditions

$$\psi_{j+1,2m+1}\left(g_{j+1,2\ell+1}\right) = \delta_{\ell,m}, \qquad \ell = 0,1,\ldots,d_j - 1. \qquad (2.4)$$

Then, we want to establish reconstruction and decomposition relation in matrix form. Therefore, we define the function vectors $\boldsymbol{\phi}_{j-1} := (\phi_{j-1,k})_{k=0}^{d_j-1}$, $\boldsymbol{\psi}_j := (\psi_{j,2k+1})_{k=0}^{d_j-1-1}$, and $\boldsymbol{P}_j\,\boldsymbol{\phi}_j := (\phi_{j,0},\phi_{j,2},\ldots,\phi_{j,d_j},\phi_{j,1},\phi_{j,3},\ldots,\phi_{j,d_j-1})^T$, the following knot evaluation and permutation matrices

$$\boldsymbol{K}_{j-1} := \left(\phi_{j-1,k}(g_{j,2\ell+1})\right)_{k=0,\,\ell=0}^{d_j-1,\,d_j-1-1}, \qquad \boldsymbol{P}_j = \left(\delta_{k,2\ell} + \delta_{k+d_{j-1}+1,2\ell+1}\right)_{k,\ell=0}^{d_j-1},$$

$$\boldsymbol{L}_j := \left(\psi_{j,2k+1}(g_{j,2\ell})\right)_{k=0,\,\ell=0}^{d_j-1-1,\,d_j-1}, \qquad \vec{\boldsymbol{P}}_{j-1} = \left(\delta_{k,d_{j-1}-\ell}\right)_{k,\ell=0}^{d_j-1-1},$$

and

$$\mathbf{C}_j^e := \left(\cos\frac{k\ell\pi}{d_j}\right)_{k,\ell=0}^{d_j}, \qquad \mathcal{E}_j := \operatorname{diag}\left(\varepsilon_{j,k}\right)_{k=0}^{d_j},$$

$$\mathbf{C}_j^o := \left(\cos\frac{(2k+1)\ell\pi}{d_j}\right)_{k,\ell=0}^{d_j-1}, \qquad \vec{\mathcal{E}}_j := \operatorname{diag}\left(\varepsilon_{j,k}\right)_{k=0}^{d_j-1}.$$

The matrices for the discrete cosine transforms are given by

$$\mathbf{C}_j^{\mathrm{I}} := \mathbf{C}_j^e\,\mathcal{E}_j \qquad : \quad \text{type-I-DCT of length } d_j + 1 \text{ (DCT-I } (d_j + 1)),$$
$$\mathbf{C}_j^{\mathrm{II}} := \mathbf{C}_j^o \qquad : \quad \text{type-II-DCT of length } d_j \text{ (DCT-II } (d_j)),$$
$$\mathbf{C}_j^{\mathrm{III}} := (\mathbf{C}_j^e)^T\,\vec{\mathcal{E}}_j \qquad : \quad \text{type-III-DCT of length } d_j \text{ (DCT-III } (d_j)).$$

They satisfy the relations

$$\mathbf{C}_j^{\mathrm{I}}\,\mathbf{C}_j^{\mathrm{I}} = \frac{d_j}{2}\,\boldsymbol{I}_{d_j+1} \qquad \text{and} \qquad \mathbf{C}_j^{\mathrm{II}}\,\mathbf{C}_j^{\mathrm{III}} = \mathbf{C}_j^{\mathrm{III}}\,\mathbf{C}_j^{\mathrm{II}} = \frac{d_j}{2}\,\boldsymbol{I}_{d_j}.$$

Fast and numerically stable algorithms for the DCT-I $(d_j + 1)$, DCT-II (d_j) and DCT-III (d_j) working in real arithmetic are described in [2, 16].

We use the notation $\boldsymbol{A} \boxplus \boldsymbol{B} := \operatorname{diag}(\boldsymbol{A}, \boldsymbol{B})$ for the direct sum of matrices.

Theorem 2.1. (i) *The two-scale or reconstruction relation is given by*

$$\begin{pmatrix} \boldsymbol{\phi}_{j-1} \\ \boldsymbol{\psi}_j \end{pmatrix} = \boldsymbol{C}_j\,\boldsymbol{\phi}_j,$$

with

$$C_j = \begin{pmatrix} I_{d_{j-1}+1} & K_{j-1} \\ \vec{\mathcal{E}}_{j-1} L_j & I_{d_{j-1}} \end{pmatrix} P_j$$
$$= \tfrac{1}{d_{j-1}} \left((C_{j-1}^{\mathrm{I}})^T \boxplus (C_{j-1}^{\mathrm{II}})^T \right) C_{j,\lambda} (C_j^{\mathrm{I}})^T .$$

Here $C_{j,\lambda}$ *is the following sparse matrix*

$$C_{j,\lambda} = \begin{pmatrix} \mathrm{diag}(\alpha_{j-1,k})_{k=0}^{d_{j-1}-1} & \mathbf{o} & \mathrm{diag}(\alpha_{j-1,k})_{k=d_{j-1}+1}^{d_j}\vec{P}_{j-1} \\ \mathbf{o} & \alpha_{j-1,d_{j-1}} & \mathbf{o} \\ \mathrm{diag}(\beta_{j,k})_{k=0}^{d_{j-1}-1} & \mathbf{o} & -\mathrm{diag}(\beta_{j,k})_{k=d_{j-1}+1}^{d_j}\vec{P}_{j-1} \end{pmatrix},$$

where

$$(\alpha_{j-1,k})_{k=0}^{d_j} := C_j^{\mathrm{I}} (\phi_{j-1}(g_{j,k}))_{k=0}^{d_j} ,$$
$$(\beta_{j,k})_{k=0}^{d_j} := C_j^{\mathrm{I}} (\psi_j(g_{j,k}))_{k=0}^{d_j} .$$

(ii) *The decomposition relation can be written as*

$$\phi_j = D_j \begin{pmatrix} \phi_{j-1} \\ \psi_j \end{pmatrix} ,$$

with

$$D_j = C_j^{-1} = \tfrac{2}{d_{j-1}} (C_j^{\mathrm{I}})^T D_{j,\lambda} \left((C_{j-1}^{\mathrm{I}})^T \boxplus (C_{j-1}^{\mathrm{III}})^T \right) .$$

Here $D_{j,\lambda}$ *denotes*

$$D_{j,\lambda} = \begin{pmatrix} -\mathrm{diag}(\tilde{\alpha}_{j-1,k})_{k=0}^{d_{j-1}-1} & \mathbf{o} & \mathrm{diag}(\tilde{\beta}_{j,k})_{k=0}^{d_{j-1}-1}\vec{P}_{j-1} \\ \mathbf{o} & \tilde{\alpha}_{j-1,d_{j-1}} & \mathbf{o} \\ \mathrm{diag}(\tilde{\alpha}_{j-1,k})_{k=d_{j-1}+1}^{d_j} & \mathbf{o} & \mathrm{diag}(\tilde{\beta}_{j-1,k})_{k=d_{j-1}+1}^{d_j}\vec{P}_{j-1} \end{pmatrix},$$

where

$$\tilde{\alpha}_{j-1,k} := \begin{cases} \dfrac{\alpha_{j-1,k}}{\alpha_{j-1,k}\beta_{j,d_j-k}+\alpha_{j-1,d_j-k}\beta_{j,k}}, & \textit{for } \ k = 0,\ldots,d_{j-1}-1, \\ & \qquad d_{j-1}+1,\ldots,d_j, \\ \dfrac{1}{\alpha_{j-1,d_{j-1}}}, & \textit{for } k = d_{j-1}, \end{cases}$$

$$\tilde{\beta}_{j,k} := \dfrac{\beta_{j,k}}{\alpha_{j-1,k}\beta_{j,d_j-k}+\alpha_{j-1,d_j-k}\beta_{j,k}}, \qquad \textit{for } k = 0,\ldots,d_j.$$

Proof : We derive the form of the reconstruction matrix C_j immediately from the interpolatory properties (2.1) and (2.4) of the scaling functions and

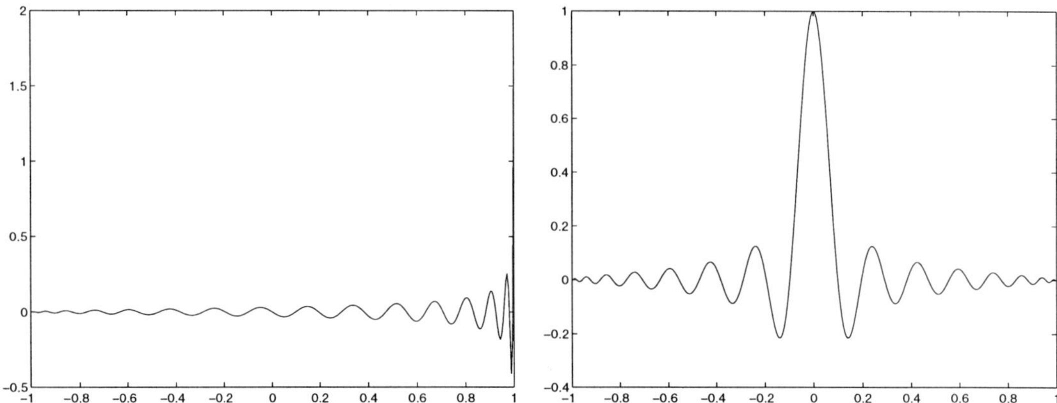

Figure 1: The polynomial scaling function ϕ_5^p and the Chebyshev shift $\phi_{5,16}^p$.

the wavelets. The factorized forms of the matrices C_j and D_j are obtained by using the results in [12].

The polynomial scaling functions and wavelets introduced in [11] and generalized in [12, 17] may serve as an example here. The scaling function is given by

$$\phi_j^p := \frac{2}{d_j} \sum_{r=0}^{d_j} \varepsilon_{j,r}\, T_r,$$

with $d = 1$. Then $V_j = \Pi_{d_j}$. The corresponding wavelet function can be written as

$$\psi_{j+1}^p := \frac{2}{d_j} \sum_{r=d_j+1}^{d_{j+1}} \varepsilon_{j+1,r}\, T_r.$$

The vectors $\boldsymbol{\alpha}_{j-1}^p$ ans $\boldsymbol{\beta}_j^p$ needed for decomposition and reconstruction have a simple form

$$\alpha_{j-1,k}^p = \begin{cases} 2, & \text{for } k = 0, \ldots, d_{j-1} - 1, \\ 1, & \text{for } k = d_{j-1}, \\ 0, & \text{otherwise,} \end{cases}$$

$$\beta_{j,k}^p = \begin{cases} 2, & \text{for } k = d_{j-1} + 1, \ldots, d_j, \\ 0, & \text{otherwise.} \end{cases}$$

As another example one can regard the fundamental spline interpolants constructed from the transformed B-splines of even order in [12].

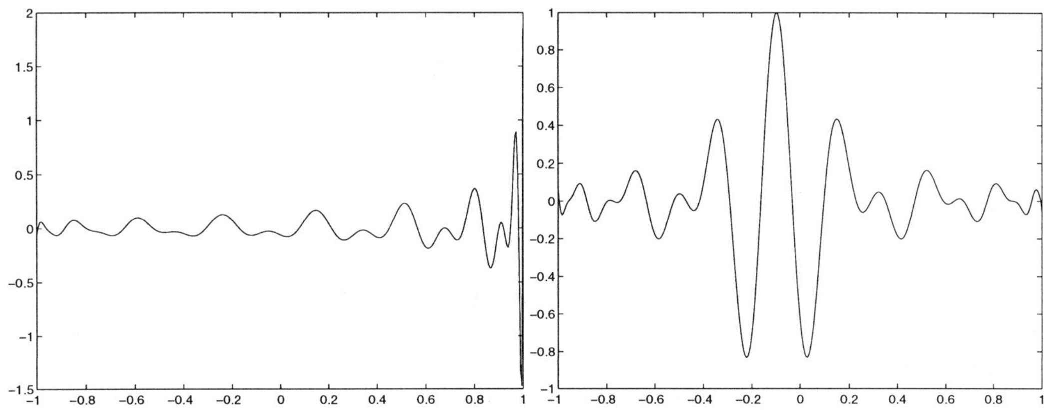

Figure 2: The polynomial wavelet ψ_5^p and the Chebyshev shift $\psi_{5,17}^p$.

3   Univariate Decomposition and Reconstruction

In this section, we want to give the matrices for complete decomposition and reconstruction of univariate functions. Functions $f_j \in V_j$ can be represented in the basis of scaling functions as well as in the basis of the wavelets contained in this sample space. If we want to change between these bases of a univariate sample space V_j we need transformations between these bases. The complete decomposition of the scaling functions into wavelets parts in the univariate case means the following. There have to exist coefficients $z_{k,\ell}^j$, such that

$$\phi_{j,k} = \sum_{\nu=0}^{d_0} z_{k,2^j\nu}^j\, \phi_{0,\nu} + \sum_{\mu=0}^{j-1}\sum_{\nu=0}^{d_\mu-1} z_{k,2^{j-\mu-1}(2\nu+1)}^j\, \psi_{\mu+1,2\nu+1}\;,$$

where the second index ℓ of the entries of $\mathbf{Z}_j := \left(z_{k,\ell}^j\right)_{k,\ell=0}^{d_j}$ corresponds to that point $g_{j,\ell}$ in which the function $\phi_{0,\nu}$ or $\psi_{\mu+1,2\nu+1}$, respectively, fulfils the interpolatory condition (2.1) or (2.4) with value 1. On the other hand, there have to be coefficients $r_{k,\ell}^j$, such that the complete reconstruction of the wavelets into terms of scaling functions of a higher level in the univariate case reads as

$$\phi_{0,\nu} \;=\; \sum_{\ell=0}^{d_j} r_{2^j\nu,\ell}^j\, \phi_{j,\ell} = \sum_{\ell=0}^{d_j} \varepsilon_{j,\ell}\, \phi_{0,\nu}(g_{j,\ell})\, \phi_{j,\ell},$$

$$\psi_{\mu+1,2\nu+1} \;=\; \sum_{\ell=0}^{d_j} r^{\,j}_{2^{j}-\mu-1\,(2\nu+1),\ell}\,\phi_{j,\ell} = \sum_{\ell=0}^{d_j} \varepsilon_{j,\ell}\,\psi_{\mu+1,2\nu+1}(g_{j,\ell})\,\phi_{j,\ell}\,,$$

where the first index k of the entries of $\boldsymbol{R}_j := (r^{j}_{k,\ell})^{d_j}_{k,\ell=0}$ corresponds to that point $g_{j,k}$ of the fine grid in which the function $\phi_{0,\nu}$ or $\psi_{\mu+1,2\nu+1}$, respectively, satisfies (2.1) or (2.4) with function value 1.

Of course, one can describe the matrices $\boldsymbol{Z}_j$ of complete decomposition and $\boldsymbol{R}_j$ of complete reconstruction by a stepwise application of the matrices $\boldsymbol{D}_r$ and $\boldsymbol{C}_r$ from Theorem 2.1, respectively. Here, we want to use discrete cosine transform in our decomposition and reconstruction algorithms. Therefore, we need some notations. We restrict ourselves to a special case. Assume here and in the sequel that $d = d_0$ is a power of 2, i.e. $d = 2^{t_d}$. Otherwise things can be done in a similar way but will become somewhat more complicated. Let $k \in \{0,\dots,d_j-1\}$ have the representation

$$k = \sum_{s=0}^{t} k_s\, 2^s\,, \qquad k_s \in \{0,1\}.$$

Then, the binary representation mirrored at $j+t_d-1$ (or in bit reversed order) is given by

$$M_j(k) := \sum_{s=0}^{t} k_s\, 2^{j+t_d-1-s}\,.$$

Additionally, we define

$$M_j(d_j) := -1 \qquad \text{and} \qquad M_j(-1) := d_j\,.$$

Now, we want to give factorizations of the matrices for complete univariate decomposition $\boldsymbol{Z}_j$ and reconstruction $\boldsymbol{R}_j$.

Theorem 3.1. *Let $j \in \mathbb{N}$ and $\boldsymbol{Q}_s$ be the permutation matrices*

$$\boldsymbol{Q}_s := \left(\delta_{k,M_s(l)}\right)^{d_s-1,\,d_s}_{k=-1,\ell=0}\,, \qquad \vec{\boldsymbol{Q}}_s := \left(\delta_{k,M_s(l)}\right)^{d_s-1}_{k,\ell=0}\,.$$

All occuring matrix products are taken from left to right.

(i) *The matrix $\boldsymbol{R}_j$ for complete univariate reconstruction can be factorized as*

$$\boldsymbol{Q}_j \boldsymbol{R}_j \boldsymbol{Q}_j^T = \tfrac{1}{d_{j-1}} \left(\boldsymbol{Q}_0 (\mathrm{C}_0^{\mathrm{I}})^T \boxplus \bigboxplus_{r=0}^{j-1} \vec{\boldsymbol{Q}}_r (\mathrm{C}_r^{\mathrm{II}})^T \right) \boldsymbol{R}_{j,\lambda} \left((\mathrm{C}_j^{\mathrm{I}})^T \boldsymbol{Q}_j^T \right)$$

with the inner sparse matrix

$$R_{j,\lambda} := \prod_{r=1}^{j} C_{r,\lambda} \boxplus I_{d_j - d_r} .$$

(ii) *The matrix Z_j for complete univariate decomposition can be factorized as*

$$Q_j Z_j Q_j^T = \tfrac{2}{d_{j-1}} \left(Q_j (C_j^{\mathrm{I}})^T\right) Z_{j,\lambda} \left((C_0^{\mathrm{I}})^T Q_0^T \boxplus \overset{j-1}{\underset{r=0}{\boxplus}} (C_r^{\mathrm{III}})^T \vec{Q}_r^T\right)$$

with the inner sparse matrix

$$Z_{j,\lambda} := \prod_{r=1}^{j} D_{r,\lambda} \boxplus I_{d_j - d_r} .$$

We skip the proof which uses the properties of the DCT-I and the orthogonality of the matrices Q_r. As we can see by this theorem univariate complete decomposition and reconstruction, respectively, for the level j cost some discrete cosine transforms of length at most d_{j-1} and a multiplication with a matrix with $\mathcal{O}(j2^j)$ non-zero entries, i.e., altogether $\mathcal{O}(j2^j)$ essential operations.

4 Sample and Wavelet Spaces for Sparse Grids

Starting from the univariate interpolation, we want to investigate bivariate interpolation on a sparse grid. The j-th order Boolean sum interpolation operator is defined by

$$B_j := \bigoplus_{r=0}^{j} L_r \otimes L_{j-r},$$

where $A \oplus B := A + B - AB$ and the representation of B_j in terms of ordinary sums is known (see [1, 6]) to be

$$B_j = \sum_{r=0}^{j} L_r \otimes L_{j-r} - \sum_{r=0}^{j-1} L_r \otimes L_{j-r-1}.$$

The range of this interpolation operator is given by

$$\mathrm{Im}\, B_j = \sum_{r=0}^{j} \mathrm{Im}\, L_r \otimes \mathrm{Im}\, L_{j-r} = \sum_{r=0}^{j} V_r \otimes V_{j-r} \subset \mathrm{Im}\, B_{j+1} . \qquad (4.1)$$

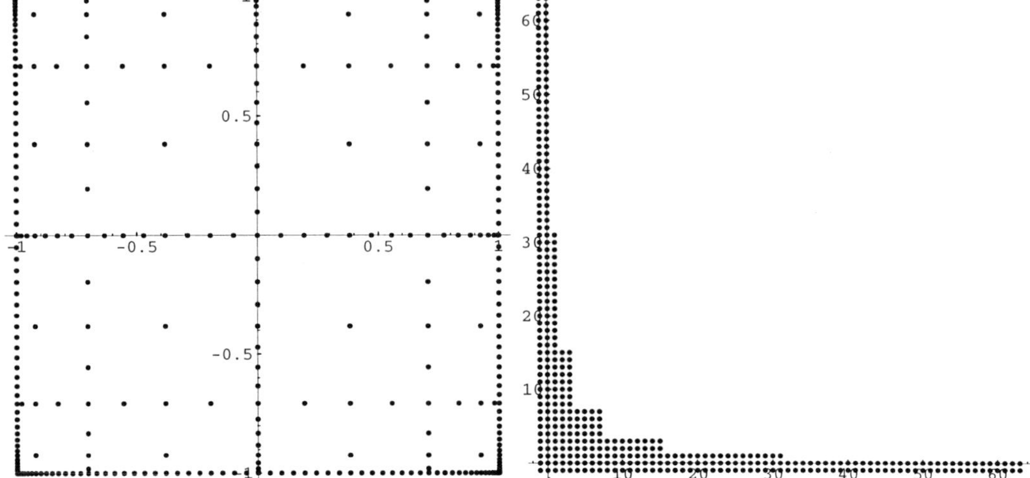

Figure 3: Sparse Gauß–Chebyshev grid (left) and the index set $M_j(\mathcal{J}_j)$ for $d = 1, j = 6$ (right).

It interpolates on the sparse grid

$$\mathcal{G}_j^B = \bigcup_{r=0}^{j} \mathcal{G}_r \times \mathcal{G}_{j-r} \subset \mathcal{G}_{j+1}^B .$$

Now we define new sample spaces of bivariate functions by

$$V_j^B := \mathrm{Im}\, B_j = \sum_{r=0}^{j} V_r \otimes V_{j-r} .$$

From that definition and (4.1) we obtain immediately

$$V_j^B \subset V_{j+1}^B .$$

In order to characterize bivariate scaling functions again by their interpolatory properties we observe the relation $\mathcal{G}_j^B \subset \mathcal{G}_j \times \mathcal{G}_j$ between the sparse and the regular interpolation grid. Thus, we can define an index set $\mathcal{J}_j$ by

$$\mathcal{J}_j := \{(k, \ell) \in \mathbb{N}_0^2 \; ; \; (g_{j,k}, g_{j,\ell}) \in \mathcal{G}_j^B\}$$

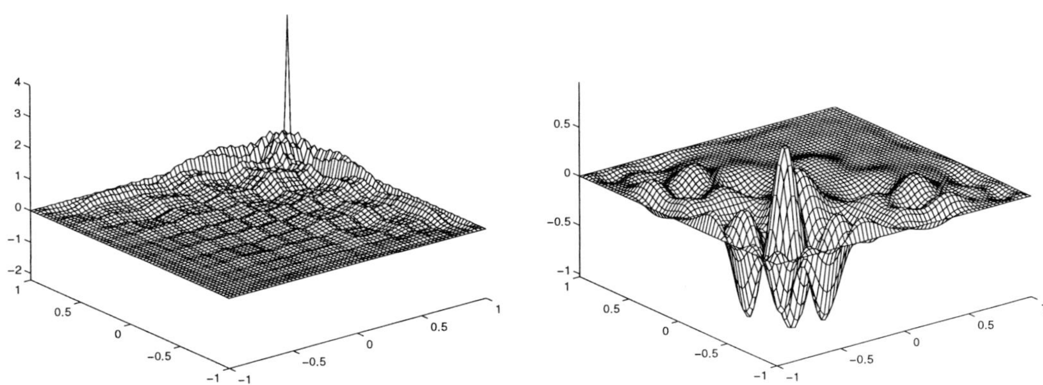

Figure 4: The polynomial scaling functions $\Phi^p_{6,0,0}$ (left) and $\Phi^p_{6,48,48}$ (right).

and choose scaling functions $\Phi_{j,k,\ell}$ as modified fundamental interpolants in V^B_j on the sparse grid $\mathcal{G}^B_j$, i.e., with the property,

$$\Phi_{j,k,\ell}(g_{j,m}, g_{j,n}) = \frac{1}{\varepsilon_{j,k}\varepsilon_{j,\ell}}\delta_{k,m}\delta_{\ell,n}, \qquad (k,\ell),(m,n) \in \mathcal{J}_j.$$

For describing these functions in detail, we first take a finer splitting of the index set $\mathcal{J}_j$

$$\mathcal{J}_j = 2\,\mathcal{J}_{j-1} \cup (\mathcal{J}_j \backslash 2\,\mathcal{J}_{j-1}), \qquad \mathcal{J}_j \backslash 2\,\mathcal{J}_{j-1} = \bigcup_{r=0}^{j} \mathcal{K}_{j,p}$$

with

$$\mathcal{K}_{0,0} = \{(\eta,\kappa)\,;\ \eta = 0,\ldots,d_0,\ \kappa = 0,\ldots,d_0\}$$

and for $j \geq 1$

$$\begin{aligned}
\mathcal{K}_{j,p} &= \{(M_j(\eta), M_j(\kappa))\,;\ \eta = d_{p-1},\ldots,d_p - 1, \\
&\qquad \kappa = d_{j-p-1},\ldots,d_{j-p} - 1\},\ p = 1,\ldots,j-1 \\
\mathcal{K}_{j,0} &= \{(M_j(\eta), M_j(\kappa))\,;\ \eta = -1,\ldots,d_0 - 1,\ \kappa = d_{j-1},\ldots,d_j - 1\} \\
\mathcal{K}_{j,j} &= \{(M_j(\eta), M_j(\kappa))\,;\ \eta = d_{j-1},\ldots,d_j - 1,\ \kappa = -1,\ldots,d_0 - 1\}.
\end{aligned}$$

Hence,

$$\mathcal{J}_j = \bigcup_{t=0}^{j} 2^{j-t} \bigcup_{r=0}^{t} \mathcal{K}_{t,r} \,. \tag{4.2}$$

For indices $(k,\ell) \in 2^{j-t}\mathcal{K}_{t,p}, 0 \le p \le t \le j$, we can describe the modified fundamental interpolants $\Phi_{j,k,\ell}$ in V_j^B by

$$\Phi_{j,k,\ell} = \sum_{s=p}^{j+p-t} \phi_{s,2^{s-j}k} \otimes \phi_{j-s,2^{-s}\ell} - \sum_{s=p}^{j+p-t-1} \phi_{s,2^{s-j}k} \otimes \phi_{j-s-1,2^{-s-1}\ell} \tag{4.3}$$

$$= \sum_{s=p}^{j+p-t} \phi_{s,M_s(M_j(k))} \otimes \phi_{j-s,M_{j-s}(M_j(\ell))}$$

$$- \sum_{s=p}^{j+p-t-1} \phi_{s,M_s(M_j(k))} \otimes \phi_{j-s-1,M_{j-s-1}(M_j(\ell))} \,.$$

The wavelet space W_{j+1}^B is defined as the orthogonal complement of the sample space on the lower level j relative to the sample space on the higher level $j+1$, i.e.,

$$V_{j+1}^B = V_j^B \oplus^\perp W_{j+1}^B$$

with respect to the inner product in $L_w^2(I^2)$

$$\langle f, g\rangle_{I^2} := \frac{4}{\pi^2} \iint_{I^2} f(x,y)g(x,y)w(x)w(y)\,dx\,dy \,.$$

The wavelet spaces can be characterized in terms of tensor products of the univariate wavelet spaces.

Theorem 4.1. *For $j \in \mathbb{N}_0$, it holds that*

$$W_{j+1}^B = \bigoplus_{r=0}^{j+1}{}^\perp W_r \otimes W_{j+1-r}$$

with the notation $W_0 := V_0$.

The proof can be found in [14].

We define for $(m,n) \in \mathcal{K}_{j+1,p}$ the wavelet function

$$\Psi_{j+1,m,n} := \psi_{p,2^{p-j-1}m} \otimes \psi_{j+1-p,2^{-p}n} \tag{4.4}$$

with $\psi_{0,r} := \phi_{0,r}$. Using the representation of the wavelet space given by Theorem 4.1, we obtain immediately bases of the sample and the wavelet spaces.

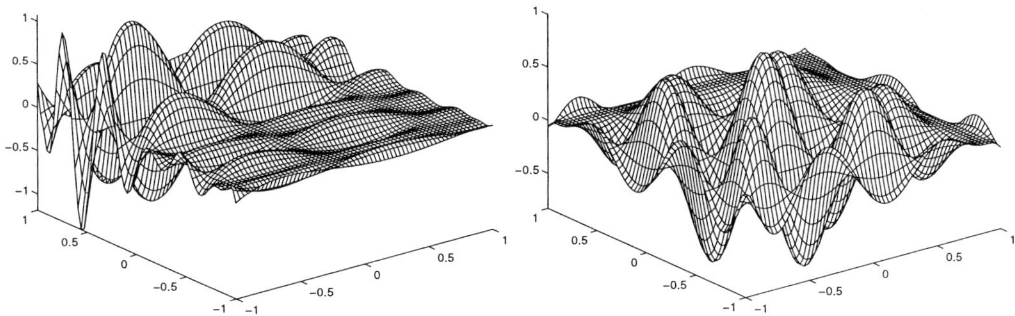

Figure 5: The polynomial wavelet functions $\Psi^p_{6,20,48}$ (left) and $\Psi^p_{6,40,40}$ (right).

Corollary 4.2. *Let $j \in \mathbb{N}_0$, then,*

(i) *the basis and the dimension of the sample space are given by*

$$
\begin{aligned}
\mathcal{B}(V_j^B) &= \{\Phi_{j,k,\ell}\,;\ (k,\ell) \in \mathcal{J}_j\}, \\
\dim V_j^B &= d^2((j-1)2^{j-1} + 2^{j+2} + 1),
\end{aligned}
$$

(ii) *the basis and the dimension of the wavelet space are given by*

$$
\begin{aligned}
\mathcal{B}(W_{j+1}^B) &= \{\Psi_{j+1,m,n}\,;\ (m,n) \in \mathcal{K}_{j+1,p},\ p = 0,\ldots,j+1\}, \\
\dim W_{j+1}^B &= d^2(j2^{j-1} + 2^{j+2}).
\end{aligned}
$$

Hence, the dimension of the constructed bivariate sample space is growing as $\mathcal{O}(j2^j)$ while the dimension of ordinary tensor products of univariate sample spaces grows as $\mathcal{O}(2^{2j})$.

5   Complete Decomposition and Reconstruction

We want to establish the relations for the complete decomposition of the fundamental interpolants on the sparse grid, i.e., the scaling functions in the bivariate case, into wavelets parts in matrix form

$$\boldsymbol{\Phi}_j = \boldsymbol{Z}_j^B \, \boldsymbol{\Psi}_j$$

and for complete reconstruction of the wavelet parts into the basis of fundamental interpolants

$$\boldsymbol{\Psi}_j = \boldsymbol{R}_j^B \, \boldsymbol{\Phi}_j$$

with

$$\boldsymbol{\Phi}_j = \left(\Phi_{j,k,\ell}\right)_{(k,\ell)\in\mathcal{J}_j},$$

where the indices follow the ordering

$$2^j \mathcal{K}_{0,0}, \; 2^{j-1}\mathcal{K}_{1,0}, \; 2^{j-1}\mathcal{K}_{1,1}, \; \ldots, \; \mathcal{K}_{j,j}$$

and

$$\boldsymbol{\Psi}_j = \left(\Psi_{u,m,n}\right)_{u=0,\ldots,j,\, q=0,\ldots,u,\, (m,n)\in\mathcal{K}_{u,q}}.$$

We want to describe the matrices $\boldsymbol{Z}_j^B$ and $\boldsymbol{R}_j^B$ in terms of matrices for the univariate relations.

Theorem 5.1. *With the permutation matrices*

$$\boldsymbol{S}_{k,s} \; := \; \left(\delta_{\eta,\lfloor(d_{k-s}+1)^{-1}r\rfloor-1}\,\delta_{\kappa,r\bmod(d_{k-s}+1)-1}\right)_{(\eta,\kappa)\in M_j(\mathcal{J}_j),\,r=0,\ldots,(d_s+1)(d_{k-s}+1)},$$

$$\boldsymbol{T}_j \; := \; \left(\delta_{M_j(k),m}\,\delta_{M_j(\ell),n}\right)_{(k,\ell)\in\mathcal{J}_j,\,(m,n)\in M_j(\mathcal{J}_j)},$$

and $\boldsymbol{Q}_s$ as in Theorem 3.1,

(i) *the matrix $\boldsymbol{Z}_j^B$ for complete decomposition has the form*

$$\boldsymbol{Z}_j^B \;=\; \boldsymbol{T}_j \left(\sum_{s=0}^{j} \boldsymbol{S}_{j,s} \left((\boldsymbol{Q}_s \boldsymbol{Z}_s \boldsymbol{Q}_s^T) \otimes (\boldsymbol{Q}_{j-s} \boldsymbol{Z}_{j-s} \boldsymbol{Q}_{j-s}^T) \right) \boldsymbol{S}_{j,s}^T \right.$$

$$\left. - \sum_{s=0}^{j-1} \boldsymbol{S}_{j-1,s} \left((\boldsymbol{Q}_s \boldsymbol{Z}_s \boldsymbol{Q}_s^T) \otimes (\boldsymbol{Q}_{j-s-1} \boldsymbol{Z}_{j-s-1} \boldsymbol{Q}_{j-s-1}^T) \right) \boldsymbol{S}_{j-1,s}^T \right) \boldsymbol{T}_j^T,$$

(ii) *the matrix $\boldsymbol{R}_j^B$ for complete reconstruction has the form*

$$\boldsymbol{R}_j^B \;=\; \boldsymbol{T}_j \left(\sum_{s=0}^{j} \boldsymbol{S}_{j,s} \left((\boldsymbol{Q}_s \boldsymbol{R}_s \boldsymbol{Q}_s^T) \otimes (\boldsymbol{Q}_{j-s} \boldsymbol{R}_{j-s} \boldsymbol{Q}_{j-s}^T) \right) \boldsymbol{S}_{j,s}^T \right.$$

$$\left. - \sum_{s=0}^{j-1} \boldsymbol{S}_{j-1,s} \left((\boldsymbol{Q}_s \boldsymbol{R}_s \boldsymbol{Q}_s^T) \otimes (\boldsymbol{Q}_{j-s-1} \boldsymbol{R}_{j-s-1} \boldsymbol{Q}_{j-s-1}^T) \right) \boldsymbol{S}_{j-1,s}^T \right) \boldsymbol{T}_j^T.$$

For the proof, we refer the reader to [15] where it has been carried out in the periodic case (only some slight modifications are necessary).

Now, we want to present the algorithms for complete decomposition and reconstruction, respectively, of a given function for our Boolean sum approach. We project a bivariate continuous (or piecewise continuous) function f into the sample space V_j. We obtain the projection f_j of f by interpolation

$$f_j := B_j f = \sum_{(k,\ell) \in \mathcal{J}_j} \varepsilon_{j,k}\,\varepsilon_{j,\ell}\,f(t_{j,k}, t_{j,\ell})\,\Phi_{j,k,\ell}$$

on the sparse Gauß–Chebyshev grid. Writing this in vector form, we obtain $f_j = \boldsymbol{v}_j^T \boldsymbol{\Phi}_j$ with the coefficient vector $\boldsymbol{v}_j$ consisting of the function values $v_{j,k,\ell} = \varepsilon_{j,k}\,\varepsilon_{j,\ell}\,f(t_{j,k}, t_{j,\ell})$, $(k,\ell) \in \mathcal{J}_j$ sampled from the sparse grid. In terms of the wavelet basis of V_j, we get another representation of the interpolant as $f_j = \boldsymbol{w}_j^T \boldsymbol{\Psi}_j$ with a coefficient vector $\boldsymbol{w}_j$. Then, we use the relations of complete decomposition and reconstruction for the basis vectors

$$\boldsymbol{\Phi}_j = \boldsymbol{Z}_j^B \boldsymbol{\Psi}_j \qquad \text{and} \qquad \boldsymbol{\Psi}_j = \boldsymbol{R}_j^B \boldsymbol{\Phi}_j$$

and obtain the relations of complete decomposition and reconstruction, respectively, for the coefficient vectors

$$\boldsymbol{w}_j = (\boldsymbol{Z}_j^B)^T \boldsymbol{v}_j \qquad \text{and} \qquad \boldsymbol{v}_j = (\boldsymbol{R}_j^B)^T \boldsymbol{w}_j. \tag{5.1}$$

The matrices involved in the Equations (5.1) are of the same structure. We have

$$\boldsymbol{y}_j = (\boldsymbol{A}_j^B)^T \boldsymbol{x}_j$$

with the letter $\boldsymbol{A}$ standing for $\boldsymbol{Z}$ in the complete decomposition relation and for $\boldsymbol{R}$ in the reconstruction relation, respectively. By Theorem 5.1, the multiplication with the matrix $(\boldsymbol{A}_j^B)^T$ reduces to the stepwise application of the algorithms in the univariate case.

Algorithm. (Complete decomposition or reconstruction)

Input vector: $\quad \boldsymbol{x}_j = (x_{j,k,\ell})_{(k,\ell) \in \mathcal{J}_j}$

Output vector: $\boldsymbol{y}_j = (y_{j,k,\ell})_{(k,\ell) \in \mathcal{J}_j}$

Help vector: $\quad \boldsymbol{z}_j = (z_{j,k,\ell})_{(k,\ell) \in \mathcal{J}_j}$

Notation: $\boldsymbol{z}_{j,k} = (z_{j,k,\ell})_{\ell \in \mathcal{L}}, \boldsymbol{z}_{j,\ell} = (z_{j,k,\ell})_{k, \in \mathcal{K}}$, with an appropriate choice of the index sets $\mathcal{K}, \mathcal{L}$ (the ranges for k, ℓ as given in the algorithm)

Initialization: $y_{j,k,\ell} := 0$ for all $(k,\ell) \in \mathcal{J}_j$

For all $s = 0, \ldots, j$ do

$$z_{j,k,\ell} := x_{j,M_j(k),M_j(\ell)} \qquad \text{for } k = -1, \ldots, d_s - 1,$$
$$\ell = -1, \ldots, d_{j-s} - 1$$

$$\boldsymbol{z}_{j,k} := \boldsymbol{Q}_{j-s}\, \boldsymbol{A}_{j-s}^T\, \boldsymbol{Q}_{j-s}^T\, \boldsymbol{z}_{j,k} \qquad \text{for } k = -1, \ldots, d_s - 1$$

$$\boldsymbol{z}_{j,\ell} := \boldsymbol{Q}_s\, \boldsymbol{A}_s^T\, \boldsymbol{Q}_s^T\, \boldsymbol{z}_{j,\ell} \qquad \text{for } \ell = -1, \ldots, d_{j-s} - 1$$

$$y_{j,M_j(k),M_j(\ell)} := y_{j,M_j(k),M_j(\ell)} + z_{j,k,\ell} \quad \text{for } k = -1, \ldots, d_s - 1,$$
$$\ell = -1, \ldots, d_{j-s} - 1$$

For all $s = 0, \ldots, j - 1$ do

$$z_{j,k,\ell} := x_{j,M_j(k),M_j(\ell)} \qquad \text{for } k = -1, \ldots, d_s - 1,$$
$$\ell = -1, \ldots, d_{j-s-1} - 1$$

$$\boldsymbol{z}_{j,k} := \boldsymbol{Q}_{j-s-1}\, \boldsymbol{A}_{j-s-1}^T\, \boldsymbol{Q}_{j-s-1}^T\, \boldsymbol{z}_{j,k} \quad \text{for } k = -1, \ldots, d_s - 1$$

$$\boldsymbol{z}_{j,\ell} := \boldsymbol{Q}_s\, \boldsymbol{A}_s^T\, \boldsymbol{Q}_s^T\, \boldsymbol{z}_{j,\ell} \qquad \text{for } \ell = -1, \ldots, d_{j-s-1} - 1$$

$$y_{j,M_j(k),M_j(\ell)} := y_{j,M_j(k),M_j(\ell)} - z_{j,k,\ell} \quad \text{for } k = -1, \ldots, d_s - 1,$$
$$\ell = -1, \ldots, d_{j-s-1} - 1$$

The products $\boldsymbol{Q}_s \boldsymbol{A}_s^T \boldsymbol{Q}_s^T$ are of the structure given in Theorem 3.1. That's why they will be realized by fast cosine transforms and multiplication with a sparse matrix.

We count the univariate complete decompositions and the univariate complete reconstructions, respectively, for the levels occuring in the algorithm. In the end we need altogether $\mathcal{O}(j^2\, 2^j)$ essential operations for the bivariate complete decomposition and the complete reconstruction, respectively, for the level j if we use the Boolean sum approach.

References

[1]   G. Baszenski, F.-J. Delvos: *A discrete Fourier transform scheme for Boolean sums of trigonometric operators*, in: *Multivariate Approximation Theory IV*, Internat. Ser, Numer. Math. **90**, C. K. Chui, W. Schempp, K. Zeller (eds.), 15–24, Birkhäuser, Basel 1989.

[2] G. Baszenski, M. Tasche: *Fast DCT-algorithms, interpolating wavelets and hierarchical bases*, in: *Wavelets, Images, and Surface Fitting*, P. J. Laurent, A. L. Méhauté, L. L. Schumaker (eds.), 37–50, AK Peters, Boston 1994.

[3] P. L. Butzer, R. L. Stens: *The operational properties of the Chebyshev transform*, Funct. Approx. Comment. Math. **5** (1977), 129–160.

[4] C. K. Chui, E. Quak: *Wavelets on a bounded interval*, in: *Numerical Methods of Approximation Theory*, D. Braess, L. L. Schumaker (eds.), 53–75, Birkhäuser, Basel 1992.

[5] A. Cohen, I. Daubechies, P. Vial: *Wavelets on the interval and fast wavelet transforms*, Appl. Comput. Harmonic Anal. **1** (1993), 54–81 .

[6] F.-J. Delvos, W. Schempp: *Boolean Methods in Interpolation and Approximation*, Pitman Research Notes in Mathematics Series, Longman Scientific & Technical, Harlow 1989.

[7] R. A. DeVore, S. V. Konyagin, V. N. Temlyakov: *Hyperbolic wavelet approximation*, Constr. Appr., to appear.

[8] R. A. DeVore, P. Petrushev, V. N. Temlyakov: *Multivariate trigonometric approximation with frequencies from the hyperbolic cross*, Mat. Zametki **56** (1994), 36–63 , English transl. in Math. Notes **56** (1994).

[9] M. Griebel, P. Oswald: *On additive Schwarz preconditioners for sparse grid discretizations*, Numer. Math. **66** (1994), 449–464.

[10] M. Griebel, P. Oswald: *Tensor product type subspace splitting and multilevel iterative methods for anisotropic problems*, Adv. Comput. Math. 4 (1995), 171–206.

[11] T. Kilgore, J. Prestin: *Polynomial wavelets on the interval*, Const. Approx. **12** (1996), 95–110.

[12] G. Plonka, K. Selig, M. Tasche: *On the construction of wavelets on a bounded interval*, Adv. in Comput. Math. 4 (1995), 357–388.

[13] E. Quak, N. Weyrich: *Decomposition and reconstruction algorithms for spline wavelets on a bounded interval*, Appl. Comput. Harmonic Anal. **1** (1994), 217–231.

[14] F. Sprengel: *Polynomial wavelets on the square*, in: *Approximation Theory VIII, Volume 2: Wavelets and Multilevel Approximation*, C. K. Chui, L. L. Schumaker (eds.), 383–390, World Scientific Publ., Singapore 1995.

[15] F. Sprengel: *Periodic interpolation and wavelets on sparse grids*, submitted (1996).

[16] M. Tasche: *Fast algorithms for discrete Chebyshev–Vandermonde transforms and applications*, Numer. Algorithms **5** (1993), 453–464.

[17] M. Tasche: *Polynomial wavelets on* $[-1, 1]$, in: S. P. Singh (Ed.): *Approximation Theory, Wavelets, and Applications*, 497–512, Kluwer, Dordrecht 1995.

[18] V. N. Temlyakov: *Approximation of functions with bounded mixed derivative*, Proc. Steklov Institute **1** (1989).

[19] V. N. Temlyakov: *Approximation of Periodic Functions*, Nova Science, New York 1993.

[20] C. Zenger: *Sparse grids*, in: *Notes on Numerical Fluid Mechanics*, W. Hackbusch (ed.), Volume 31, 297–301, Parallel Algorithms for Partial Differential Equations: Procceedings of the Sixth GAMM-Seminar, Vieweg, Braunschweig 1991.

Address:

FRAUKE SPRENGEL
Department of Mathematics
University of Rostock
D–18051 Rostock
Germany

The General Structure of Multivariate Affine Frames

Joachim Stöckler

Abstract

We present three general results on the structure of multivariate affine frames. First, the well–known estimates for the frame bounds are extended to the most general case of an arbitrary scaling matrix. Then we describe the characterization of tight frames, which was recently obtained in [14]. Thirdly we find new results on the comparison of affine frames and wavelet bases, which are based on operator algebra techniques introduced in [5].

1 Introduction and Notations

The concept of frames in a separable Hilbert space $\mathcal{H}$ was introduced in connection with sampling of band–limited functions in [9, 18]. Since then, frames have found important applications in sampling theory, signal and image analysis, noise reduction and quantum mechanics, see [7] for a short survey. In this paper, we deal with a certain type of frames, which are closely related to multivariate wavelet bases. Our aim is to give some important results on the structure of these frames. The chosen topics are rather abstract in nature, but some hints to possible applications and explicit constructions of frames are included, too.

It is useful to have the general definition of frames available, since it avoids some of the technical appearance of special types of frames. The definitions and elementary concepts can be found in [9, 11, 18].

Definition 1.1. *Let $\mathcal{H}$ be a separable Hilbert space. A countable family $X = (x_i; \ i \in I)$ is called a* frame *of $\mathcal{H}$, if there exist constants $A, B > 0$ such that*

$$A\|h\|^2 \le \sum_{i \in I} |\langle h, x_i \rangle|^2 \le B\|h\|^2 , \qquad h \in \mathcal{H}. \tag{1.1}$$

The constants A, B are called frame bounds *of X, and X is* tight, *if the condition (1.1) holds with $A = B > 0$.*

Multivariate Approximation: Recent Trends and Results; W. Haußmann, K. Jetter and M. Reimer (eds.)
Mathematical Research, Vol. 101, pp. 287–302, ISBN 3-05-501770-6
© Akademie–Verlag, Berlin 1997

Frequently used operators for the description of a frame X are

$$\mathcal{T}_X : \mathcal{H} \to \ell_2(I), \qquad \mathcal{T}_X h = (\langle h, x_i \rangle)_{i \in I}\ , \tag{1.2}$$

$$\mathcal{T}_X^* : \ell_2(I) \to \mathcal{H}, \qquad \mathcal{T}_X^*(c_i)_{i \in I} = \sum_{i \in I} c_i x_i\ . \tag{1.3}$$

These are called *analysis* and *synthesis operator* in the literature. Both are bounded operators, with $\mathcal{T}_X^*$ the adjoint of $\mathcal{T}_X$, and condition (1.1) is equivalent to

$$A \operatorname{id}_{\mathcal{H}} \leq \mathcal{T}_X^* \mathcal{T}_X \leq B \operatorname{id}_{\mathcal{H}}\ . \tag{1.4}$$

By means of the generalized inverse of $\mathcal{T}_X^*$ we can find a representation

$$h = \mathcal{T}_X^*(c_i)_{i \in I} \qquad \text{where} \quad (c_i)_{i \in I} = \mathcal{T}_X(\mathcal{T}_X^* \mathcal{T}_X)^{-1} h\ ,$$

for any $h \in \mathcal{H}$, and this frame decomposition of h is stable in the sense that $B^{-1}\|h\|^2 \leq \|(c_i)_{i \in I}\|^2_{\ell_2(I)} \leq A^{-1}\|h\|^2$. Therefore, frames are redundant families, which allow stable representations of all elements in the Hilbert space. In order to compare frames with Riesz bases, we mention that X is a Riesz basis of $\mathcal{H}$ iff X is complete and

$$\widetilde{A} \operatorname{id}_{\ell_2(I)} \leq \mathcal{T}_X \mathcal{T}_X^* \leq \widetilde{B} \operatorname{id}_{\ell_2(I)} \tag{1.5}$$

holds for some constants $\widetilde{A}, \widetilde{B} > 0$, which are called Riesz bounds of X.

The interest in wavelet bases over the past two decades has naturally extended to a new type of frames, which are called *wavelet frames* or *affine frames*. These are families in $L_2(\mathbb{R}^d)$, which are generated from a finite subset $\Psi = \{\psi_i;\ i \in I\}$ of $L_2(\mathbb{R}^d)$ by means of dilation and translation in the following way. Two real and invertible $d \times d$ matrices M, N are given, and M is supposed to be expansive, i.e. all eigenvalues of M have absolute value larger than 1. The equations

$$Df := |\det M|^{1/2} f(M\cdot) \qquad \text{and} \qquad T_k f = f(\cdot - k), \quad k \in N\mathbf{Z}^d,$$

define unitary operations on $L_2(\mathbb{R}^d)$. Then the family

$$\Psi_{M,N} := (D^j T_k \psi_i;\ j \in \mathbf{Z},\ k \in N\mathbf{Z}^d,\ i \in I) \tag{1.6}$$

is called the *affine frame relative to* $(M, N\mathbf{Z}^d)$ *generated by* Ψ, if it is a frame of $L_2(\mathbb{R}^d)$, hence (1.1) holds where $\mathcal{H} = L_2(\mathbb{R}^d)$ and $X = \Psi_{M,N}$. Likewise, Ψ generates a wavelet basis relative to $(M, N\mathbf{Z}^d)$, if $\Psi_{M,N}$ is a Riesz basis of

$L_2(\mathbb{R}^d)$. The operator in (1.2) will be denoted by $\mathcal{T}_\Psi$, since no confusion will occur by suppressing M and N in this notation.

There are several reasons for our interest in affine frames. First, the construction of families with compactly supported generators ψ_i, which have further desirable properties like symmetry and smoothness, has been more successful for frames than for wavelet bases until now, see e.g. [2, 15]. Secondly, the redundance of affine frames can be useful for deletion of noise and translation invariance of the coefficient functionals in the frame decomposition, see e.g. [8, Chapter 3] and [1].

Let us give a short outline of the paper. We present results which come from three different techniques for the study of affine frames. In section 2 we extend the well–known estimates for the frame bounds of univariate affine frames [3, 8] to the multivariate case. Unlike earlier work by Chui and Shi [4], no restrictions apply to the scaling matrix M. This is due to a new technique introduced in the proof, which uses a lemma by Edwards and Gaudry [10] in order to circumvent arguments which are based on the Lebesgue set of multivariate functions, see Lemma 2.2. Section 3 gives a short account of recent results by Ron and Shen [14], who obtained a characterization of *tight* affine frames, which are subject to some mild assumptions on the decay of the Fourier transforms at ∞. In section 4 we use methods from operator algebras, which were introduced by Dai and Larson [5] to the theory of orthonormal wavelets. The local commutant plays an important role for affine frames, too. We can thus generalize a negative result on the existence of affine frames from our earlier work [17] to a much larger class of generators ψ whose Fourier transform is supported in a union of so–called wavelet sets.

Our notation for Fourier transforms is $\widehat{f}(\xi) = \int_{\mathbb{R}^d} f(x)e^{-ix\cdot\xi}\,dx$. The open cube $(-r, r)^d$ is denoted by Q_r. For matrices we let $\|M\|_\infty = \sup_{x\neq 0}\|Mx\|_\infty/\|x\|_\infty$.

2   Estimates for the Frame Bounds

One of the main structural theorems about affine frames describes a relation of the frame bounds in (1.1) and the dilation–invariant function

$$F_{\Psi,M}(\xi) := \sum_{i\in I}\sum_{j\in\mathbb{Z}}\left|\widehat{\psi}_i\left((M^t)^j\xi\right)\right|^2, \qquad \xi \in \mathbb{R}^d. \tag{2.1}$$

We present the result in its most general form as obtained in [16]. A slightly shortened proof is also given in this section.

Theorem 2.1. *If Ψ generates an affine frame of $L_2(\mathrm{IR}^d)$ relative to $(M, N\mathbf{Z}^d)$, with frame bounds $0 < A \le B < \infty$, then*

$$A \le \frac{1}{|\det N|} F_{\Psi,M}(\xi) \le B, \qquad for\ a.a.\ \xi \in \mathrm{IR}^d. \tag{2.2}$$

Earlier results in the literature deal with the univariate case [3, 7] or require a special form of the scaling matrix M [4]. A new technique for the proof is used here. It is based on the following result of [10, Lemma 2.1.4] which might be of interest in other studies of measure–theoretic type, too.

Lemma 2.2. *Let $(U_j)_{j \ge 0}$ be a decreasing sequence of relatively compact measurable subsets of IR^d, which define a basis of neighborhoods of 0. Then, for any $f \in L_1(\mathrm{IR}^d)$, the sequence*

$$f_j(\xi) := |U_j|^{-1} \int_{U_j} f(\xi + \eta)\, d\eta \tag{2.3}$$

converges to f in $L_1(\mathrm{IR}^d)$.

Here, $|U|$ means the Lebesgue measure of U. Consequently, in the situation described in the lemma, there exists a subsequence $(f_{j_k})_{k \ge 0}$ which converges pointwise a.e. to f. This fact can be used in our situation of Theorem 2.1 in the following way. Chui and Shi [4, Theorem 1] prove that $F_{\Psi,M}$ is essentially bounded, if Ψ generates an affine frame relative to $(M, \mathbf{Z}^d)$, with M an arbitrary expansive matrix. Their proof immediately extends to the case of a general lattice $N\mathbf{Z}^d$. Therefore,

$$f := F_{\Psi,M} \cdot \chi_Q \in L_1(\mathrm{IR}^d)$$

for any open cube Q which is centered around 0. Since $F_{\Psi,M}$ is dilation-invariant with respect to M^t and M^t is expansive, the essential ranges of $F_{\Psi,M}$ and of f coincide. Hence, for any $y \in \mathrm{IR}$, which is an element of the essential range of $F_{\Psi,M}$, and any $\epsilon > 0$, there exists a point $\xi_0 \in Q \setminus \{0\}$ and a subsequence $(f_{j_k})_{k \ge 0}$ of integral means (2.3) such that

$$|y - f_{j_k}(\xi_0)| = \left| y - |U_{j_k}|^{-1} \int_{U_{j_k}} F_{\Psi,M}(\xi_0 + \eta)\, d\eta \right| < \varepsilon, \qquad k \ge k_0(\varepsilon, y). \tag{2.4}$$

Note that equality of the integral means of f and $F_{\Psi,M}$ holds if U_{j_k} is small enough. This describes how one can substitute arguments which rely on the Lebesgue set of $F_{\Psi,M}$ by the result of Lemma 2.2.

The sets U_j of interest for our study are defined as follows. Since M is expansive, there is an invertible matrix S such that $M^t = S\widetilde{M}S^{-1}$ and $\|\widetilde{M}^{-1}\|_\infty < 1$. With $Q_\pi = (-\pi, \pi)^d$, we let

$$U_0 := SQ_\pi, \quad U_j := (M^t)^{-j} U_0, \quad j \geq 1. \tag{2.5}$$

Then $(U_j)_{j\geq 0}$ is a decreasing sequence and defines a basis of neighborhoods of 0, as required by Lemma 2.2. We can now give the proof of the theorem.

Proof of Theorem 2.1 : Since all norms of vectors and matrices are $\|.\|_\infty$, we can drop the index without confusion. Let $y \in \mathrm{IR}$ be an element of the essential range of $F_{\Psi,M}$ and $\epsilon > 0$ be given. We choose several parameters and denote their dependency on M, ϵ etc. in their definitions. Our final aim is to find a function $f \in L_2(\mathrm{IR}^d)$ with $\|f\|_2 = 1$, which gives

$$\left| \, |\det N| \cdot \|\mathcal{T}_\Psi f\|_{\ell_2}^2 - y \right| < \text{const} \cdot \epsilon, \tag{2.6}$$

where the constant does not depend on ϵ. Then Theorem 2.1 is established by letting $y = \operatorname{ess\,sup} F_{\Psi,M}$ or $y = \operatorname{ess\,inf} F_{\Psi,M}$.

First we can find $\xi_0 = \xi_0(y, \epsilon, \Psi, M) \in Q_\pi \setminus \{0\}$ and an appropriate subsequence $(U_{j_k})_{k\geq 0}$, such that (2.4) holds. For sufficiently large $K = j_k$ in the subsequence we let $\xi_K := (M^t)^K \xi_0$; then a simple substitution and the dilation invariance of $F_{\Psi,M}$ give

$$f_K(\xi_0) = |U_K|^{-1} \int_{U_K} F_{\Psi,M}(\xi_0 + \eta)\, d\eta = |U_0|^{-1} \int_{U_0} F_{\Psi,M}(\xi_K + \eta)\, d\eta. \tag{2.7}$$

Second we pick $r = r(\epsilon, \Psi) > 0$, such that

$$\int_{\mathrm{IR}\setminus Q_r} \sum_{i\in I} \left|\widehat{\psi}_i(\xi)\right|^2 d\xi < \epsilon.$$

Then $K = K(y, \epsilon, \Psi, M, N)$ can be taken as large as needed in order to satisfy (2.4) and the relations

$$\inf_{\eta\in U_0} \|\xi_K + \eta\| > \pi \,, \tag{2.8}$$

$$\inf_{\eta\in U_0} \|\xi_K + \eta\| > r\|(M^t)^j\| \quad \text{for all } j \text{ with} \quad \|(M^t)^{-j}\| > 1/(\|N^t\| \cdot \|S\|). \tag{2.9}$$

Note that the latter is a condition for small $j \in \mathbf{Z}$. These choices are performed in order to find appropriate bounds for certain integrals which will appear later in the proof.

We are now ready to define a function $f \in L_2(\mathrm{IR}^d)$ with $\|f\|_2 = 1$, which gives (2.6). The Fourier transform of $f = f(y, \epsilon, \Psi, M, N)$ is given by

$$\widehat{f}(\xi) := \left((2\pi)^d/|U_0|\right)^{1/2} \cdot \chi_\Omega, \qquad \text{where} \quad \Omega = \xi_K + U_0,$$

and χ_Ω denotes the characteristic function of the set Ω. By the standard transformations as in [8, p. 67], but with our normalization of the Fourier transform,

$$|\det N| \cdot \|\mathcal{T}_\Psi f\|_{\ell_2}^2 = (2\pi)^{-d} \left\{ \int_{\mathrm{IR}^d} |\widehat{f}(\xi)|^2 \, F_{\Psi,M}(\xi) \, d\xi + \sum_{i \in I} \sum_{j \in \mathsf{Z}} R_{i,j} \right\}, \quad (2.10)$$

where

$$|R_{i,j}| \leq \sum_{\ell \in 2\pi \mathsf{Z}^d \setminus \{0\}} \int_{\mathrm{IR}^d} \left| \widehat{f}(\xi) \, \widehat{f}\left(\xi + (M^t)^j (N^t)^{-1}\ell\right) \times \right. \\ \left. \widehat{\psi}_i\left((M^t)^{-j}\xi\right) \, \widehat{\psi}_i\left((M^t)^{-j}\xi + (N^t)^{-1}\ell\right) \right| d\xi. \quad (2.11)$$

Due to the special form of f, the first integral in (2.10) gives

$$(2\pi)^{-d} \int_{\mathrm{IR}^d} |\widehat{f}(\xi)|^2 \, F_{\Psi,M}(\xi) \, d\xi = |\Omega|^{-1} \int_\Omega F_{\Psi,M}(\xi) \, d\xi = f_K(\xi_0)$$

by (2.7). Since K is assumed to be so large that (2.4) is satisfied, it remains to show that

$$\sum_{i \in I} \sum_{j \in \mathsf{Z}} |R_{i,j}| < \text{const} \cdot \epsilon.$$

The integration and summation in (2.11) extend only over compact sets. More precisely, let

$$L_j := \left\{ \ell \in \mathsf{Z}^d \setminus \{0\}; \; \widehat{f}(\xi) \, \widehat{f}\left(\xi + 2\pi(M^t)^j (N^t)^{-1}\ell\right) \neq 0 \right\}.$$

Since U_0 is balanced and convex, we have

$$L_j \subset \left(\pi^{-1} N^t (M^t)^{-j} U_0\right) \cap \left(\mathsf{Z}^d \setminus \{0\}\right) =: L_j^\wedge.$$

It follows immediately, that L_j is empty, if $\pi^{-1} N^t (M^t)^{-j} U_0 \subset Q_1$, especially if

$$\|(M^t)^{-j}\| \leq 1/(\|N^t\| \cdot \|S\|). \quad (2.12)$$

On the other hand, a counting argument works for $L_j^\wedge$, if $Q_1 \subset \pi^{-1}N^t(M^t)^{-j}U_0$, especially if

$$\|(M^t)^j\| \le 1/\left(\|(N^t)^{-1}\| \cdot \|S^{-1}\|\right). \tag{2.13}$$

Then a simple result [16, Lemma 2.17] gives

$$\#L_j \le \#L_j^\wedge \le 2^{2d}|\det(N^t(M^t)^{-j}S)| = \text{const}\,|\det M|^{-j}.$$

Note that the constant only depends on d, N and S, but not on ϵ. The remaining case, where neither (2.12) nor (2.13) are satisfied, defines a finite range of values for $j \in \mathbf{Z}$ which is determined by M, N, and S. This means that there exists a constant κ, which depends on d, M, N, and S, such that

$$\#L_j \le \kappa\,|\det M|^{-j}, \quad \text{for all } j \text{ such that} \quad \|(M^t)^{-j}\| > 1/(\|N^t\| \cdot \|S\|).$$

In the sequel we write "$j \in J$" for the above condition, with suitable $J = J(M,N,S) \subset \mathbf{Z}$. This is the same condition, which we used in (2.9) for our choice of K.

Finally, by using the above counting argument, we can find estimates for the sum of the terms $|R_{i,j}|$ in (2.11). The Cauchy–Schwarz inequality yields

$$\begin{aligned}
|R_{i,j}| &\le \sum_{\ell \in L_j} \int_{\mathbb{R}^d} \left|\widehat{f}(\xi)\,\widehat{\psi}_i\left((M^t)^{-j}\xi\right)\right|^2 d\xi \\
&= \#L_j \cdot (2\pi)^d/|U_0| \int_{U_0} \left|\widehat{\psi}_i\left((M^t)^{-j}(\xi_K + \eta)\right)\right|^2 d\eta.
\end{aligned}$$

The choice of K in (2.9) implies that each set $(M^t)^{-j}(\xi_K + U_0)$, $j \in J$, is disjoint from the cube Q_r. Furthermore, only a constant number $\kappa_2 = \kappa_2(M)$ of these sets may overlap, since M^t is expansive and $\xi_K + U_0$ lies in a certain annulus by (2.8). Combined with the previous discussion of L_j this leads to the final estimates

$$\begin{aligned}
\sum_{i \in I}\sum_{j \in \mathbf{Z}} |R_{i,j}| &\le \kappa_1 \sum_{i \in I}\sum_{j \in J} |\det M|^{-j} \int_{U_0} \left|\widehat{\psi}_i\left((M^t)^{-j}(\xi_K + \eta)\right)\right|^2 d\eta \\
&\le \kappa_1\kappa_2 \sum_{i \in I} \int_{\mathbb{R}^d \setminus Q_r} \left|\widehat{\psi}_i(\xi)\right|^2 d\xi \ < \ \kappa_1\kappa_2 \cdot \epsilon\,,
\end{aligned}$$

and the constants κ_1, κ_2 do not depend on ϵ. This completes the proof of Theorem 2.1. $\qquad\qquad\square$

As a direct consequence of Theorem 2.1, for any tight affine frame the function $F_{\Psi,M}$ is constant a.e. Estimates of the frame bounds in the other directions can be found, if one assumes certain decay conditions of the Fourier transforms of the ψ_i, $i \in I$, see [3, 4, 7, 8, 16].

3 A Characterization of Tight Affine Frames

In a series of papers, Ron and Shen study various aspects of multivariate frames. In [14] they give a characterization of *tight* affine frames, which we wish to include here as one of the main structural theorems. In their study M is an integer matrix, which is expansive, and the Fourier transforms of the ψ_i, $i \in I$, satisfy a mild decay condition, namely

$$\sum_{k=0}^{\infty} \operatorname*{ess\,sup}_{\xi \in Q_\pi} \left(\sum_{\substack{\alpha \in 2\pi \mathbf{Z}^d \\ |\alpha| > 2^k}} |\widehat{\psi}_i(\cdot + \alpha)|^2 \right) < \infty . \tag{3.1}$$

Besides these restrictions, they obtain a complete characterization by means of two relatively simple criteria. The first is the necessary condition on the function $F_{\Psi,M}$ in (2.1) being constant a.e., which is also implied by our Theorem 2.1. The second condition, which is of a similar nature, makes this an equivalence.

Theorem 3.1. ([14]) *Let a finite family* $\Psi = (\psi_i; \ i \in I)$ *be given, such that* (3.1) *holds. Then* Ψ *generates a tight affine frame relative to* $(M, \mathbf{Z}^d)$, *with* M *an integer matrix and with frame bounds* $A = B$, *if and only if the following two conditions are satisfied:*

(a) $$\sum_{i \in I} \sum_{j \in \mathsf{Z}} \left| \widehat{\psi}_i \left((M^t)^j \xi \right) \right|^2 = A \qquad a.e.\ in\ \mathrm{I\!R}^d,$$

(b) *For any* $\alpha \in 2\pi(\mathbf{Z}^d \setminus M^t \mathbf{Z}^d)$,

$$\sum_{i \in I} \sum_{j \geq 0} \widehat{\psi}_i \left((M^t)^j \xi \right) \overline{\widehat{\psi}_i \left((M^t)^j (\xi + \alpha) \right)} = 0 \qquad a.e.\ in\ \mathrm{I\!R}^d.$$

Based on this result, the authors were able in [15] to describe an explicit construction of tight affine frames in any dimension $\mathrm{I\!R}^d$, where the generating set Ψ consists of box–splines. This method works, if the scaling matrix M has a power which is a multiple of the identity matrix. Important examples are the constructions of bivariate tight affine frames with scaling matrix $M = \left(\begin{smallmatrix} 1 & 1 \\ -1 & 1 \end{smallmatrix} \right)$. A feature which is immanent to this work is the polynomial growth of the number of generators with the degree of smoothness of the box–spline. A different result in [2] shows, that (non–tight) affine frames for this scaling matrix exist with a single generator, which is a four–directional box–spline of an arbitrary degree of smoothness. For further information on this subject we refer the reader to [2, 15].

4 Wavelet Sets and Algebraic Techniques

Several new aspects of wavelet analysis were recently addressed by Dai and Larson [5], Hernandez et. al. [12], and others. They investigate algebraic and topological properties of the set of all "mother wavelets" $\psi \in L_2(\mathbb{R})$, which generate an orthonormal wavelet basis. A new and surprising result was the construction of *wavelet sets* in $\mathbb{R}^d$ in [6], thus ascertaining the existence of Shannon–type wavelets for an arbitrary expansive matrix $M \in \mathbb{R}^{d \times d}$. More precisely, it is proved that a measurable set $E \subset \mathbb{R}^d$ exists with the following properties:

$$E + 2\pi k, \ k \in \mathbf{Z}^d, \quad \text{are disjoint,} \tag{4.1}$$

$$(M^t)^j E, \ j \in \mathbf{Z}, \quad \text{are disjoint, and} \tag{4.2}$$

$$\mathbb{R}^d \setminus \bigcup_{k \in \mathbf{Z}^d} (E + 2\pi k) \quad \text{and} \quad \mathbb{R}^d \setminus \bigcup_{j \in \mathbf{Z}} (M^t)^j E \quad \text{are null sets.} \tag{4.3}$$

Consequently, the function $\psi_E \in L_2(\mathbb{R}^d)$, with Fourier transform $\widehat{\psi}_E = \chi_E$, is the generator of an orthonormal wavelet basis relative to $(M, \mathbf{Z}^d)$. This shows that only one generator is needed, whereas any construction by multiresolution analysis requires $|\det M| - 1$ generators (where $\det M$ must be an integer).

The techniques in [5] need not be restricted to orthonormal wavelet bases. We first give an extension of [5, Proposition 1.3] to Riesz bases and frames. As a central tool, we define for $\psi \in L_2(\mathbb{R}^d)$ as in [5] the *local commutant*

$$C_\psi(D,T) := \{X \in \mathcal{B}(L_2(\mathbb{R}^d)); \ X D^j T_k \psi = D^j T_k X \psi \text{ for all } j \in \mathbf{Z}, \ k \in \mathbf{Z}^d\}\,.$$

More generally, we let $C_\Psi(D,T) := \bigcap_{i=1}^n C_{\psi_i}(D,T)$ for a finite collection $\Psi = (\psi_i; 1 \le i \le n) \in L_2(\mathbb{R}^d)^n$. Obviously, the classical commutant $\{D,T\}'$ of $\{D,T\}$ is a subspace of the local commutant. Several further properties are described in [5, Lemmas 1.1 and 3.1]. It is not difficult to see that any operator

$$m_h : L_2(\mathbb{R}^d) \to L_2(\mathbb{R}^d), \quad (m_h f)^\wedge = h\widehat{f} \tag{4.4}$$

with $h \in L_\infty(\mathbb{R}^d)$ and $h(M^t \cdot) = h$ is an element of $\{D,T\}'$. Hence, the local commutant is non–trivial for any Ψ. On the other hand, $C_\Psi(D,T)$ is not closed under multiplication of operators (so it is no algebra); especially it often occurs that X is in $C_\Psi(D,T)$, but X^2 is not. The following result gives a straightforward extension of [5, Proposition 1.3] to Riesz bases and multiple generators in the multivariate case. For $n \in \mathbb{N}$ we use the notation

$$\mathcal{R}_n := \left\{ \Psi \in L_2(\mathbb{R}^d)^n \ \middle| \ \begin{array}{l} (D^j T_k \psi_i; \ j \in \mathbf{Z}, \ k \in \mathbf{Z}^d, \ 1 \le i \le n) \\ \text{is a Riesz basis of } L_2(\mathbb{R}^d) \end{array} \right\}\,.$$

Proposition 4.1. *Let $\Psi \in L_2(\mathbb{R}^d)^n$ generate an orthonormal wavelet basis relative to $(M, \mathbf{Z}^d)$, and let $V\Psi := (V\psi_i; 1 \le i \le n)$ for any operator $V \in \mathcal{B}(L_2(\mathbb{R}^d))$. Then*

$$\mathcal{R}_n = \{V\Psi \mid V \in C_\Psi(D, T) \text{ and } V \text{ is bounded and boundedly invertible}\}.$$

Furthermore, the correspondence $V \to V\Psi$, with $V \in C_\Psi(D, T)$, is one-to-one, and the operator bounds of V^*V are the exact Riesz bounds of the basis which is generated by $V\Psi$.

Proof: If $(\eta_i; 1 \le i \le n) \in \mathcal{R}_n$, then the operator V on $L_2(\mathbb{R}^d)$ with $V D^j T_k \psi_i = D^j T_k \eta_i$ is bounded and boundedly invertible, since it maps an orthonormal basis into a Riesz basis. Moreover, $\eta_i = V\psi_i$, and therefore $V \in C_\Psi(D, T)$. Conversely, let $V \in C_\Psi(D, T)$ be bounded and boundedly invertible. Defining $\eta_i = V\psi_i$, $1 \le i \le n$, we obtain $V D^j T_k \psi_i = D^j T_k \eta_i$. The properties of the operator V assure, that the family $(D^j T_k \eta_i; \ j \in \mathbf{Z}, \ k \in \mathbf{Z}^d, \ 1 \le i \le n)$ is a Riesz basis of $L_2(\mathbb{R}^d)$, hence $V\Psi \in \mathcal{R}_n$. This proves the first statement of the proposition. Since V defines the correspondence of elements of the orthonormal basis and the Riesz basis, the exact Riesz bounds are the operator bounds of V^*V. The fact that the mapping $V \to V\Psi$ is injective follows, since $V\psi_i = 0$ for all $1 \le i \le n$ implies $V D^j T_k \psi_i = 0$ for all elements of the wavelet basis generated by Ψ, since $V \in C_\Psi(D, T)$, hence $V = 0$.   □

A similar application of the local commutant gives a parameterization of all affine frames which are generated by n functions.

Theorem 4.2. *Let the assumptions of Proposition* 4.1 *hold. Then the family $\Phi = (\phi_i; 1 \le i \le n)$ generates an affine frame relative to $(M, \mathbf{Z}^d)$ with frame bounds $A, B > 0$, if and only if there is an operator $V \in C_\Psi(D, T)$, such that $\Phi = V\Psi$ and*

$$A \operatorname{id}_{L_2(\mathbb{R}^d)} \le V V^* \le B \operatorname{id}_{L_2(\mathbb{R}^d)}. \tag{4.5}$$

Proof: Let a family Φ be given as in the theorem, which generates an affine frame relative to $(M, \mathbf{Z}^d)$. As before, we define a linear operator on $L_2(\mathbb{R}^d)$ by $V D^j T_k \psi_i = D^j T_k \phi_i$. Since the frame operator T_Φ^* in (1.3) satisfies $\|T_\Phi^*\| \le \sqrt{B}$, the operator V is bounded by the same constant. Moreover, its definition implies $V \in C_\Psi(D, T)$. For any $f \in L_2(\mathbb{R}^d)$ we obtain

$$\langle f, \, D^j T_k \phi_i \rangle = \langle V^* f, \, D^j T_k \psi_i \rangle.$$

Hence the frame condition (1.1) for the affine frame generated by Φ is equivalent to (4.5), since Ψ generates an orthonormal basis. This gives one direction of the assertion. The converse direction is shown in an analogous way.   □

Let us now develop a new application of the previous theorem which is related to wavelet sets in (4.1)-(4.3). Let E, F be two wavelet sets in $\mathbb{R}^d$, which share the same M, and ψ_E, ψ_F the corresponding Shannon–type wavelets. There is a measure preserving map

$$\sigma_E^F(\xi) := \begin{cases} \xi + 2\pi k \in F \quad \text{with} \quad k \in \mathbf{Z}^d, \quad \text{if } \xi \in E, \\ (M^t)^j \, \sigma_E^F\left((M^t)^{-j}\xi\right), \qquad \text{if } \xi \in (M^t)^j E, \ j \neq 0, \end{cases} \tag{4.6}$$

which is one–to–one and defined a.e. in $\mathbb{R}^d$. We will understand all identities modulo null sets in the sequel. The map $\sigma = \sigma_E^F$ was defined in [5, Chapter 5] for the univariate case. By its definition, σ is homothetic with respect to scaling by M^t, i.e. $\sigma(M^t\xi) = M^t\sigma(\xi)$.

Next we define the unitary operator

$$U_\sigma : L_2(\mathbb{R}^d) \to L_2(\mathbb{R}^d), \quad (U_\sigma f)^\wedge = \widehat{f} \circ \sigma^{-1}.$$

Without change, the same argument as in [5] shows that U_σ is in the local commutant $C_{\psi_E}(D,T)$. We impose the analogous conditions as in [5, p. 55], namely

$$\sigma^K = \mathrm{id}_{\mathbb{R}^d} \text{ modulo a null set, for some } K \in \mathbb{N}, \tag{4.7}$$

$$E_\kappa := \sigma^\kappa E \text{ is a wavelet set for all } 0 \leq \kappa \leq K - 1. \tag{4.8}$$

Such sets for $K = 2$ (when σ is an involution) and $K = 3$ have been found in the univariate case in [5]. In order to avoid some technical difficulties, we further assume that the sets E_κ are mutually disjoint for $0 \leq \kappa \leq K - 1$.

By (4.7) and (4.8) the cyclic group generated by U_σ lies in $C_{\psi_E}(D,T)$, too. This implies that any operator of the form

$$V = \sum_{\kappa=0}^{K-1} m_{h_\kappa} U_\sigma^\kappa, \tag{4.9}$$

with $m_{h_\kappa} \in \{D,T\}'$ as in (4.4), belongs to the local commutant at ψ_E. Note that the space of all such operators is a von Neumann algebra, since $U_\sigma m_h = m_{h\circ\sigma^{-1}} U_\sigma$ and $h \circ \sigma^{-1}$ is dilation–invariant relative to M^t. Note also that

$$(V \psi_E)^\wedge = \sum_{\kappa=0}^{K-1} h_\kappa \, \widehat{\psi_E}(\sigma^{-\kappa}\cdot) = \sum_{\kappa=0}^{K-1} h_\kappa \, \chi_{E_\kappa}. \tag{4.10}$$

This describes a one–to–one correspondence between V in (4.9) and the dilation–invariant functions h_κ, $0 \leq \kappa \leq K - 1$.

Dai and Larson [5] introduce a *coefficient criterion* for V which we wish to apply to our situation. With the above restrictions on the sets E_κ, $0 \le \kappa \le K - 1$, there is a $*$-isomorphism of operators V in (4.9) and the algebra of $K \times K$ matrix functions

$$H(\xi) = \Big(H_{ij}(\xi) \Big)_{0 \le i,j \le K-1} \qquad \text{where} \quad H_{ij} = h_{(j-i) \bmod K} \circ \sigma^{-i} . \qquad (4.11)$$

By its definition, $H(\xi)$ is dilation–invariant relative to M^t. It is a simple exercise to show that these matrices, with entries in $L_\infty(\mathbb{R}^d)$, define a von Neumann algebra. E.g., for the involutive case, when $K = 2$, the matrix has the form

$$H = \begin{pmatrix} h_0 & h_1 \\ h_1 \circ \sigma & h_0 \circ \sigma \end{pmatrix} .$$

In order to describe the mapping $V \mapsto H$ more precisely, we define the function vector

$$F(\xi) := \Big(\widehat{f}(\xi), \widehat{f}(\sigma^{-1}(\xi)), \dots, \widehat{f}(\sigma^{1-K}(\xi)) \Big)^t , \qquad f \in L_2(\mathbb{R}^d) , \qquad (4.12)$$

which is an element of $L_2(\mathbb{R}^d)^K$. The fact, that σ is measure preserving, gives

$$\int_{\mathbb{R}^d} \|F(\xi)\|^2 \, d\xi = (2\pi)^d K \, \|f\|^2_{L_2(\mathbb{R}^d)} . \qquad (4.13)$$

The next results are essential for the understanding of the coefficient criterion.

Lemma 4.3. *For a.a. $\xi \in \mathbb{R}^d$ we have*

$$(Vf)^\wedge(\sigma^{-\nu}(\xi)) = (H(\xi) \cdot F(\xi))_\nu , \qquad 0 \le \nu \le K - 1 , \qquad (4.14)$$

where the subscript ν denotes the ν-th entry of the vector function. Moreover, the mapping $V \mapsto H$ is a $*$-algebra isomorphism of operators V in (4.9) and matrix functions H in (4.11). The relation

$$A \, \mathrm{id}_{L_2(\mathbb{R}^d)} \le V^* V \le B \, \mathrm{id}_{L_2(\mathbb{R}^d)} \qquad (4.15)$$

is equivalent to

$$A \, I_K \le H(\xi)^* \, H(\xi) \le B \, I_K \qquad a.e. \ in \ E , \qquad (4.16)$$

where A, B are constants and I_K denotes the identity matrix.

Proof : Parts of the proof are only sketched for space limitations. The first relation and the properties of an isomorphism can be checked algebraically. For the equivalence of the operator bounds we find

$$(2\pi)^d K \|V f\|^2_{L_2(\mathbb{R}^d)} = \int_{\mathbb{R}^d} \|H(\xi) \cdot F(\xi)\|^2 \, d\xi$$

by (4.12)-(4.14). Hence (4.16) implies (4.15), by pointwise estimates in the right integral and another application of (4.13). We show the converse direction by contradiction. Here we only sketch the proof for the lower bound, since the upper bound is treated in the same way. Let $\epsilon > 0$ and a set Ω with positive Lebesgue measure be given, such that the minimal eigenvalue of $H(\xi)^* H(\xi)$ is less than $A - \epsilon$ for all $\xi \in \Omega$. Since $H(\xi)$ is dilation–invariant, we can assume that $\Omega \subset E$. The matrix $H^* H$ is hermitian and its entries are measurable functions. Thus an application of [13, Lemma 2.3.5] gives a measurable function of eigenvectors $s(\xi)$, $\xi \in \Omega$, with

$$\|s(\xi)\| = 1 \qquad \text{and} \qquad \|H(\xi) s(\xi)\|^2 \leq A - \epsilon, \qquad \xi \in \Omega.$$

Now let $f \in L_2(\mathbb{R}^d)$ be the function with $F(\xi) = s(\xi)$ for $\xi \in \Omega$, and $\widehat{f}(\xi) = 0$ outside of $\bigcup_{\kappa=0}^{K-1} \sigma^\kappa(\Omega)$. Note that f is well–defined, since the sets $\sigma^{-\nu}(\Omega)$ are mutually disjoint by our assumptions on $E_{-\nu}$, $0 \leq \nu \leq K - 1$. Then some straightforward calculation, which we suppress here, shows that $\|V f\|^2 \leq (A - \epsilon) \|f\|^2$. This shows that, by contradiction, (4.15) implies (4.16). The lemma is proved. $\square$

Remark. There is an oversight in [5, p. 55]. The above mapping $V \mapsto H$ is not well–defined, if the sets E_κ overlap. (There exist different representations (4.9) for the same operator V.) The difficulty can be resolved by using a certain compression of the matrix $H(\xi)$ with respect to a subspace of $\mathbb{C}^K$, which contains all vectors of the form $F(\xi)$. Since the argument which assures measurability of the mapping is quite technical, we leave its discussion to a forthcoming paper.

We are now ready to develop an application of the coefficient criterion to affine frames. This gives a generalization of our result in [17] which was obtained by different methods, see also Example 4.5 below.

Let an arbitrary function ψ with Fourier transform in $L_\infty(\mathbb{R}^d)$ and

$$\widehat{\psi}(\xi) = 0 \quad \text{for a.a.} \quad \xi \notin \bigcup_{\kappa=0}^{K-1} \sigma^\kappa(E) \tag{4.17}$$

be given. Then ψ has the form $\psi = V\psi_E$ with an operator V of type (4.9), where $h_\kappa(\xi) = \widehat{\psi}(\xi)$ for all $\xi \in E_\kappa$, see (4.10).

Now let ψ with (4.17) generate an affine frame relative to $(M, \mathbf{Z}^d)$. Then $\widehat{\psi} \in L_\infty(\mathrm{I\!R}^d)$ by Theorem 2.1. The results of Theorem 4.2 and Lemma 4.3 show, that

$$A\, I_K \le H(\xi)\, H(\xi)^* \le B\, I_K \qquad \text{a.e. in } E. \qquad (4.18)$$

By [13, Lemma 2.3.5] there is a measurable matrix function $S(\xi)$, $\xi \in E$, where $S(\xi)$ is a unitary matrix and $S^* H H^* S$ is a diagonal matrix a.e. in E. The bounds A, B in (4.18) are the essential infimum (supremum) of the smallest (resp. largest) eigenvalue of $H(\xi)\, H(\xi)^*$. Since $H(\xi)$ is a quadratic matrix, the diagonalization of $H(\xi)^* H(\xi)$ leads to the same operator bounds A, B. In view of the equivalent estimates in Lemma 4.3, we obtain both relations

$$A\, \mathrm{id}_{L_2(\mathrm{I\!R}^d)} \le V\, V^* \le B\, \mathrm{id}_{L_2(\mathrm{I\!R}^d)}, \qquad A\, \mathrm{id}_{L_2(\mathrm{I\!R}^d)} \le V^*\, V \le B\, \mathrm{id}_{L_2(\mathrm{I\!R}^d)}.$$

Hence V is bounded and boundedly invertible. Consequently, $\psi = V\psi_E$ generates a wavelet basis of $L_2(\mathrm{I\!R}^d)$ by Proposition 4.1. We have thus proved the following theorem.

Theorem 4.4. *Let a wavelet set E and a map σ be given as above, and let $\psi \in L_2(\mathrm{I\!R}^d)$ satisfy (4.17). Then the following statements are equivalent.*

(a) *ψ generates an affine frame relative to $(M, \mathbf{Z}^d)$.*

(b) *$\psi \in \mathcal{R}_1$, i.e. ψ generates a wavelet basis of $L_2(\mathrm{I\!R}^d)$ relative to $(M, \mathbf{Z}^d)$.*

(c) *The matrix function $H(\xi)$ in (4.11), with dilation–invariant entries $h_\kappa|_{E_\kappa} = \widehat{\psi}|_{E_\kappa}$, satisfies (4.18).*

Recall that we still assumed that the sets E_κ in (4.8) are mutually disjoint. We conjecture that the result also holds in the general case. This would shed some light on the method of Dai and Larson [5] of interpolation of wavelets. It shows that situations as in (4.7), (4.8) cannot happen if the union of the sets E_κ is too large, which would admit frames which are no Riesz bases.

Example 4.5. The sets

$$E = [-8\pi/3, -4\pi/3) \cup [2\pi/3, 4\pi/3), \qquad F = [-4\pi/3, -2\pi/3) \cup [4\pi/3, 8\pi/3)$$

are wavelet sets with respect to scaling by 2, and σ_E^F is involutive. Hence Theorem 4.4 applies to any function $\psi \in L_2(\mathrm{I\!R})$ with supp $\widehat{\psi} \subset E \cup F$. Due to the simple form of σ_E^F, it is sufficient to check the coefficient criterion (4.18) on the interval $(2\pi/3, 4\pi/3)$ only. On this interval the matrix H has the form

$$H(\xi) = \begin{pmatrix} \widehat{\psi}(\xi) & \widehat{\psi}(2\xi) \\ \widehat{\psi}(\xi - 2\pi) & \widehat{\psi}(2\xi - 4\pi) \end{pmatrix}, \qquad \xi \in [2\pi/3, 4\pi/3).$$

Since the four entries belong to disjoint subsets of the support of $\widehat{\psi}$, we have a complete characterization of all such wavelet bases by the coefficient criterion, which also gives the exact Riesz bounds. The same result was obtained in [17, Theorem 4] by a different method.

References

[1] J. J. Benedetto and S. Li: *Subband coding and noise reduction in multiresolution analysis frames,* in: *SPIE Proceedings on Wavelet Applications in Signal and Image Processing,* 154–165, San Diego 1994.

[2] C. K. Chui, K. Jetter and J. Stöckler: *Wavelets and frames on the four–directional mesh,* in: *Wavelets: Theory, Algorithms and Applications,* C. K. Chui, L. Montefusco and L. Puccio (eds.), 215–230, Academic Press, San Diego 1994.

[3] C. K. Chui and X. Shi: *Inequalities of Littlewood–Paley type for frames and wavelets,* SIAM J. Math. Anal. **24** (1993), 263–277.

[4] C. K. Chui and X. Shi: *Inequalities on matrix–dilated Littlewood–Paley energy functions and oversampled affine operators,* SIAM J. Math Anal. **28** (1997), 213–232.

[5] X. Dai and D. R. Larson: *Wandering vectors for unitary systems and orthogonal wavelets,* Memoirs of the American Mathematical Society, to appear.

[6] X. Dai, D. R. Larson and D. M. Speegle: *Wavelet sets in* $\mathbb{R}^n$, preprint, 1995.

[7] I. Daubechies: *The wavelet transform, time–frequency localization and signal analysis,* IEEE Trans. Inform. Theory **36** (1990), 961–1005.

[8] I. Daubechies: *Ten Lectures on Wavelets,* CBMS–NSF Reg. Conf. Series in Appl. Math. **61**, SIAM, Philadelphia 1992.

[9]   R. J. Duffin and A. C. Schaeffer: *A class of nonharmonic Fourier series,* Trans. Amer. Math. Soc. **72** (1952), 341–366.

[10]   R. E. Edwards and G. I. Gaudry: *Littlewood–Paley and Multiplier Theory,* Springer, Berlin–Heidelberg–New York 1977.

[11]   C. Heil and D. Walnut: *Continuous and discrete wavelet transforms,* SIAM Review **31** (1989), 628–666.

[12]   E. Hernandez, X. Wang and G. Weiss: *Smoothing minimally supported frequency (MSF) wavelets. Part I,* J. of Fourier Analysis and Appl. **2** (1996), 329–340.

[13]   A. Ron and Z. Shen: *Frames and stable bases for shift–invariant subspaces of $L_2(\mathrm{I\!R}^d)$,* Canadian J. Math. **47** (1995), 1051–1094.

[14]   A. Ron and Z. Shen: *Affine systems in $L_2(\mathrm{I\!R}^d)$: the analysis of the analysis operator,* J. Funct. Anal., to appear.

[15]   A. Ron and Z. Shen: *Compactly supported tight affine spline frames in $L_2(\mathrm{I\!R}^d)$,* Math. Comp., to appear.

[16]   J. Stöckler: *Multivariate affine Frames,* Habilitationsschrift, University of Duisburg, 1995.

[17]   J. Stöckler: *A Laurent operator technique for multivariate frames and wavelet bases,* in: *Advanced Topics in Multivariate Approximation,* F. Fontanella, K. Jetter and P.–J. Laurent (eds.), 339–354 World Scientific Publ., Singapore 1996.

[18]   R. M. Young: *An Introduction to Nonharmonic Fourier Series,* Academic Press, New York 1980.

Address:

JOACHIM STÖCKLER
Institut für Angewandte Mathematik und Statistik
Universität Hohenheim
D–70593 Stuttgart
Germany

Spectral Synthesis with Wavelet Methods

G. Zimmermann

Abstract

The aim of this paper is to demonstrate how spectral synthesis on $\mathbb{R}^d$ can be done with wavelet methods. The standard approach to generate a wavelet basis from a Multiresolution Analysis does not always work in Banach spaces. We exhibit some of the problems one may encounter and also suggest methods to avoid them. Then we apply these methods to the space $\boldsymbol{A}(\mathbb{R}^d)$ and construct an explicit example of a wavelet basis for spectral synthesis.

Introduction

Spectral synthesis is the approximation of elements in $\boldsymbol{L}^\infty$ by trigonometric polynomials in the weak*–sense, i.e., as linear functionals on $\boldsymbol{L}^1$. Equivalently, the problem can be stated on the Fourier side as approximation in $\boldsymbol{A}'$, the space of *pseudo-measures*, by discrete measures with finite support. A good introduction to the problem is given in [2].

We want to approach this problem with wavelet methods. I.e., we want to construct a weak* basis for $\boldsymbol{A}'(\mathbb{R}^d)$ of wavelet type consisting of discrete measures, or equivalently, construct a wavelet for $\boldsymbol{A}(\mathbb{R}^d)$ with a discrete measure as dual wavelet.

The usual approach to construct wavelets in a Hilbert space is the use of a *Multiresolution Analysis* (MRA), a doubly infinite sequence of closed subspaces with certain properties making it possible to generate a wavelet basis. This strategy is not always successful in Banach spaces, as examples in $\boldsymbol{L}^1(\mathbb{R}^d)$ show. Thus we give a short overview over a modification of this strategy which uses a *projective MRA*, a sequence of uniformly bounded projection operators rather than closed subspaces with corresponding properties, a concept which has been developed and successfully applied to closed subspaces of $\boldsymbol{L}^1$ with finite codimension in [11]. Using this method we show how to reach our goal. Finally, we construct an explicit example.

Multivariate Approximation: Recent Trends and Results; W. Haußmann, K. Jetter and M. Reimer (eds.)
Mathematical Research, Vol. 101, pp. 303–320, ISBN 3-05-501770-6
© Akademie–Verlag, Berlin 1997

Notation and Preliminaries.

We define the *Fourier transform* $\mathcal{F}f = \widehat{f}$ of an integrable function on $\mathbb{R}^d$ by

$$\widehat{f}(\xi) = \int_{\mathbb{R}^d} f(x)\, e^{-2\pi i x \cdot \xi}\, dx\,.$$

The image of $\boldsymbol{L}^1(\mathbb{R}^d)$ under $\mathcal{F}$ is the Banach space $\boldsymbol{A}(\mathbb{R}^d)$, its norm being defined by postulating that $\mathcal{F}$ be an isometry, i.e., $\|\widehat{f}\|_{\boldsymbol{A}} = \|f\|_{\boldsymbol{L}^1}$. Thus the Fourier transform can be extended via $\mathcal{F} = \mathcal{F}^{*-1}$ to an isometry between $\boldsymbol{L}^\infty$ and $\boldsymbol{A}'$. The fact that $\mathcal{F}$ is a unitary operator on $\boldsymbol{L}^2$ (Parseval–Plancherel) ensures that the two definitions agree on $\boldsymbol{L}^1 \cap \boldsymbol{L}^\infty$.

On the quotient group $\mathbb{T}^d = \mathbb{R}^d/\mathbb{Z}^d$, we define the Fourier transformations as usual by $\mathcal{F}(f) = \widehat{f} = (\widehat{f}[n])_{n\in\mathbb{Z}^d}$ where

$$\widehat{f}[n] = \int_{\mathbb{T}^d} f(x)\, e^{-2\pi i x \cdot n}\, dx\,.$$

Consequently, the Banach space $\boldsymbol{A}(\mathbb{Z}^d)$ and $\boldsymbol{A}'(\mathbb{Z}^d)$ are the images of $\boldsymbol{L}^1(\mathbb{T}^d)$ and $\boldsymbol{L}^\infty(\mathbb{T}^d)$, respectively, under $\mathcal{F}$.

For $y \in \mathbb{R}^d$, the *translation operator* T_y is defined by $(T_y f)(x) = f(x-y)$, and for $\lambda > 0$, we denote by D_λ the *dilation operator* given by $(D_\lambda f)(x) = f(\lambda x)$.

For the *closed span* of a family $\{x_\alpha : \alpha \in I\}$ of vectors in a topological vector space $\boldsymbol{X}$, we write $\overline{\mathrm{sp}}_{\boldsymbol{X}}\{x_\alpha : \alpha \in I\}$.

A *basis* of a topological vector space $\boldsymbol{X}$ is a family of vectors $\{x_\alpha\}_{\alpha\in I}$ together with a certain ordering such that each $x \in \boldsymbol{X}$ can be written uniquely as a series $x = \sum_{\alpha\in I} c_\alpha(x)\, x_\alpha$, where the limit of the series is to be taken according to the ordering of the vectors. The coefficient maps $x \mapsto c_\alpha(x)$ are linear functionals on $\boldsymbol{X}$. The basis is a *Schauder basis*, if each of these functionals is continuous. Thus it makes sense to consider a Schauder basis not only as a family in $\boldsymbol{X}$, but as a biorthogonal system in $(\boldsymbol{X}, \boldsymbol{X}')$. Note that this implies that a Schauder basis for $\boldsymbol{X}$ is — after reversing the order of the biorthogonal system — also a basis for $\boldsymbol{X}'$ with the weak topology induced by $\boldsymbol{X}$. In case $\boldsymbol{X}$ is a Banach space, this topology is usually referred to as the *weak*–topology* on $\boldsymbol{X}'$. For details, e.g., see [9].

A biorthogonal system $\{x_\alpha, x_\alpha^*\}_{\alpha\in I}$ in $(\boldsymbol{X}, \boldsymbol{X}')$ is a *projection basis* for a closed subspace $\boldsymbol{Y}$ of $\boldsymbol{X}$, if it is a basis for $\boldsymbol{Y}$ with the property that for all $x \in \boldsymbol{X}$, the series $Px = \sum_{\alpha\in I} \langle x, x_\alpha^* \rangle\, x_\alpha$ converges. [9], Theorem IV.1.1 states that under these assumptions, the map $x \mapsto Px$ is a bounded projection $\boldsymbol{X} \twoheadrightarrow \boldsymbol{Y}$.

1 Wavelet Bases and Multiresolution

Definition. Let $\boldsymbol{X}(\mathbb{R}^d)$ be a translation and dilation invariant Banach space of functions or distributions on $\mathbb{R}^d$. A *wavelet basis* for $\boldsymbol{X}$ with scaling factor k ($k \geq 2$, integer) is a basis of the form

$$\left\{ \psi_{m,n}^{(j)} = k^{m\alpha} D_{k^m} T_n \psi^{(j)} : m \in \mathbb{Z}, n \in \mathbb{Z}^d, j \in J \right\},$$

generated by the *wavelets* $\psi^{(j)}$, with dual basis

$$\left\{ \psi_{m,n}^{(j)*} = k^{m\alpha'} D_{k^m} T_n \psi^{(j)*} : m \in \mathbb{Z}, n \in \mathbb{Z}^d, j \in J \right\},$$

generated by the *dual wavelets* $\psi^{(j)*}$, where α and α' have to be chosen appropriately. The order of summation is given by

$$f = \sum_{m=-\infty}^{+\infty} \sum_{\substack{n \in \mathbb{Z}^d \\ j \in J}} \langle f, \psi_{m,n}^{(j)*} \rangle\, \psi_{m,n}^{(j)},$$

where for each $m \in \mathbb{Z}$, we assume the inner series to converge unconditionally.

Given a wavelet basis for a space $\boldsymbol{X}(\mathbb{R}^d)$, we can define

$$\boldsymbol{W}_m = \overline{\mathrm{sp}}_{\boldsymbol{X}} \left\{ \psi_{m,n}^{(j)} : n \in \mathbb{Z}^d, j \in J \right\}$$

and obtain a sequence of closed subspaces $(\boldsymbol{W}_m)_{m \in \mathbb{Z}}$ of $\boldsymbol{X}$ with the following properties:

$$\bigoplus_{m=-\infty}^{+\infty} \boldsymbol{W}_m = \boldsymbol{X}(\mathbb{R}^d), \tag{W1,2}$$

$$\forall m \in \mathbb{Z}, \qquad \boldsymbol{W}_m = D_{k^m} \boldsymbol{W}_0, \tag{W3}$$

$$\forall n \in \mathbb{Z}^d, \qquad T_n \boldsymbol{W}_0 = \boldsymbol{W}_0, \tag{W4}$$

$$\left\{ T_n \psi^{(j)}, T_n \psi^{(j)*} : n \in \mathbb{Z}^d, j \in J \right\} \text{ is a projection basis for } \boldsymbol{W}_0, \tag{W5}$$

where (W1,2) means that for each $f \in X$, there are unique $g_m \in W_m$ such that

$$f = \lim_{M_1, M_2 \to \infty} \sum_{m=-M_1}^{+M_2} g_m .$$

Now we can define

$$V_m = \bigoplus_{\mu=-\infty}^{m-1} W_\mu$$

and obtain a sequence of closed subspaces $(V_m)_{m\in\mathbb{Z}}$ of X with the following properties:

$$\forall m\in\mathbb{Z}, \qquad V_m \subseteq V_{m+1}, \tag{V1}$$

$$\bigcap_{m\in\mathbb{Z}} V_m = \{0\} \text{ (a)} \quad \text{and} \quad \overline{\bigcup_{m\in\mathbb{Z}} V_m} = X \text{ (b)}, \tag{V2}$$

$$\forall m\in\mathbb{Z}, \qquad V_m = D_{k^m} V_0, \tag{V3}$$

$$\forall n\in\mathbb{Z}^d, \qquad T_n V_0 = V_0. \tag{V4}$$

If we furthermore postulate that there exist $\varphi \in X$ and $\varphi^* \in X'$ such that

$$\{ T_n\varphi, T_n\varphi^* : n\in\mathbb{Z}^d \} \text{ is a projection basis for } V_0 , \tag{V5}$$

then $(V_m)_{m\in\mathbb{Z}}$ is a *Multiresolution Analysis* (MRA) which has become the standard tool for the construction of wavelets. Since $\varphi \in V_0 \subseteq V_1$, this function has to be *refinable*, i.e., it has to satisfy a scaling equation of the form

$$\varphi = \sum_{n\in\mathbb{Z}^d} h[n]\, \varphi_{1,n}$$

where $h[n] = \langle \varphi, \varphi^*_{1,n} \rangle$. For duality reasons, φ^* has to be refinable also. Therefore, φ is called the *scaling function* and φ^* the *dual scaling function* generating the MRA.

The usual strategy for the construction of wavelet bases in a Hilbert space is to begin with an MRA. For the case $k = 2$, there are well–established algorithms to obtain the $\psi^{(j)}$ and $\psi^{(j)*}$ from φ and φ^*, e.g., see [8]. The problem is that in a Banach space, these algorithms may fail for the following reasons. Given an MRA $(\boldsymbol{V}_m)_{m\in\mathbb{Z}}$, we have to complement $\boldsymbol{V}_m$ in $\boldsymbol{V}_{m+1}$, i.e., we have to find a sequence of closed subspaces $\boldsymbol{W}_m$ such that

$$\forall m\in\mathbb{Z}, \qquad \boldsymbol{V}_m \oplus \boldsymbol{W}_m = \boldsymbol{V}_{m+1}\,. \qquad\qquad (\text{VW})$$

Because of (V3) and (W3), it suffices to satisfy this condition for $m = 0$. In a Hilbert space, $\boldsymbol{W}_0$ is usually chosen to be the orthogonal complement of $\boldsymbol{V}_0$ in $\boldsymbol{V}_1$, and then (V1) and (V2) are equivalent to (W1,2). In a Banach space, two problems arise. Firstly, it is not clear whether the complementation is possible at all, and if so, there is no canonical way of doing it. Secondly, if the complementation is done, (V1) and (V2) do not necessarily imply (W1,2), as can be seen from the following example in $\boldsymbol{L}^1(\mathbb{R})$. If we use $k = 2$, $\varphi = \varphi^* = \chi_{[0,1)}$ which yields, e.g., $\psi = \psi^* = \chi_{[0,1/2)} - \chi_{[1/2,1)}$, we obtain the so–called Haar system, a well–known wavelet basis for $\boldsymbol{L}^2(\mathbb{R})$. In $\boldsymbol{L}^1(\mathbb{R})$, though, we obtain an MRA, but not a wavelet basis. Therefore it becomes necessary to take a different approach.

Note that in a Hilbert space, we have a canonical equivalence between closed subspaces and orthogonal projections. Thus it makes no difference whether we study sequences of closed subspaces or of orthogonal projections. For a closed subspace of a Banach space, there exist bounded projection operators onto it if and only if it is complementable (see [10], Corollary 5.16), and then there is no canonically distinguished projection in general. To construct wavelets in Banach spaces, we should therefore consider an MRA as a sequence of bounded projection operators rather than of closed subspaces.

2 Projective Decomposition and Multiresolution

Given a wavelet basis $\left\{\psi_{m,n}^{(j)}, \psi_{m,n}^{(j)*}\right\}_{m\in\mathbb{Z},n\in\mathbb{Z}^d,j\in J}$ in a Banach space $\boldsymbol{X}(\mathbb{R}^d)$, we can use the wavelet expansion to express each $f \in \boldsymbol{X}$ as

$$f = \sum_{\substack{m=-\infty \\ j\in J}}^{+\infty} \sum_{n\in\mathbb{Z}^d} \langle f, \psi_{m,n}^{(j)*}\rangle\, \psi_{m,n}^{(j)} =: \sum_{m=-\infty}^{+\infty} g_m\,.$$

By [9], Theorem IV.1.1, the maps $Q_m : f \mapsto g_m$ are bounded projections $\boldsymbol{X} \twoheadrightarrow \boldsymbol{W}_m$. They have the following properties:

$$\forall m_1 \neq m_2, \qquad Q_{m_1} Q_{m_2} = 0, \tag{Q1}$$

$$\sum_{m=-\infty}^{+\infty} Q_m = Id, \tag{Q2}$$

$$\forall m \in \mathbb{Z}, \qquad Q_m = D_{k^m} Q_0 D_{k^{-m}}, \tag{Q3}$$

$$\forall n \in \mathbb{Z}^d, \qquad T_n Q_0 = Q_0 T_n, \tag{Q4}$$

$$Q_0 f = \sum_{\substack{n \in \mathbb{Z}^d \\ j \in J}} \langle f, T_n \psi^{(j)*} \rangle \, T_n \psi^{(j)}. \tag{Q5}$$

We denote a sequence of uniformly bounded projection operators satisfying (Q1) and (Q2) a *projective decomposition.*

Observation. Conditions (Q1) and (Q2) are equivalent to condition (W1,2), since an analytic direct sum is defined in this way.

Letting

$$P_m = \sum_{\mu=-\infty}^{m-1} Q_\mu,$$

we obtain a sequence of uniformly bounded projection operators $(P_m)_{m \in \mathbb{Z}}$ in $\boldsymbol{X}$ with the following properties.

$$\forall m_1, m_2, \qquad P_{m_1} P_{m_2} = P_{\min\{m_1, m_2\}}, \tag{P1}$$

$$\lim_{m \to -\infty} P_m = 0 \text{ (a)} \quad \text{and} \quad \lim_{m \to +\infty} P_m = Id \text{ (b)}, \tag{P2}$$

$$\forall m \in \mathbb{Z}, \qquad P_m = D_{k^m} P_0 D_{k^{-m}}, \tag{P3}$$

$$\forall n \in \mathbb{Z}^d, \qquad T_n P_0 = P_0 T_n. \tag{P4}$$

where the limits in (P2) are taken in the strong operator topology. If in addition, there exist $\varphi \in \boldsymbol{X}$ and $\varphi^* \in \boldsymbol{X}'$ such that $\{T_n \varphi, T_n \varphi^*\}_{n \in \mathbb{Z}^d}$ is a projection basis with

$$P_0 f = \sum_{n \in \mathbb{Z}^d} \langle f, T_n \varphi^* \rangle \, T_n \varphi, \tag{P5}$$

then $(P_m)_{m \in \mathbb{Z}}$ is a *projective Multiresolution Analysis* (pMRA).

Vice versa, given $(P_m)_{m\in\mathbb{Z}}$ satisfying (P1)–(P5), we can define $(Q_m)_{m\in\mathbb{Z}}$ via

$$\forall m\in\mathbb{Z}, \qquad P_m + Q_m = P_{m+1} \qquad \text{(PQ)}$$

which immediately yields (Q1)–(Q4). Furthermore, we can try and use the standard methods for the construction of the $\psi^{(j)}$ and the $\psi^{(j)*}$ from φ and φ^* to obtain (Q5). We denote a sequence of projection operators satisfying (P1) and (P2) a *double projective approximation*.

Proposition 2.1. ([12], Theorem 3.1) *Double projective approximations and projective decompositions are equivalent.*

Lemma 2.2. *Let $(P_m)_{m\in\mathbb{Z}}$ be a pMRA in $\boldsymbol{X}(\mathbb{R}^d)$. Then the following hold.*

(i) *The images $\boldsymbol{V}_m = P_m(\boldsymbol{X})$ form an MRA.*

(ii) *The sequence of adjoint operators $(P_m^*)_{m\in\mathbb{Z}}$ is a pMRA for $\boldsymbol{X}'(\mathbb{R}^d)$ with the weak*–topology.*

(iii) *The images of the adjoint operators $\boldsymbol{V}'_m = P_m^*(\boldsymbol{X}')$ are isometric representations of the dual spaces of the $\boldsymbol{V}_m$.*

Proof: (i) and (ii) are obvious. For (iii), see [12], Lemma 2.4. $\square$

Consequently, $(\boldsymbol{V}'_m)_{m\in\mathbb{Z}}$ is an MRA for $\boldsymbol{X}'$ with the weak*–topology, and it seems natural to refer to properties (V1)–(V5) for these spaces as (V1')–(V5').

We saw that (W1,2) is equivalent to (Q1) and (Q2), which in turn are equivalent to (P1) and (P2). Therefore the problems we mentioned that we encounter in Banach spaces must stem from (P1) and (P2) not being equivalent to (V1) and (V2). This is confirmed by the following result.

Theorem 2.3. ([12], Theorems 3.2 and 3.3) *Let $(P_m)_{m\in\mathbb{Z}}$ be a sequence of projections in $\boldsymbol{X}(\mathbb{R}^d)$ with images $(\boldsymbol{V}_m)_{m\in\mathbb{Z}}$. Then*

(i) *(P1) is equivalent to (V1) and (V1').*

Assuming that (P1) is satisfied, we have

(ii) *(P2b) is equivalent to (V2b) and uniform boundedness of $(P_m)_{m>0}$.*

(iii) *(P2a) is equivalent to (V2a') and uniform boundedness of $(P_m)_{m<0}$.*

Remark. Note that if X is a Hilbert space and we assume the P_m to be orthogonal projections, then $P_m^* = P_m$, thus $V'_m = V_m$ and (P1) and (P2) are equivalent to (V1) and (V2), as one would intuitively expect.

Construction of a pMRA.

We assume that dilation can be defined in $X(\mathbb{R}^d)$ in such a way that it becomes an isometry. Note, e.g., that in $L^p(\mathbb{R}^d)$, we have

$$\left\| \lambda^{d/p} D_\lambda f \right\|_{L^p} = \|f\|_{L^p} .$$

In C_b, C_0, or $A(\mathbb{R}^d)$, we can use D_λ by itself, while in M_b or $A'(\mathbb{R}^d)$, we can use $\lambda^d D_\lambda$. Under this assumption, we can construct a pMRA as follows.

Find $\varphi \in X$ and $\varphi^* \in X'$ such that if we define $V_0 = \overline{\mathrm{sp}}_X\{T_n\varphi : n\in\mathbb{Z}^d\}$ and $V_m = D_{k^m} V_0$, analogously V'_m in X' with the weak*–topology, the following hold:

i) φ and φ^* are refinable.

ii) $(T_n\varphi, T_n\varphi^*)_{n\in\mathbb{Z}^d}$ is a projection basis in (X, X').

iii) (V2a′) and (V2b) hold.

Then $(P_m)_{m\in\mathbb{Z}}$ defined by (P5) and (P3) form a pMRA, by the following argument. By construction, P_0 is a bounded projection, and so are all the P_m. (P4) is obviously satisfied. Refinability of φ and φ^* implies (V1) and (V1′), respectively, so (P1) holds. The isometric dilation property of X guarantees that the P_m are uniformly bounded, and thus (V2a′) and (V2b) imply (P2).

Remark. It is well known that there are no wavelet bases for $L^1(\mathbb{R}^d)$. The reason that the above construction fails is that condition (V2a′) can not be satisfied. In [11], it has been shown that it is necessary to have $\varphi^* \in L^1\cap L^\infty$, and it is a result from [6] that any refinable function $\varphi^* \in L^1(\mathbb{R}^d)$ with stable integer translates satisfies $\sum_{n\in\mathbb{Z}^d} 1 \cdot T_n\varphi^* \equiv \widehat{\varphi^*}(0) \neq 0$. This implies that the constant functions are contained in V'_0 and thus by (V3′) in V'_m for all m, which contradicts (V2a′).

3 A Wavelet Basis for $A(\mathbb{R}^d)$

We want to find a wavelet basis for $A(\mathbb{R}^d)$ with finite discrete measures as dual wavelet. Therefore we first construct a pMRA with a finite discrete measure as dual scaling function.

The choice of φ^*. Using $\varphi^* = \delta$ will not work, since $\lambda^d D_\lambda \delta = \delta$ for all $\lambda > 0$, and thus $\delta \in V_0'$ implies $\delta \in \bigcap_m V_m' \neq \{0\}$. Instead, we can use an appropriate translate of δ; for symmetry reasons, let us consider $\varphi^* = \delta_{1/2}$ where we let $1/2 = (\frac{1}{2}, \ldots, \frac{1}{2}) \in \mathbb{R}^d$ for short. For scaling factor $k = 2$, this φ^* is not refinable, though; but we can use $k = 3$ (or any odd integer > 1).

To show (V2a′), we split the projection

$$
\begin{aligned}
P_0 : \quad & A(\mathbb{R}^d) && \to && V_0 \subseteq A(\mathbb{R}^d) \\
& f && \mapsto && \sum_{n \in \mathbb{Z}^d} \langle f, T_n \varphi^* \rangle \, T_n \varphi
\end{aligned}
$$

into the two parts

$$
\begin{aligned}
S_{\varphi^*} : \quad & A(\mathbb{R}^d) \to Y(\mathbb{Z}^d) \\
& f \quad \mapsto \left(\langle f, T_n \varphi^* \rangle \right)_{n \in \mathbb{Z}^d}
\end{aligned}
$$

and

$$
\begin{aligned}
R_\varphi : \quad & Y(\mathbb{Z}^d) && \to && A(\mathbb{R}^d) \\
& (c[n])_{n \in \mathbb{Z}^d} && \mapsto && \sum_{n \in \mathbb{Z}^d} c[n] \, T_n \varphi
\end{aligned}
$$

and we have to find out what $Y(\mathbb{Z}^d)$ is. To see this, we need the Poisson Summation Formula in the following form.

Definition. For a function or distribution $\widehat{f}$ on $\mathbb{R}^d$, we define its *periodization* to be $\widehat{f}^\circ = \sum_{n \in \mathbb{Z}^d} T_n \widehat{f}$. For pointwise well–defined functions f on $\mathbb{R}^d$, we define the *sampling operator* $S : f \mapsto (f(n))_{n \in \mathbb{Z}^d}$.

Proposition 3.1. (Poisson Summation)

   (i)   *Periodization is a bounded linear map from $L^1(\mathbb{R}^d)$ onto $L^1(\mathbb{T}^d)$.*

   (ii)   *Periodization is the Fourier transform of sampling in the sense that $(f(n))_{n \in \mathbb{Z}^d}$ are the Fourier coefficients of $\widehat{f}^\circ$.*

   (iii)   *Sampling is a bounded linear map from $A(\mathbb{R}^d)$ onto $A(\mathbb{Z}^d)$.*

Proof: (i) and (ii) are straightforward calculations, and (iii) is a direct consequence. $\qquad\square$

Remark. Note that the above implies for $f \in \boldsymbol{A}(\mathbb{R}^d)$ that for $x \in \mathbb{R}^d$,

$$\sum_{n\in\mathbb{Z}^d} f(n+x)\, e^{-2\pi i n\cdot\xi} = \sum_{n\in\mathbb{Z}^d} \widehat{f}(\xi+n)\, e^{2\pi i(\xi+n)\cdot x} \qquad\text{(PSF)}$$

in the sense that the left hand side is the Fourier series of the right hand side. Because of the translation invariance of $\boldsymbol{A}(\mathbb{R}^d)$, this implies that we have for the map S_{φ^*} from above

$$S_{\varphi^*}: \quad \boldsymbol{A}(\mathbb{R}^d) \quad \twoheadrightarrow \quad \boldsymbol{A}(\mathbb{Z}^d)$$
$$f \quad \mapsto \quad \big(f(n+1/2)\big)_{n\in\mathbb{Z}^d}.$$

Consequently, its adjoint

$$S_{\varphi^*}^*: \quad \boldsymbol{A}'(\mathbb{Z}^d) \quad \hookrightarrow \quad \boldsymbol{A}'(\mathbb{R}^d)$$
$$(c[n])_{n\in\mathbb{Z}^d} \mapsto \sum_{n\in\mathbb{Z}^d} c[n]\, \delta_{n+1/2}$$

is a weak*–continuous linear map with closed range. Therefore we may conclude that

$$\boldsymbol{V}'_0 = \overline{\mathrm{sp}}_{\boldsymbol{A}'}\big\{\delta_{n+1/2}\big\}$$
$$= \mathrm{range}\,(S_{\varphi^*}^*) = \left\{ \sum_{n\in\mathbb{Z}^d} c[n]\, \delta_{n+1/2} : (c[n])_{n\in\mathbb{Z}^d} \in \boldsymbol{A}'(\mathbb{Z}^d) \right\},$$

which implies in particular that $\bigcap_m \boldsymbol{V}'_m = \{0\}$ because of (V3).

Conditions on φ. Apart from refinability, we know by the above that φ has to be chosen such that

$$R_\varphi: \quad \boldsymbol{A}(\mathbb{Z}^d) \quad \hookrightarrow \quad \boldsymbol{A}(\mathbb{R}^d)$$
$$(c[n])_{n\in\mathbb{Z}^d} \mapsto \sum_{n\in\mathbb{Z}^d} c[n]\, T_n\varphi$$

and equivalently, on the Fourier side,

$$\widehat{R}_\varphi: \quad \boldsymbol{L}^1(\mathbb{T}^d) \quad \hookrightarrow \quad \boldsymbol{L}^1(\mathbb{R}^d)$$
$$C \quad \mapsto \quad C\,\widehat{\varphi},$$

are bounded linear maps.

Proposition 3.2. ([11, 13]) *The operator* $\widehat{R}_\varphi \,:\, C \to C\widehat{\varphi}$ *is a bounded linear map* $L^1(\mathbb{T}^d) \to L^1(\mathbb{R}^d)$ *if and only if* $\widehat{\varphi} \in L^{1,\infty}(\mathbb{Z}^d, \mathbb{T}^d)$ *(the mixed norm space defined in* [1]*), i.e.*

$$\Big\| \big\| (\widehat{\varphi}(\xi+n))_{n \in \mathbb{Z}^d} \big\|_{\ell^1(\mathbb{Z}^d)} \Big\|_{L^\infty(\mathbb{T}^d)} < \infty \,.$$

Remark. In [6], where one direction of the above result has been stated already, the symbol $\mathcal{L}_p(\mathbb{R}^d)$ has been introduced for the space $L^{1,p}(\mathbb{Z}^d, \mathbb{T}^d)$. For systematic reasons, we prefer the notation from [1], since it has become necessary to consider $L^{p_1,p_2}(\mathbb{Z}^d, \mathbb{T}^d)$ for $p_1 > 1$ (see e.g. [11, 13]).

Biorthogonality of $(T_n\varphi, T_n\varphi^*)_{n \in \mathbb{Z}^d}$ means

$$\varphi(n+1/2) = \begin{cases} 1 & \text{for } n = 0, \\ 0 & \text{for } n \in \mathbb{Z}^d \setminus \{0\}. \end{cases}$$

This condition together with $\widehat{\varphi} \in L^{1,\infty}(\mathbb{Z}^d, \mathbb{T}^d)$ will thus ensure that P_0 is a bounded projection operator in $A(\mathbb{R}^d)$.

It remains to satisfy (V2b) or, equivalently, (P2b). The expression

$$P_0(f) = \sum_{n \in \mathbb{Z}^d} f(n+1/2)\, T_n\varphi$$

becomes on the Fourier side (by (PSF))

$$\widehat{P_0 f}(\xi) = \left(\sum_{n \in \mathbb{Z}^d} \widehat{f}(\xi+n)\, e^{i\pi(\xi+n)} \right) \widehat{\varphi}(\xi) \,.$$

(To avoid bulky notation, we define $e^{i\pi\eta} := e^{i\pi(\eta_1 + \cdots + \eta_d)}$ for $\eta \in \mathbb{R}^d$, which makes the equations above and below a little easier to read.)
Using the dilation relation (P3), we obtain

$$\widehat{P_m f}(\xi) = \left(\sum_{n} \widehat{f}(\xi+k^m n)\, e^{i\pi(k^{-m}\xi+n)} \right) \widehat{\varphi}(k^{-m}\xi) \,.$$

Thus we can make the estimate

$$
\begin{aligned}
\left\| \widehat{P_m f} - \widehat{f} \right\|_{L^1} &\leq \int_{\mathbb{R}^d} \left| \widehat{f}(\xi) \left(e^{i\pi k^{-m}\xi} \, \widehat{\varphi}(k^{-m}\xi) - 1 \right) \right| d\xi \\
&\quad + \sum_{n\neq 0} \int_{\mathbb{R}^d} \left| \widehat{f}(\xi + k^m n) \, \widehat{\varphi}(k^{-m}\xi) \right| d\xi \\
&= \int_{\mathbb{R}^d} \left| \widehat{f}(\xi) \right| \left(\left| \widehat{\varphi}(k^{-m}\xi) - e^{-i\pi k^{-m}\xi} \right| + \sum_{n\neq 0} \left| \widehat{\varphi}(k^{-m}\xi + n) \right| \right) d\xi \\
&= \int_{\mathbb{R}^d} \left| k^{md} \widehat{f}(k^m \xi) \right| \left(\sum_{n\in\mathbb{Z}^d} \left| \widehat{\varphi}(\xi + n) - e^{-i\pi\xi} \delta_{o,n} \right| \right) d\xi . \qquad (*)
\end{aligned}
$$

If we define the vector valued function

$$
\vec{\widehat{\varphi}} : \xi \mapsto \left(\widehat{\varphi}(\xi + n) \right)_{n\in\mathbb{Z}^d} ,
$$

we observe that $\widehat{\varphi} \in L^{1,\infty}(\mathbb{Z}^d, \mathbb{T}^d)$ is equivalent to $\vec{\widehat{\varphi}} \in L^\infty(\mathbb{R}^d, \ell^1(\mathbb{Z}^d))$, and that

$$
\sum_{n\in\mathbb{Z}^d} \left| \widehat{\varphi}(\xi + n) - e^{-i\pi\xi} \delta_{o,n} \right| = \left\| \vec{\widehat{\varphi}}(\xi) - e^{-i\pi\xi} \delta \right\|_{\ell^1(\mathbb{Z}^d)} .
$$

Now recall that for $\widehat{f} \in L^1(\mathbb{R}^d)$ with $\| \widehat{f} \|_{L^1} = 1$ the family $\left(|k^{md} \widehat{f}(k^m \xi)| \right)_{m\in\mathbb{N}}$ forms an approximative identity. Thus we see that the right hand side of $(*)$ converges to 0 for all $\widehat{f} \in L^1(\mathbb{R}^d)$ if and only if φ satisfies the condition

$$
\xi = 0 \text{ is a Lebesgue point of } \vec{\widehat{\varphi}} \text{ with } \vec{\widehat{\varphi}}(0) = \delta .
$$

That this condition is also necessary to satisfy (P2b) can be seen from the fact that for compactly supported $\widehat{f}$, the inequality in $(*)$ becomes sharp for sufficiently large m.

For practical purposes, it should be noted that this condition is certainly satisfied if $\vec{\widehat{\varphi}}$ is continuous at $\xi = 0$.

An explicit example. A straightforward example which should be easy to work with is given by the d–fold tensor product of the linear cardinal spline shifted by $1/2$, i.e.,

$$\varphi(x) = \prod_{k=1}^{d} \varphi_1(x_k)$$

where $\varphi_1(x) = \max\{0, 1 - |x - \tfrac{1}{2}|\}$.

Letting $\sin \eta = \prod_{k=1}^{d} \sin \eta_k$ and $\operatorname{sinc} \eta = \prod_{k=1}^{d} \frac{\sin \eta_k}{\eta_k}$ for $\eta \in \mathbb{R}^d$, we have

$$\widehat{\varphi}(\xi) = e^{-i\pi\xi} \operatorname{sinc}^2(\pi\xi).$$

It is well known that $\sum_{n \in \mathbb{Z}^d} |\widehat{\varphi}(\xi + n)| = 1$ for all $\xi \in \mathbb{R}^d$, which implies that $\widehat{\varphi} \in L^{1,\infty}(\mathbb{Z}^d, \mathbb{T}^d)$. Furthermore, we have for $\|\xi\| < \tfrac{1}{2}$

$$\|\vec{\widehat{\varphi}}(\xi) - \vec{\widehat{\varphi}}(0)\|_{\ell^1} = |\widehat{\varphi}(\xi) - 1| + \sum_{n \neq 0} |\widehat{\varphi}(\xi+n)|$$

$$= |e^{-i\pi\xi}\operatorname{sinc}^2(\pi\xi) - 1| + \sum_{n \neq 0} \prod_{k=1}^{d} \frac{\sin^2 \pi(\xi_k + n_k)}{\pi^2(\xi_k + n_k)^2} \to 0 \quad (\xi \to 0),$$

i.e., $\vec{\widehat{\varphi}}$ is continuous at $\xi = 0$ with $\vec{\widehat{\varphi}}(0) = \delta$, as desired.

To obtain the wavelet basis, we use the standard strategy of constructing wavelets for the one–dimensional case first and then forming tensor products. For $d = 1$, the refinement equations for φ and φ^* are

$$\varphi_{0,0} = \sum_{n \in \mathbb{Z}} h[n]\, \varphi_{1,n} = \tfrac{1}{3}\varphi_{1,-1} + \tfrac{2}{3}\varphi_{1,0} + 1\,\varphi_{1,1} + \tfrac{2}{3}\varphi_{1,2} + \tfrac{1}{3}\varphi_{1,3}$$

$$\varphi_{0,0}^* = \sum_{n \in \mathbb{Z}} h^*[n]\, \varphi_{1,n}^* = \qquad\qquad\qquad 1\,\varphi_{1,1}^*$$

Since $k = 3$, we need to construct two wavelets $\psi^{(1)}$, $\psi^{(2)} \in V_1$ and two dual wavelets $\psi^{*(1)}$, $\psi^{*(2)} \in V_1'$. Thus we can write

$$\psi^{(j)} = \sum_{n \in \mathbb{Z}} g_j[n]\, \varphi_{1,n} \quad \text{and} \quad \psi^{*(j)} = \sum_{n \in \mathbb{Z}} g_j^*[n]\, \varphi_{1,n}^* , \quad j = 1, 2,$$

where we have to choose the coefficients in such a way that

$$\left\{ T_n \varphi,\, T_n \psi^{(1)},\, T_n \psi^{(2)};\, T_n \varphi^*,\, T_n \psi^{*(1)},\, T_n \psi^{*(2)} \right\}_{n \in \mathbb{Z}}$$

is a projection basis for V_1 in $\big(A(\mathbb{R}), A'(\mathbb{R})\big)$. Equivalently, the family of sequences

$$\left\{ T_{3n} h,\, T_{3n} g_1,\, T_{3n} g_2;\, T_{3n} h^*,\, T_{3n} g_1^*,\, T_{3n} g_2^* \right\}_{n \in \mathbb{Z}}$$

has to be a biorthogonal system in $\big(A(\mathbb{Z}), A'(\mathbb{Z})\big)$.

A possible choice is shown in the following table.

n	$\ldots$	-3	-2	-1	0	1	2	3	4	5	$\ldots$
h	$\ldots$	0	0	$\frac{1}{3}$	$\frac{2}{3}$	1	$\frac{2}{3}$	$\frac{1}{3}$	0	0	$\ldots$
h^*	$\ldots$	0	0	0	0	1	0	0	0	0	$\ldots$
g_1	$\ldots$	0	0	$\frac{2}{3}$	$\frac{4}{3}$	0	0	0	0	0	$\ldots$
g_1^*	$\ldots$	0	0	$-\frac{1}{2}$	1	$-\frac{1}{2}$	0	0	0	0	$\ldots$
g_2	$\ldots$	0	0	0	0	0	$\frac{4}{3}$	$\frac{2}{3}$	0	0	$\ldots$
g_2^*	$\ldots$	0	0	0	0	$-\frac{1}{2}$	1	$-\frac{1}{2}$	0	0	$\ldots$

The graphs of the functions and measures are shown on the next page.

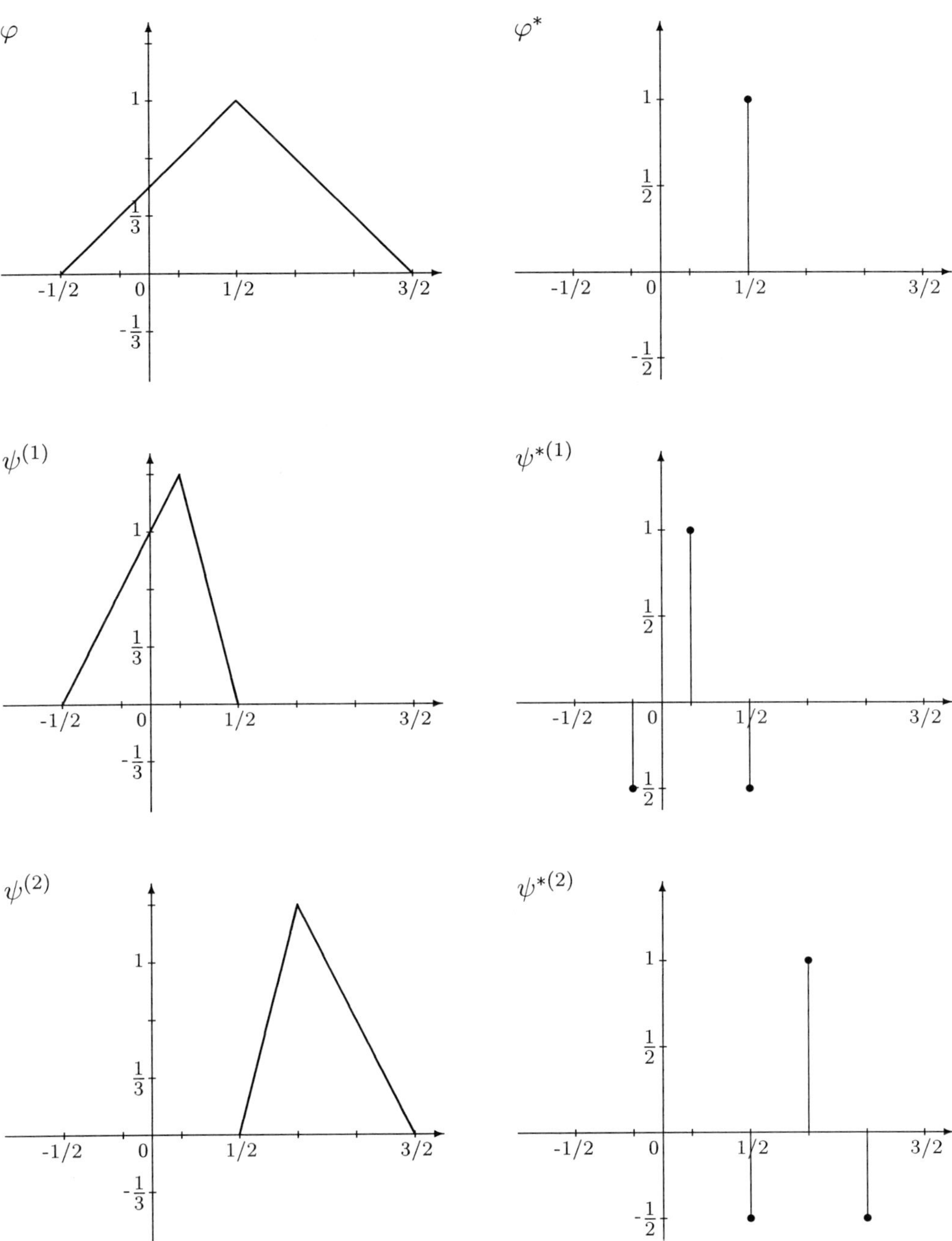

The scaling function, the wavelets, and their duals.

Another possible solution based on the above one can be obtained by following the suggestion made in [3] and constructing a symmetric and an antisymmetric wavelet and dual wavelet each, which can easily be done by considering $\psi^{(1)} \pm \psi^{(2)}$ and $(\psi^{*(1)} \pm \psi^{*(2)})/2$.

The wavelets for the d–dimensional case are constructed by forming tensor products. To simplify notation, we define $\psi^{(0)} = \varphi$ and $\psi^{*(0)} = \varphi^*$, and consider

$$\psi^{(\vec{j})} = \psi^{(j_1)} \otimes \cdots \otimes \psi^{(j_d)} \quad \text{and} \quad \psi^{*(\vec{j})} = \psi^{*(j_1)} \otimes \cdots \otimes \psi^{*(j_d)}$$

for $\vec{j} \in \{0,1,2\}^d$. Then we obtain the scaling function on $\mathbb{R}^d$ and its dual as $\psi^{(\vec{0})} = \varphi \otimes \cdots \otimes \varphi$ and $\psi^{*(\vec{0})} = \varphi^* \otimes \cdots \otimes \varphi^*$, respectively, and the desired family of wavelets and dual wavelets is given by $\{\psi^{(\vec{j})}, \psi^{*(\vec{j})} : \vec{j} \neq \vec{0}\}$.

Remark. It is worth noting that $\psi^{(1)}$ and $\psi^{(2)}$ are no "wavelets" in the original sense of the word, since they do not have zero mean. Neither $\psi^{(1)}$ and $\psi^{(2)}$ nor $\psi^{*(1)}$ and $\psi^{*(2)}$ satisfy the *admissibility condition* (see e.g. [5])

$$\int_{\mathbb{R}} |\widehat{\psi}(\xi)|^2 \, \frac{d\xi}{|\xi|} < \infty \, .$$

We do have though that

$$\int_{\mathbb{R}} |\widehat{\psi^{(j)}}(\xi) \, \widehat{\psi^{*(j)}}(\xi)| \, \frac{d\xi}{|\xi|} < \infty$$

for $j = 1,2$, which appears to be a natural adaptation of the admissibility condition to the biorthogonal approach.

4   Outlook and Acknowledgements

We have shown that it is possible to construct wavelet bases for $\boldsymbol{A}(\mathbb{R}^d)$ with finite discrete measures as dual wavelets, and thus, by considering the dual space $\boldsymbol{A}'(\mathbb{R}^d)$ with the weak*–topology, to do spectral synthesis on $\mathbb{R}^d$ with wavelets. A strong advantage of this approach is that it is constructive. Since the spectral synthesis problem as such (e.g., see [2]) is yet unsolved, the question is of interest to which subsets of $\mathbb{R}^d$ this method can be transferred. Therefore it seems worthwhile to try and connect our results to those known about wavelets on subsets of $\mathbb{R}^d$.

The author would like to use this opportunity to express his thanks to his thesis adviser, Professor John J. Benedetto, for inspiring him to work on these problems; and to the organizers of the 2nd International Conference on Multivariate Approximation for their invitation to participate in the conference.

References

[1] A. Benedek, R. Panzone: *The spaces L^P, with mixed norm*, Duke Math. J. **28** (1961), 301–324.

[2] J. J. Benedetto: *Spectral Synthesis*, B. G. Teubner, Stuttgart 1975.

[3] C. K. Chui, J. Lian: *Construction of compactly supported symmetric and antisymmetric orthonormal wavelets with scale = 3*, Appl. Comput. Harmonic Anal. **2** (1995), 21–51.

[4] W. Dahmen: *Some remarks on multiscale transformations, stability, and biorthogonality*, in: *Wavelets, Images and Surface Fitting*, P. J. Laurent et al. (eds.), 157–188, A. K. Peters, Wellesley 1994.

[5] I. Daubechies: *Ten Lectures on Wavelets*, CBMS–NSF Series in Applied Mathematics **61**, SIAM, Philadelphia 1992.

[6] R.-Q. Jia, C. A. Micchelli: *Using the refinement equations for the construction of prewavelets II : Powers of two*, in: *Curves and Surfaces*, P. J. Laurent et al. (eds.), 209–246, Academic Press, Boston 1991.

[7] P. G. Lemarié–Rieusset: *Projection operators in multiresolution analysis*, in: *Different Perspectives on Wavelets*, Proc. Sympos. Appl. Math. **47**, I. Daubechies (ed.), 59–76, Amer. Math. Soc., Providence 1993.

[8] S. G. Mallat: *Multiresolution approximations and wavelet orthonormal bases of $L^2(\mathbb{R})$*, Trans. Amer. Math. Soc. **315** (1989), 69–87.

[9] J. T. Marti: *Introduction to the Theory of Bases*, Springer Tracts in Natural Philosophy **18**, Springer Verlag, New York 1969.

[10] W. Rudin: *Functional Analysis*, International Series in Pure and Applied Mathematics, McGraw–Hill, New York 1991.

[11] G. Zimmermann: *Projective Multiresolution Analysis and Generalized Sampling,* PhD thesis, University of Maryland, College Park 1994.

[12] G. Zimmermann: *Double projective approximation and projective decomposition in the construction of wavelet bases,* submitted, 1996.

[13] G. Zimmermann: *Generalized sampling and quasi-interpolation,* in preparation, 1997.

Address:

GEORG ZIMMERMANN
Institut für Mathematik
Universität Wien
Strudlhofgasse 4
A–1090 Wien
Austria

87 **ICIAM 95**
Proceedings of the Third International
Congress on Industrial and Applied
Mathematics held in Hamburg, Germany,
July 3–7, 1995
Edited by Klaus Kirchgässner,
Oskar Mahrenholtz,
and Reinhard Mennicken
488 pages, 1996
ISBN 3-05-501682-3

88 **Asymptotics and Extrapolation**
By Guido Walz
330 pages, 1996
ISBN 3-05-501732-3

89 **Numerical Methods and Error Bounds**
Proceedings of the IMACS-GAMM
International Symposium on Numerical
Methods and Error Bounds held in
Oldenburg, Germany, July 9–12, 1995
Edited by Götz Alefeld
and Jürgen Herzberger
303 pages, 1996
ISBN 3-05-501696-3

90 **Scientific Computing
and Validated Numerics**
Proceedings of the International
Symposium on Scientific Computing,
Computer Arithmetic and Validated
Numerics SCAN-95 held in Wuppertal,
Germany, September 26–29, 1995
Edited by Götz Alefeld,
Andreas Frommer, and Bruno Lang
340 pages, 1996
ISBN 3-05-501737-4

91 **Transient Tunnel Effect
and Sommerfeld Problem**
Waves in Semi-Infinite Structures
By Felix Ali Mehmeti
ca. 200 pages, 1996
ISBN 3-05-501707-2

92 **Global Hypoellipticity
and Spectral Theory**
By Paolo Boggiatto, Ernesto Buzano,
and Luigi Rodino
187 pages, 1996
ISBN 3-05-501724-2

93 **The Degenerate Oblique
Derivative Problem for Elliptic
and Parabolic Equations**
By Dian Palagachev
and Peter Popivanov
ca. 140 pages, 1997
ISBN 3-05-501757-8

94 **Pseudo-Differential Operators
and Markov Processes**
By Niels Jacob
207 pages, 1996
ISBN 3-05-501731-5

95 **Maslov Classes, Metaplectic
Representation
and Lagrangian Quantization**
By Maurice de Gosson
186 pages, 1996
ISBN 3-05-501714-5

96 **Parcella 96**
Proceedings of the VII. Internatinal
Workshop on Parallel Processing by
Cellular Automata and Arrays
held in Berlin, September 16–20, 1996
Edited by Roland Vollmar,
Erhard Werner, and Vesselin Jossifov
341 pages, 1996
ISBN 3-05-501750-1

97 **Parameter-Free Iterative
Linear Solvers**
By Rüdiger Weiss
217 pages, 1996
ISBN 3-05-501763-3

98 **Spectral Theory of Indefinite
Krein-Feller Differential Operators**
By Andreas Fleige
133 pages, 1996
ISBN 3-05-501742-0

99 **Uncertainty: Models and Measures**
Proceedings of the International
Workshop held in Lambrecht, Germany,
July 22–24, 1996
Edited by H. Günther Natke and
Yakov Ben-Haim
276 pages, 1997
ISBN 3-05-501740-4